Industry 4.0 and Hyper-Customized Smart Manufacturing Supply Chains

S.G. Ponnambalam
University Malaysia Pahang, Malaysia

Nachiappan Subramanian
University of Sussex, UK

Manoj Kumar Tiwari
Indian Institute of Technology Kharagpur, India

Wan Azhar Wan Yusoff
University Malaysia Pahang, Malaysia

A volume in the Advances in Logistics, Operations, and Management Science (ALOMS) Book Series

IGI Global
DISSEMINATOR OF KNOWLEDGE

Published in the United States of America by
IGI Global
Business Science Reference (an imprint of IGI Global)
701 E. Chocolate Avenue
Hershey PA, USA 17033
Tel: 717-533-8845
Fax: 717-533-8661
E-mail: cust@igi-global.com
Web site: http://www.igi-global.com

Library of Congress Cataloging-in-Publication Data

Names: Ponnambalam, S. G. (Sivalinga Govinda), editor.
Title: Industry 4.0 and hyper-customized smart manufacturing supply chains / S.G. Ponnambalam [and three others] editors.
Description: Hershey, PA : Business Science Reference, [2020]
Identifiers: LCCN 2018056090| ISBN 9781522590781 (hardcover) | ISBN 9781522590804 (ebook) | ISBN 9781522590798 (softcover)
Subjects: LCSH: Manufacturing industries--Technological innovations. | Production engineering. | Business logistics.
Classification: LCC HD9720.5 .I525 2020 | DDC 658.5--dc23 LC record available at https://lccn.loc.gov/2018056090

This book is published in the IGI Global book series Advances in Logistics, Operations, and Management Science (ALOMS) (ISSN: 2327-350X; eISSN: 2327-3518)

British Cataloguing in Publication Data
A Cataloguing in Publication record for this book is available from the British Library.

All work contributed to this book is new, previously-unpublished material.
The views expressed in this book are those of the authors, but not necessarily of the publisher.

For electronic access to this publication, please contact: eresources@igi-global.com.

Advances in Logistics, Operations, and Management Science (ALOMS) Book Series

ISSN:2327-350X
EISSN:2327-3518

Editor-in-Chief: John Wang, Montclair State University, USA

MISSION

Operations research and management science continue to influence business processes, administration, and management information systems, particularly in covering the application methods for decision-making processes. New case studies and applications on management science, operations management, social sciences, and other behavioral sciences have been incorporated into business and organizations real-world objectives.

The **Advances in Logistics, Operations, and Management Science** (ALOMS) Book Series provides a collection of reference publications on the current trends, applications, theories, and practices in the management science field. Providing relevant and current research, this series and its individual publications would be useful for academics, researchers, scholars, and practitioners interested in improving decision making models and business functions.

COVERAGE

- Information Management
- Political Science
- Computing and information technologies
- Organizational Behavior
- Finance
- Services management
- Operations Management
- Marketing engineering
- Decision analysis and decision support
- Risk Management

IGI Global is currently accepting manuscripts for publication within this series. To submit a proposal for a volume in this series, please contact our Acquisition Editors at Acquisitions@igi-global.com or visit: http://www.igi-global.com/publish/.

The Advances in Logistics, Operations, and Management Science (ALOMS) Book Series (ISSN 2327-350X) is published by IGI Global, 701 E. Chocolate Avenue, Hershey, PA 17033-1240, USA, www.igi-global.com. This series is composed of titles available for purchase individually; each title is edited to be contextually exclusive from any other title within the series. For pricing and ordering information please visit http://www.igi-global.com/book-series/advances-logistics-operations-management-science/37170. Postmaster: Send all address changes to above address.

Titles in this Series

For a list of additional titles in this series, please visit:
https://www.igi-global.com/book-series/advances-logistics-operations-management-science/37170

Strategic Thinking, Planning, and Management Practice in the Arab World
Fayez Albadri (Middle East University, Jordan) and Yacoub Adel Nasereddin (Middle East University, Jordan)
Business Science Reference • ©2019 • 312pp • H/C (ISBN: 9781522580485) • US $225.00

Global Supply Chains and Multimodal Logistics Emerging Research and Opportunities
Deepankar Sinha (Indian Institute of Foreign Trade, India)
Business Science Reference • ©2019 • 220pp • H/C (ISBN: 9781522582984) • US $165.00

Agile Approaches for Successfully Managing and Executing Projects in the Fourth ...
Hür Bersam Bolat (Istanbul Technical University, Turkey) and Gül Tekin Temur (Bahçeşehir University, Turkey)
Business Science Reference • ©2019 • 424pp • H/C (ISBN: 9781522578659) • US $235.00

Technology Optimization and Change Management for Successful Digital Supply Chains
Ehap Sabri (KPMG LLP, USA & University of Texas at Dallas, USA)
Business Science Reference • ©2019 • 323pp • H/C (ISBN: 9781522577003) • US $210.00

Strategy and Superior Performance of Micro and Small Businesses in Volatile Economies
João Conrado de Amorim Carvalho (Unidade de Ensino Superior Dom Bosco, Brazil) and Emmanuel M.C.B. Sabino (FORUM (Centro de Formação, Estudos e Pesquisas), Brazil)
Business Science Reference • ©2019 • 389pp • H/C (ISBN: 9781522578888) • US $225.00

Logistics and Transport Modeling in Urban Goods Movement
Jesus Gonzalez-Feliu (Ecole des Mines de Saint-Etienne, France)
Business Science Reference • ©2019 • 273pp • H/C (ISBN: 9781522582922) • US $185.00

Hierarchical Planning and Information Sharing Techniques in Supply Chain Management
Atour Taghipour (Normandy University, France)
Business Science Reference • ©2019 • 287pp • H/C (ISBN: 9781522572992) • US $185.00

For an entire list of titles in this series, please visit:
https://www.igi-global.com/book-series/advances-logistics-operations-management-science/37170

IGI Global
DISSEMINATOR OF KNOWLEDGE

701 East Chocolate Avenue, Hershey, PA 17033, USA
Tel: 717-533-8845 x100 • Fax: 717-533-8661
E-Mail: cust@igi-global.com • www.igi-global.com

Editorial Advisory Board

Table of Contents

Detailed Table of Contents

The literature concerning supply chain management (SCM) and blockchain integration is scarce, and organizations and decision makers still have a limited knowledge about blockchain solution diffusion in SCM. Under these circumstances, studies aimed to improving the understanding of the main blockchain diffusion factors are fundamental. This chapter is a contribution to bridging such gaps, as it aims to unlock the driving factors of blockchain diffusion in the SCM environment. Based on the emerging literature on blockchain, supply chain management, expectancy theory, and diffusion of innovations, a model was developed and validated, considering data from India's supply chain management professionals. The data was analyzed using partial least square structural equation modeling (PLS-SEM). The results indicated that IT deployment capability, compatibility, and trading partner pressure are factors that affect blockchain diffusion significantly. Also, the results bring essential managerial and theoretical implications regarding the blockchain diffusion in the SCM.

This chapter aims to explore the feasibility of using blockchain in the beef supply chain to reduce waste. A mono-method, qualitative, inductive, single case study approach was taken on a cross-sectional scale from June 2018 to August 2018, with two individuals interviewed: a beef and a blockchain expert. The case study also involved observations, a field visit, and other secondary source data. Beef is a high demand, valuable food product with a limited shelf life. By using blockchain in conjunction with RFID and sensor technologies, farming and processing stages in the beef supply chain can be streamlined. Firstly, using the technology to monitor the animals on the farm and during transportation can reduce the amount of water and energy wasted. Secondly, blockchain can be used to establish exactly when and where the meat is cut and packaged, improving the accuracy of information between supply chain entities, resulting in improved inventory management, specifically more accurate delivery times and lengthened product shelf lives.

While the new retail is revolutionizing retailing in China, the supply chain of the new retail has a few problems or potential risks which will decrease the customer satisfaction level. However, the implementation of one of the cutting-edge

environmental objectives simultaneously. The objectives considered are minimizing the total supply chain cost and minimizing CO2 emission of the supply chain network. The proposed model falls into NP-hard category. Multi-objective genetic algorithm is proposed to solve the proposed model and illustration is provided to explain the use of the proposed model. A procedure that could be followed to find the best possible solution based on user's choice among the Pareto front solutions is also explained.

This chapter aims to investigate the potential economic and environmental sustainability outcomes of additive manufacturing (AM) for spare parts logistics. System dynamic simulation was conducted to analyze the sustainability of producing a spare part used in a railways subsystem using a particular additive manufacturing (AM) technology (i.e., selective laser sintering [SLS]) compared to producing it using injection molding. The results of the simulation showed that using SLS for the chosen part is superior to the conventional one in terms of total variable costs as well as for carbon footprint. Compared to the conventional supply chain, for the AM supply chain, the costs of the supplier reduces by 46%, that of the railways company reduces by 71%, while the overall supply chain costs reduce by 61.9%. The carbon emissions in the AM supply chain marginally reduces by 2.89% compared to the conventional supply chain.

Rapid technological advances have revolutionized the industrial sector. In the global market, it is necessary to consider the new paradigm of Industry 4.0 that presents a lot of features in the industrial logistics application. It has been seen through literature that innovation management practices enable companies to compete within the autonomous and connected vehicle market and is considered as an emerging and competitive differentiator towards the growth of the product and that of meeting customer demands within the changing markets. The first case study explores the integration of GPS and GLONASS signals in AGV for localization and navigation of customer destination and materials in the indoor and outdoor environment. The second case study implemented in obstacle environment that recognized the

Preface

Next generation industrial revolution depends on the pillars of smart manufacturing which is characterized by manufacturing technology and processes, materials, data, predictive engineering, sustainability and resource sharing and networking. Smart on a wider note is connected with concepts of cyber-physical systems spearheaded by the internet of things, cloud computing, service-oriented computing, artificial intelligence and data science. Smart manufacturing is an advancement of autonomous and flexible manufacturing systems including integration of sensors, computing platforms, communication technology, control, simulation, data intensive modelling and predictive analytics. Typical example of smart manufacturing is 3D printing, which is an evolving manufacturing process as per contextual factors and instant delivers of products. In addition, next generation supply chain also revolves around smart manufacturing processes and personalized customization of products and services. Attaining hyper-customization is a huge challenge in the Industry 4.0 era, where listening and reacting to customers at a rapid rate without any concerns is going to be a demanding requirement. In addition, there are several challenges connected to the digital supply chain are investigated in this book. Considering these, the book unfolds original contributions including all aspects of innovation and operational issues in smart manufacturing and digital supply chain management in the era of Industry 4.0.

Research in the area of hyper-customized smart supply chains is garnering vital importance in the era of Industry 4.0. Several stakeholders including government and big multinational firms are evolving with several strategic plans that accounts the key issues to achieve flexible and reconfigurable manufacturing processes in order to address dynamic and global market. Recent advancements also benefits small and medium-sized enterprises (SMEs) other than large companies. To convert all resources as intelligent objects, which can sense, act, and behave as per the environment is challenging and need the support of multiple stakeholders including researchers thought process to progress. Despite promised benefits, however, this area is under-researched. Certainly, we believe the topics covered in this book will advance theoretical and practical knowledge and serves as a foundation for further development.

The book consists of eleven chapters in the three major themes such as digital supply chain, smart manufacturing and supply chain and Industry 4.0 of eleven chapters covering

THEME 1: DIGITAL SUPPLY CHAIN

This theme includes the majority of chapters of this book, which consists of blockchain implementation and security threats in supply chain contexts in emerging and developed economies.

The first chapter investigates the potential improvement blockchain can offer to supply chain that will enable to achieve multiple benefits such as reduction of costs, improved efficiency in processes, enhanced security and better transparency in the operations. Authors' conclusions are based on 30 studies that was published during 2017-2018. Certainly, this chapter will shed lights on open research question in application of blockchain which may kindle further insights and research opportunity.

Chapter 2 determines the factors that enable blockchain diffusion in supply chain. The authors developed the conceptual model and validated using empirical data collected from the information technology-oriented people in India. The outcome of the study such as IT deployment capability, compatibility, and trading partner pressure are perceived to impact blockchain diffusion significantly in the emerging economy.

The next two chapters investigates the application of blockchain in two specific sectors such as food supply chain from the perspective of supplier and retailers. Chapter 3 studies how blockchain application in a beef supply chain could potentially reduce waste in future. The study encompasses waste surrounding beef supply chain and processing including meat, water and energy waste. The authors also proposed mitigation strategies to overcome waste in the end-to-end beef supply chain. Typically, the study highlights the theoretical contributions made to ignorance theory.

Chapter 4 studied the potential application of blockchain in the Chinese new retail business model. The authors studied the benefits of blockchain to the new retail format such as security and efficiency using IBM based tool. The study uncovers potential strategies that companies can consider to implement blockchain in the emerging economies. On continuation of this, Chapter 5 analyses the cybersecurity challenges in Malaysian companies. The study is a descriptive to understand the emerging digital threats, which is ubiquitous, and need further solution and proactive action plan in the future.

THEME 2: SMART MANUFACTURING AND SUPPLY CHAIN

Three chapters are included in this theme which covers the broader spectrum including supply chain, manufacturers and logistics services. Chapter 6 develops a multi-objective optimisation model for a three-echelon supply chain considering both economic and environmental aspects. The mathematical model's objective is to reduce the total supply chain cost and CO_2 emission of the supply chain network. The authors explained the model using an illustrative example. The next chapter's (Chapter 7) focus is on additive manufacturing where authors simulated the case of spare parts production for a railway subsystem to assess economic and environmental impacts. The additive manufacturing (AM) technology considered in this study is selective laser sintering (SLS) compared to conventional injection moulding where substantial reduction in cost and carbon foot print is possible. The final study in this theme is discussed in Chapter 8, where tracking of movement of goods in logistics services is explained using multiple case studies. The first case study explores the integration of GPS and GLONASS signals in AGV for localization and navigation of customer destination and materials in the indoor and outdoor environment. The second case study understands the obstacle in environment such as robot movement including dimension of obstacle such as size, length, width, circumference, height, and distance from a robot

THEME 3: INDUSTRY 4.0

Under this theme the book, include three distinctive chapters with the focus on real time dynamic scheduling, assessment of Industry 4.0 principles in coal industry and the outlook of research issues in Industry 4.0. Chapter 9 proposes smart make to order production in a flow shop environment for Industry 4.0 with the specific focus on scheduling procedures for flow shop production environment. In this procedure, customers place orders into the system dynamically where decision maker make a choice whether to accept the order considering the available production capacity. If accepted, then the researchers address how to schedule the jobs of an accepted order in short computation time. Influential factors that could affect coal sect in industry 4.0 is discussed in Chapter 10. The study uses Total Interpretive Structural Modeling (TISM) keeping in mind the potential ways to minimize the impacts of environment with the help of Industry 4.0 principles. The study's major concern is handling of environmental sustainability in era of Industry 4.0. Chapter 11, the final chapter in this theme and the book sheds light on causes and mitigation strategies during

3rd industrial revolution and the futuristic issues in managing supply and demand in Industry 4.0. Specially the study analyses the literature appeared in a popular production research journal. The major findings of the study are dependence on technology will simplify tracking of transit inventory, improving transparency of real-time sharing and continuous updating will enable to match supply and demand in the future.

Overall, the book's three themes will enlighten the research thought process and provokes additional studies to the readers. We believe this book will benefit students, researchers and practitioners. As all of us know Industry 4.0 is an emerging topic and we made a first attempt to link hyper customisation and smart manufacturing under the overarching Industry 4.0 terminology. We know it is not possible to cover all aspects, still believe there is a long way to go and to get additional studies related to IoT and supply chain, supply chain management with Web 2.0 and Enterprise 2.0 and Management 2.0.

Acknowledgment

The editors would like to acknowledge the help of all the people involved in this project and, more specifically, to the authors and reviewers that took part in the review process. Without their support, this book would not have become a reality.

First, the editors would like to thank each one of the authors for their contributions. Our sincere gratitude goes to the chapter's authors who contributed their time and expertise to this book.

Second, the editors wish to acknowledge the valuable contributions of the reviewers regarding the improvement of quality, coherence, and content presentation of chapters. Most of the authors also served as referees; we highly appreciate their double task.

Finally, we are grateful to the IGI Global staff for their unstinted support and patience during the production of this book. Special mention goes to Professor John Wang, Advances in Logistics, Operations, and Management Science (ALOMS) series editor and the executive board for approving our book proposal. We are also very much grateful to Jan Travers, Director of Intellectual Property and Contracts; Marianne Caesar, Assistant Development Editor-Books; and Josephine Dadeboe, Assistant Development Editor for their support in bringing out this book in time.

S. G. Ponnambalam
University Malaysia Pahang, Malaysia

Nachiappan Subramanian
University of Sussex, UK

Manoj Kumar Tiwari
Indian Institute of Technology Kharagpur, India

Wan Azhar Wan Yusoff
University Malaysia Pahang, Malaysia

Section 1
Digital Supply Chain

Chapter 1
New Era in the Supply Chain Management With Blockchain:
A Survey

Jesus Alvarado
Charles Sturt University, Australia

Malka N. Halgamuge
https://orcid.org/0000-0001-9994-3778
The University of Melbourne, Australia

ABSTRACT

The results show that transparency and auditability, security and indelibility, and distribution and sustainability are the key attributes of blockchain-based solutions in 56% of the articles reviewed. These three aspects represent the foundation of blockchain technologies which may contribute positively to improve supply management processes. Moreover, immutability, tracking and tracing, and smart contracts are also included in nearly a third of the cases. Moreover, efficiencies and costs through this technology would reduce the costs in payment of intermediaries, reduce paperwork, and help in the shipment of physical documents. Supply chain plays a critical role in the global trade and urgently needs to reassess its models in searching for greater efficiencies. Moreover, better results in visibility across the chain will increase trust for the customers and all interested parties. Secure transactions, strong security mechanisms that prevent fraud and illegal practices, could be achieved through the blockchain.

DOI: 10.4018/978-1-5225-9078-1.ch001

INTRODUCTION

The term blockchain first actualized by Nakamoto (2008) to exchange money between peers known as Bitcoin. It utilizes blocks to store the data and implement specific cryptography strategies to keep up the privacy of the data. In the last decades, the role played by supply chain management in the business world has become critical. World trade is evolving at an accelerated pace, and this requires more demands from the organizations for more rapid responses and efficient solutions according to the expectations of the market (Xu et al., 2017). Supply chain in addition to its elemental functions, such as ordering/receipt of goods or services, handling physical cargos (Nakasumi, 2017), acts at the heart of the business and is responsible for managing information flows and supporting core activities such as planning and forecasting, procurement, customer service and performance measurement (Xu et al., 2017). Furthermore, information is the central axis around which all supply chains work (Khadraoui, 2018). In the supply chain, information flow is the key element required between stakeholders to trigger new processes or visualize their current status. In addition, with each movement of products, a record of entry and exit times is made as well as the location of the goods, legal documentation and other procedures which are inherently necessary at each stage (Khadraoui, 2018). Therefore, interested parties can obtain information about their products and services throughout the process.

The supply chain is an essential component in the business structure for modern organizations to deliver customer value (Kshetri, 2017). Previously, companies produced their own goods, but globalization and the increasingly challenging demands of the market have caused drastic changes in the way they operate. As a consequence, today large manufacturing companies outsource a significant part of their processes and concentrate exclusively on the critical activities of their business (Nakasumi, 2017). Under these new scenarios, different companies are responsible for providing various kinds of goods or services within supply chain, which requires them to incorporate their own business strategies at both operational and strategic level to meet the demands that the market imposes and achieve their own benefits. Supply chain management has been an area of research and constant evolution over time; numerous applications have been developed, making it an exciting market for software companies. On the other hand, some research proposes incorporating new technologies such as GPS or RFID sensors into the logistics chain (Xu et al., 2017).

However, technological efforts and innovations have helped alleviate inefficiencies; business operations remain inefficient, expensive and vulnerable within the supply chain (IBM Institute for Business Values, 2017). Some issues affect the current supply chain systems. First, the significant number of third parties involved in the life cycle of a product means divided interests are higher (Xu, 2017; Chen et al.,

2017). Second, there is likely to be a technological gap between the diverse parties and their capacity to collect, process, analyze and share information following the real needs of the supply chain. This is reflected in inefficiencies, over costs, lack of transparency and confidence along the supply chain (Korpela & Hallikas, 2017; Pilkington, 2017). Third, interoperability of the current supply management system is another issue, as a majority of solutions may not offer end-to-end tracking and information sharing. When information is centralized, issues regarding the integrity of the information may rise (Khadraoui, 2018). In addition, lack of trust is evident since the information shared is not immutable. Data integration and non-repudiation properties cannot be assured. Companies which possess sensitive information may not be prepared to reveal and share it with others due to the fear of opportunistic behavior of its partners generating uncertainties (Xu, 2017; Khadraoui, 2018; Kshetri, 2017). Fourth, like any other IT system, supply chain management system is prone to new threats like cyber-attacks, which leads to security breaches of the integrity of the information. This may result in fraud, economic losses and breach of commercial agreements (Xu et al., 2017; IBM Institute for Business Value, 2017). Finally, when there are conflicts of interest between the parties involved, the quality of information is affected by the given inputs along the process. Consequently, the outcomes at the end are entirely different as initially expected. It seems that many actors of the supply chain are still tied to ineffective and obsolete procedures that do not allow real progress in this field (Chen et al., 2017).

Recently, a new technological innovation has emerged with great force in the business environment. With its origins in the financial industry, blockchain has been considered one of the top technological mega-trends with the capacity to revolutionize global trade (Kshetri, 2017; O'Leary, 2017). Blockchain offers the opportunity for transformation so that companies can reinvent their processes and integrate better with participants in the supply chain. This can achieve benefits such as traceability from the end-to-end in the supply chain, high transparency for stakeholders in the chain, strengthening trust relationships among all stakeholders, mitigating business risk, cost reduction, process optimization and profit increase (Khadraoui, 2018; Hackius, 2017). In simple terms, the blockchain represents a chain of blocks, and its main core is a distributed ledger technology (Law, 2017). Blockchain consists of a distributed database where each party or user can access the entire database; the input of new information represents a steady cycle, which is stored in its respective blocks (Edwards, 2017). A block refers to all related transactions which are transmitted to the network every 10 minutes (based on Bitcoin protocol) (Xu et al., 2017; Nakasumi, 2017). Moreover, the consensus mechanisms require prior agreement through the network to guarantee the security and adhesion of new blocks (Lin, 2017).

Blockchain incorporates a series of technical features which offer enormous possibilities to take the global supply chain into a new state. The key attributes underlying the blockchain technology will be described next. First, blockchain constitutes an autonomous and decentralized structure, where every single user has the same rights to access the whole database (Iansiti, 2017). Communication happens in a peer-to-peer way instead of the central entity which results in the reduction of third party involvement (Law, 2017; Liu et al., 2017; Lo, 2017). This will lead in providing greater agility, transparency, and cost-effectiveness along the process (Nakasumi, 2017; Iansiti, 2017). Second, operations are performed in real-time and each operation constitutes a new record which is disseminated to the entire network where each user can visualize new information added according to their security privileges (Nakasumi, 2017). Moreover, anonymity is secured throughout all process as each user has a unique alphanumeric identifier (Iansiti, 2017). Additionally, the users have their own private key, and access to the public key infrastructure (PKI), which is shared between all users and public notifications are sent about transactions in progress (Korpela & Hallikas, 2017). Third, all transactions are verified and authenticated through cryptographic methods which ensure that nobody can access information that is not relevant to them (Nakasumi, 2017; IBM Institute for Business Values, 2017). Furthermore, the transactions are verified before being recorded in the public ledger and all of them are linked with their respective IDs which remain visible for everyone in the system. There are no additional permissions required to audit any record within the network which promotes high transparency and easy auto tracking processes (Xu et al., 2017). Likewise, blockchain may provide a tamperproof history across the process in the supply chain (Pilkington, 2017).

Fourth, blockchains are highly secure due to their design which ensures data immutability and non-repudiation. This is because once the data has been accepted by the entire network, introducing any change represents a difficult task because every transaction is linked with its predecessor and requires consensus and approval by all users (Iansiti, 2017; Lo, 2017). In addition, blockchain-based models operate in distributed form, and each user keeps its own copy of the ledger which guarantees that even if an individual account is breached, multiple copies are stored and connected through the network (IBM Institute for Business Values, 2017). Importantly, blockchain provides a clear comparison to the concept of "security through obscurity", a prevalent mechanism in large industries, which stands for keeping security systems secret. This is because the latter approach is vulnerable as a simple breach point ends up compromising the entire network. For example, the Swift system, which operates in financial transactions, has been the target of repeated attacks in recent times (Nakasumi, 2017; Kshetri, 2017).

Finally, smart contracts represent an enormous potential for the supply chain when implemented through blockchain technologies. Furthermore, smart contracts ensure compliance with business rules as they are self-executing and self-enforcing mechanism embedded through a digital code in the distributed ledger (Liu et al., 2017; Francisco & Swanson, 2018). Many supply chain processes could be transformed or even eliminated using smart contracts. For example, the automation of payments, once the agreed conditions in the contract have been met, through the transfer of funds between the buyer's accounts to the provider (Iansiti, 2017). As a result, essential benefits for the companies in the supply chain could be reached such as decrease of intermediaries, reduce losses due to fraud, data tamper-proof, and automated solutions leading to decreased costs. (Korpela & Hallikas, 2017; Bocek, 2017)

This research aims to analyze new alternatives and possible solutions that contribute to improving supply chain management. Bringing solutions to the supply chain constitutes a significant challenge for business, governments, organizations and the academy. Studies from different industries show significant figures about how businesses lose millions or even trillions of dollars every year because of inefficient operations and a lack of visibility evidenced through the supply chain (Chen et al., 2017; Value, 2018). Despite being in the most revolutionary era of computer science and emerging technologies, the results in this field are still in a developing phase.

With the goal of exemplifying how blockchain technologies may impact in supply chain management, this chapter describes the key principles behind it as well as the critical issues within the supply chain that can be addressed through this technology. The rest of the chapter is organized as follows; the materials and methods are presented in Section 2. Then, results are presented in Section 3. Discussions and implications are presented in section 4. Section 5 provides the conclusions and comments.

MATERIALS AND METHODS

This chapter covers an extensive review of blockchain studies and applications in emerging areas with a focus in the supply chain management data extracted from the 30 peer-reviewed scientific publications (2017–2018). Different approaches utilized by both supporters and critics of blockchain technologies will be analyzed to identify the most significant benefits as well as the weaknesses of this new technology in its present application in supply chain management.

Table 1 shows a summary of the 30 studies that were analyzed the specific application within the blockchain technology, industries involved, the fundamental principles behind blockchain as well as the main issues in Supply Chain Management addressed through the chapter.

Table 1. Blockchain in supply Chain, areas of application, key factors and issues. Data extracted from 30 review studies (2017-2018)

Study	Year	Application Processes Blockchain	Application Areas Blockchain	Key Principles of Blockchain	Issues Addressed to the Blockchain	Impact	Proposed- Solution
1	2017	Quality & Strategic Management	Supply Chain Management	Smarts contracts	Counter-feit products	Economic Losses	Create a framework and system architecture for blockchain-based supply chain quality management.
				Tracking and Tracing	Quality and Compliance	Loss market share	
				Distributed and Sustainable	Centralized structures	Cost in Pollution controls	
				Immutable	Lack of trust / Intermediaries		
2	2017	Optimization Supply Chain	Manufacturing	Secure and indelible	Information security and privacy	Economic losses	Blockchain scheme for information sharing combining blockchain with homomorphic encryption solution
				Distributed and Sustainable	Lack of trust / Intermediaries		
				Transparent and auditable			
3	2017	Product Safety & Authenticity	Food	Transparent and auditable	Information security and privacy	Loss of human beings	Implement a solution based on Blockchain technology to solve the problem of Agricultural food traceability in China.
				Tracking and Tracing	Lack of transparency and traceability	Economic losses and distrust on the part of consumers	
				Secure and indelible			
				Distributed and Sustainable			
				Immutable	Lack of trust / Intermediaries		

continued on following page

Table 1. Continued

Study	Year	Application Processes Blockchain	Application Areas Blockchain	Key Principles of Blockchain	Issues Addressed to the Blockchain	Impact	Proposed- Solution
4	2017	Product Safety & Authenticity	Health & Pharma	Tracking and Tracing	Restrictive Regulations	Reduce number of intermediaries in supply chain process	Introduce a new technology based on IoT sensors with blockchain technology to guarantee data integrity and avoid record manipulation
				Immutable	Counter-feit products	Reduce manual interventions	
				Smarts contracts	Information security and privacy	Decrease operational expenses and manipulation risks	
				Consensus-based and Transactional	Transaction Costs		
				Transparent and auditable	Lack of trust / Intermediaries		
5	2017	Customer Engagement	Various Industries	Secure and indelible	Lack of transparency and traceability	Recover visibility along the supply chain	This study explores and presents different approaches regarding the adoption of blockchain technologies in a way to improve customer experiences along different industries. According to that study, it seems to be a consensus among all consumer industry executives about opportunities and barriers that this new technology could present.
				Tracking and Tracing	Information security and privacy	Eliminates data uncertainty making data available in real time.	
				Consensus-based and Transactional	Restrictive regulations	Detect bottlenecks and trigger immediate actions to respond to unexpected changes in a variable environment	
				Distributed and Sustainable	Lack of trust / Intermediaries		

continued on following page

Table 1. Continued

Study	Year	Application Processes Blockchain	Application Areas Blockchain	Key Principles of Blockchain	Issues Addressed to the Blockchain	Impact	Proposed- Solution
6	2017	Optimization Supply Chain	Various Industries	Distributed and Sustainable	Centralized structures	Creation of a strong system that promotes transparency through a decentralized platform for massive sharing.	this executive report examines the potential of Blockchain Technologies to eliminate the frictions that are a drag on organizations. Nine frictions were identified and the impact of blockchain might have in them.
				Secure and indelible	Information security and privacy	No intermediaries mean the more efficient process	
				Transparent and auditable	Lack of transparency and traceability	Creation of new services to be distributed through the blockchain networks	
				Consensus-based and Transactional	Lack of trust / Intermediaries		
					Transaction Costs		
					Restrictive Regulations		
				Flexible			

continued on following page

Table 1. Continued

Study	Year	Application Processes Blockchain	Application Areas Blockchain	Key Principles of Blockchain	Issues Addressed to the Blockchain	Impact	Proposed- Solution
7	2017	Optimization Supply Chain	Supply Chain Management	Consensus-based and Transactional	Lack of trust / Intermediaries	Achieve a B2B integration within digital supply chain	This article investigates how blockchain technology can support digital supply chain integration in a B2B case. Requirements and functionalities for a supply chain integration are investigated. One new approach is presented based on the blockchain response and support that this technology can offer to accelerate such integration.
				Smarts contracts	Transaction Costs	Minimizing use of third parties involved in the transactions lead to cost-effective services	
				Distributed and Sustainable	Information security and privacy	Enhance security, speed and cost transactions	
				Secure and indelible			
8	2017	Product Safety & Authenticity	Health & Pharma	Transparent and auditable	Information security and privacy	Improve patient outcomes through a new healthcare digital solution	Develop a next generation of healthcare digital solutions to improve patient solutions while decrease costs and address the main concerns in e-health like a security, reliability, efficiency and flexibility.
				Immutable	Centralized structures	Decrease cost	
				Secure and indelible	Transaction Costs	Improve security and reliability	
				Efficiency and cost effectiveness	Lack of trust / Intermediaries	Promotes integration of all actors in the healthcare industry	

continued on following page

Table 1. Continued

Study	Year	Application Processes Blockchain	Application Areas Blockchain	Key Principles of Blockchain	Issues Addressed to the Blockchain	Impact	Proposed- Solution
9	2017	Optimization Supply Chain	Manufacturing	Distributed and Sustainable	Lack of transparency and traceability	Transformation of traditional systems into more secure, reliable, transparent and collaborative system	A decentralized distributed system that uses Blockchain technologies to collect, store and manage key product information of each individual product throughout its life cycle is proposed
				Transparent and auditable	Lack of trust / Intermediaries		
				Immutable	Centralized structures		
				Secure and indelible			
10	2017	Optimization Supply Chain	Supply Chain Management	Smarts contracts	Lack of transparency and traceability	Increase Transparency, traceability and efficiency	Integration of Smart contracts in supply chain management with the objective of determine provenance of goods, tracking the flow's goods, execution payments upon fulfillment of criteria, evaluate a reputation of stakeholders
				Transparent and auditable	Lack of trust / Intermediaries	Tracking flow and provenance of goods	
				Immutable		Reduce time and cost inefficiencies	
				Efficiency and cost effectiveness	Information security and privacy		

continued on following page

Table 1. Continued

Study	Year	Application Processes Blockchain	Application Areas Blockchain	Key Principles of Blockchain	Issues Addressed to the Blockchain	Impact	Proposed- Solution
11	2017	Product Safety & Authenticity	Food	Distributed and Sustainable	Information security and privacy	Enhance food safety and rebuild consumer's confidence in the food industry	Build a food supply chain traceability system for real-time food tracing based on HACCP (Hazard Analysis and Critical Control Points), blockchain and Internet of things, which could provide an information platform for all the supply chain members with openness, transparency, neutrality, reliability and security.
				Transparent and auditable	Lack of trust / Intermediaries		
				Immutable	Centralized structures		
					Lack of transparency and traceability		
12	2017	Product Safety & Authenticity	Supply Chain Management	Secure and indelible	Information security and privacy	Reduce losses caused by the counter-feits products in the post supply chain	Develop a novel product ownership management system (POMS) of RFID-attached products for anti-counterfeits that can be used in the post supply chain.
				Transparent and auditable	Quality and compliance		
13	2017	Optimization Supply Chain	Supply Chain Management	Transparent and auditable	Lack of transparency and traceability	Increase transparency and accountability along SCM	Provide a framework that considers how Blockchain can helps firms meet key SCM objectives (cost, quality, speed, dependability, risk reduction, sustainability and flexibility)
				Immutable	Lack of trust / Intermediaries		
				Tracking and Tracing	Information security and privacy		
					Transaction Costs		
					Counter-feit products		
					Restrictive Regulations		

continued on following page

Table 1. Continued

Study	Year	Application Processes Blockchain	Application Areas Blockchain	Key Principles of Blockchain	Issues Addressed to the Blockchain	Impact	Proposed- Solution
14	2017	Transaction Process	Supply Chain Management	Distributed and Sustainable	Centralized structures	Although blockchain appears as an emerging technology, its impact is not yet clearly perceived since there are already several applications that comply with what blockchain tries to solve.	Provide some of the potential blockchain configurations as well as different approaches with advantages and disadvantages (private vs public, decentralized vs centralized, peer-to-peer vs cloud). The relationship between blockchain and other existing technologies also is examined
			Various Industries	Consensus-based and Transactional	Lack of transparency and traceability		
				Immutable	Transaction Costs		
				Transparent and auditable			
15	2017	Optimization Supply Chain	Supply Chain Management	Transparent and auditable	Lack of trust / Intermediaries	Improved benefits for the stakeholders. Privacy concerns can be managed through private ledgers while data that require a high-level trust can be stored in a public ledger.	Introduces a framework for online tracking information to all stakeholders during distribution phase of Supply chain. Two distributed ledgers support this framework: a public and private ledger
				Secure and indelible	Information security and privacy	Achieve a greater level of visibility during the physical distribution in SC	
				Tracking and Tracing	Transaction Costs		
				Distributed and Sustainable	Lack of transparency and traceability		

continued on following page

Table 1. Continued

Study	Year	Application Processes Blockchain	Application Areas Blockchain	Key Principles of Blockchain	Issues Addressed to the Blockchain	Impact	Proposed- Solution
16	2017	Product Safety & Authenticity	Supply Chain Management	Secure and indelible	Information security and privacy	Blockchain is hyped like a new trend technology which can support the highest security and privacy solution levels.	Presents an approach of how cybersecurity and protecting privacy can be improved and strengthened through blockchain-based solutions
				Distributed and Sustainable	Quality and compliance		
				Tracking and Tracing	Lack of Lack of trust / Intermediaries / Intermediaries		
17	2017	Transaction Process	Various Industries	Secure and indelible	Lack of transparency and traceability	Blockchain can support and improve the Building Information Modelling technology helping to overcome the organizational and legal issues arisen from that technology	It is proposed to implement a hybrid blockchain model in a BIM-based construction process
				Smarts contracts	Lack of trust / Intermediaries		
				Immutable	Data Ownership		
				Distributed and Sustainable			

continued on following page

Table 1. Continued

Study	Year	Application Processes Blockchain	Application Areas Blockchain	Key Principles of Blockchain	Issues Addressed to the Blockchain	Impact	Proposed- Solution
18	2017	Quality & Strategic Management	Various Industries	Distributed and Sustainable	Lack of trust / Intermediaries	Blockchain appears as new technology able to revolutionize the way of doing business. However, to reach that level requires a massive transformation along the whole ecosystem	Proposes a framework that maps innovation against two main dimensions: Novelty and complexity.
				Transparent and auditable	Lack of transparency and traceability		
				Immutable	Information security and privacy		
				Secure and indelible			
19	2017	Optimization Supply Chain	Supply Chain Management	Secure and indelible	Lack of transparency and traceability	Improves the Supply chain management in the consumer electronic (CE) industry through its two main features: Transparency & immutability	Analyses the winds of change in the consumer electronics industry (CE) and the revolutionary impact that Blockchain technology could have on SC management
				Transparent and auditable	Counter-feit products	Blockchain provides digital identity tool to ensure physical property and packing purposes.	
				Immutable	Information security and privacy		
				Distributed and Sustainable	Transaction Costs		

continued on following page

Table 1. Continued

Study	Year	Application Processes Blockchain	Application Areas Blockchain	Key Principles of Blockchain	Issues Addressed to the Blockchain	Impact	Proposed- Solution
20	2017	Optimization Supply Chain	Supply Chain Management	Transparent and auditable	Lack of transparency and traceability	The greater impact of Blockchain is to provide transparency in the sharing information between all parties in the supply chain, leading to improve efficiencies and relationships across the participants	This thesis establish how Blockchain technologies can revolutionize Supply chain and logistics. The most common challenges were analysed and how the main features of Blockchain can sort out those issues.
				Tracking and Tracing	Information security and privacy		
				Secure and indelible			
				Immutable			

continued on following page

Table 1. Continued

Study	Year	Application Processes Blockchain	Application Areas Blockchain	Key Principles of Blockchain	Issues Addressed to the Blockchain	Impact	Proposed- Solution
21	2017	Optimization Supply Chain	Supply Chain Management	Transparent and auditable	Transaction Costs	Despite Blockchain technologies still are in an early stage; it is widely recognised that it has an enormous potential to disrupt economic and will change the ways of doing business.	Offers a perspective of Blockchain technologies incorporated in Supply chain and logistics processes. The document provides real examples carried out by some industries as well as the potential benefits in the future of Supply chain.
				Efficiency and cost effectiveness	Quality and compliance	Specifically, in Supply chain & logistics, blockchain offer huge perspectives of ensure transparency through the end-to-end supply chain, which can lead to dramatic decreasing of time delays, cost and human error among other benefits.	
				Secure and indelible	Lack of transparency and traceability	Regulatory processes across-borders could be managed more efficiently	
					Innovation frictions		

continued on following page

Table 1. Continued

Study	Year	Application Processes Blockchain	Application Areas Blockchain	Key Principles of Blockchain	Issues Addressed to the Blockchain	Impact	Proposed- Solution
22	2017	Optimization Supply Chain	Supply Chain Management	Efficiency and cost effectiveness	Lack of trust / Intermediaries	Multiple benefits are expected for the Australia's supply chain sector for e.g.: digitization of SC lead to eliminate paper-transactions & decentralized financial processes.	Presents a view of the potential applications of Blockchain-based technologies in Australia
				Transparent and auditable	Transaction Costs	Blockchain can trigger automatic and immediate payments while the transference of goods is happening in the Supply chain	
				Distributed and Sustainable	Lack of transparency and traceability		

continued on following page

Table 1. Continued

Study	Year	Application Processes Blockchain	Application Areas Blockchain	Key Principles of Blockchain	Issues Addressed to the Blockchain	Impact	Proposed- Solution
23	2018	Optimization Supply Chain	Supply Chain Management	Transparent and auditable	Lack of transparency and traceability	One of the most important effects of Blockchain technologies lies in its ability to generate trust due to its distributed nature and data integrity as well.	Develops a conceptual model based on **the unify theory of acceptance and use of technology (UTAUT),** to explore the adoption of users of Blockchain technologies in Supply chain traceability
				Secure and indelible	Information security and privacy	Blockchain facilitates organizations to evaluate and mitigate supply chain risks by bringing means to carry out tracking and tracing goods source and processes.	
				Immutable	Lack of trust / Intermediaries		
				Tracking and Tracing			
24	2017	Product Safety & Authenticity	Various Industries	Secure and indelible	Information security and privacy	Despite expected benefits of Blockchain technologies, some security issues concern about its implementation for example when someone reach 51% of computing power in the network and fork problems	Discusses different applications of the blockchain, its main features and the security issues and challenges that need to be overcome

continued on following page

Table 1. Continued

Study	Year	Application Processes Blockchain	Application Areas Blockchain	Key Principles of Blockchain	Issues Addressed to the Blockchain	Impact	Proposed- Solution
25	2018	Optimization Supply Chain	Health & Pharma	Tracking and Tracing	Counter-feit products	By means of blockchain technologies greater impacts are expected in security, integrity, data provenance and functionality of supply chain in the healthcare sector.	Provides an overview regarding to opportunities and challenges of Blockchain technologies when implemented in the healthcare supply chain sector; mostly in pharmaceutical supply, medical devices & supplies, IoT and public health.
				Transparent and auditable	Information security and privacy	Comply with law regulations and procedures needed in the healthcare industry	
				Efficiency and cost effectiveness	Lack of transparency and traceability		
					Lack of trust / Intermediaries		
					Transaction Costs		

continued on following page

Table 1. Continued

Study	Year	Application Processes Blockchain	Application Areas Blockchain	Key Principles of Blockchain	Issues Addressed to the Blockchain	Impact	Proposed- Solution
26	2017	Optimization Supply Chain	Supply Chain Management	Distributed and Sustainable	Information security and privacy	The two-construction block method will reduce the work load and latency. Users can request enough block reservations to the system, which can be used in the future when they really need to introduce real data in the blockchain.	Introduces a hybrid model called CoC for supply chain management which consist of the following features: 1) Public ledger acting as a platform to transaction and share information. 2) Two step block generation method to fix problems with low latency and storage. 3) Identity management and data protection schemes to address security issues.
				Secure and indelible	Quality and compliance	This new model seeks to extend the benefits of SC management system, promoting a unified platform which is convenient to sharing information and ensuring security along the SC	
				Immutable			
27	2017	Optimization Supply Chain	Supply Chain Management	Transparent and auditable	Lack of trust / Intermediaries	Blockchain offers a great potential of improving business processes and enhancing business models in supply chain.	Through this research an international survey was carried out within logistics industry with the aim of investigate which blockchain applications could be considered more suitable in Supply chain & Logistics and how Blockchain should be analysed in logistics and SC, as a treat or like a trick?
				Tracking and Tracing	Lack of transparency and traceability		
					Transaction Costs		
					Counter-feit products		

continued on following page

Table 1. Continued

Study	Year	Application Processes Blockchain	Application Areas Blockchain	Key Principles of Blockchain	Issues Addressed to the Blockchain	Impact	Proposed- Solution
28	2017	Quality & Strategic Management	Various Industries	Secure and indelible	Lack of trust / Intermediaries	Decentralized all transaction across multiple organizations-users	Proposes a framework based on several industrial trails which can be used to evaluate the suitability of implementing blockchain-based solutions.
				Distributed and Sustainable	Centralized structures	Increase transparency, allowing everyone in the SC be aware about next steps need to be executed	
				Transparent and auditable		Tracing back from the origin of the items allow to identify all key information which can be useful to solve any claim-issues.	
				Immutable			
				Tracking and Tracing			

continued on following page

Table 1. Continued

Study	Year	Application Processes Blockchain	Application Areas Blockchain	Key Principles of Blockchain	Issues Addressed to the Blockchain	Impact	Proposed- Solution
29	2018	Transaction Process	Supply Chain Management	Transparent and auditable	Centralized structures	Real traceability and resilience of goods during the transportation process. Furthermore, implementations of "smart contracts" technologies into the business models using Blockchain can improve the process towards new automatized models in supply chain.	Provides a conceptual solution based on Blockchain technology to improve the information sharing and traceability process in the transportation of dangerous goods.
				Immutable	Lack of trust / Intermediaries		
				Secure and indelible	Lack of transparency and traceability		
				Distributed and Sustainable	Information security and privacy		
				Smarts contracts			

continued on following page

Table 1. Continued

Study	Year	Application Processes Blockchain	Application Areas Blockchain	Key Principles of Blockchain	Issues Addressed to the Blockchain	Impact	Proposed- Solution
30	2017	Quality & Strategic Management	Various Industries	Distributed and Sustainable	Quality and compliance	Establish real conditions under which blockchain technologies can fit and impact positively in the overall system performance.	The solution proposed consist on develop a taxonomy that captures major and relevant features of diverse Blockchain and its support for different quality attributes. The proposed taxonomy comprises a variety of sources ranging from academics, government agencies, industrial applications among others. The taxonomy is structured as follows: Level of decentralization, storage & computation, blockchain configuration and other design choices such as anonymity and incentives
				Immutable	Information security and privacy		
				Transparent and auditable	Centralized structures		
				Consensus-based and Transactional			

RESULTS

Figure 1, based on Table 1, represents the key attributes of blockchain technologies when applied to supply chain processes. As the graph shows, transparency and auditability, security and indelibility, distribution and sustainability are the key attributes of blockchain-based solutions in 56% of the articles reviewed. These three aspects represent the foundation of blockchain technologies which may contribute positively to improved supply management processes. Moreover, immutability, tracking and tracing, and smarts contracts are also included in 33% of reviewed studies. These attributes of blockchain offer significant alternatives for the supply chain. For example, blockchain provides a tamper-proof repository of studies, once the data is accepted by consensus in the network; blockchain ensures its complete integrity and non-repudiation. In addition, the information remains unchanged, and anyone can verify the history of any transaction. Thus, high transparency and trust end to end in the supply chain may be achieved. Tracking and tracing of goods allow all interested parties to reach real visibility along the chain. Finally, smart contracts could lead to more efficient results, decrease human error as well as enforcing the compliance of agreements and commercial clauses between the parties involved.

The consensus-based and transactional, efficiency and cost-effectiveness, and flexible accounts were observed as 11% of the reviewed studies. These attributes of blockchain offer essential alternatives for the supply chain. For instance,

Figure 1. Key principles of blockchain technology

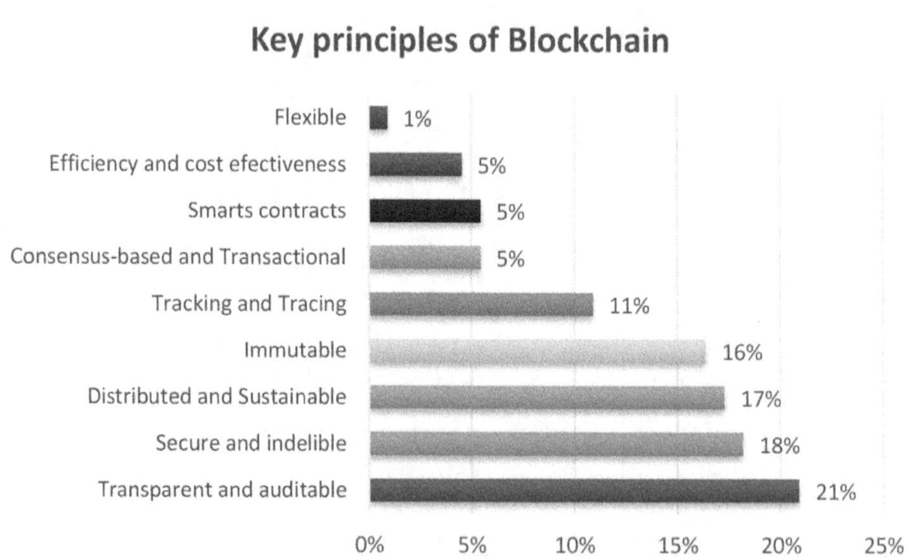

significant cost reduction, elimination of intermediaries, streamlining procedures and documentation related to shipments could be handled efficiently.

Figure 2, based on Table 1, presents the main issues in the supply chain which can be addressed through blockchain technologies. As the graph shows, lack of trust, intermediation and information security and privacy represent the issues that cause the most significant impact on the supply chain, appearing in 48% of the reviewed studies. Blockchain may be used as a system to ensure trust and transparency in the supply chain. The immutability of the information stored in blockchain is one of the most essential characteristics since it increases the trust between the parties because the information will remain unaltered through time. This resulted in eliminating the need to carry out information audit processes or mandatory supervision by a third party. Furthermore, blockchain guarantees that business rules or market agreements are respected without the need for compulsory audits. Besides, any evidence necessary to solve disputes between the parties can be extracted from the system. Blockchain provides a chronological and reliable report of events that have occurred throughout of process. In terms of information security, blockchain works as a distributed database which indicates that security falls over all participants who are responsible for the integrity and custody of the data. The information that does not fall on a single element or party, difference happens with centralized systems which are responsible for the information security, becoming a simple point of failure, which may lead to the collapse of a whole network if the system is breached. Moreover,

Figure 2. Main issues in supply chain management addressed by blockchain technologies

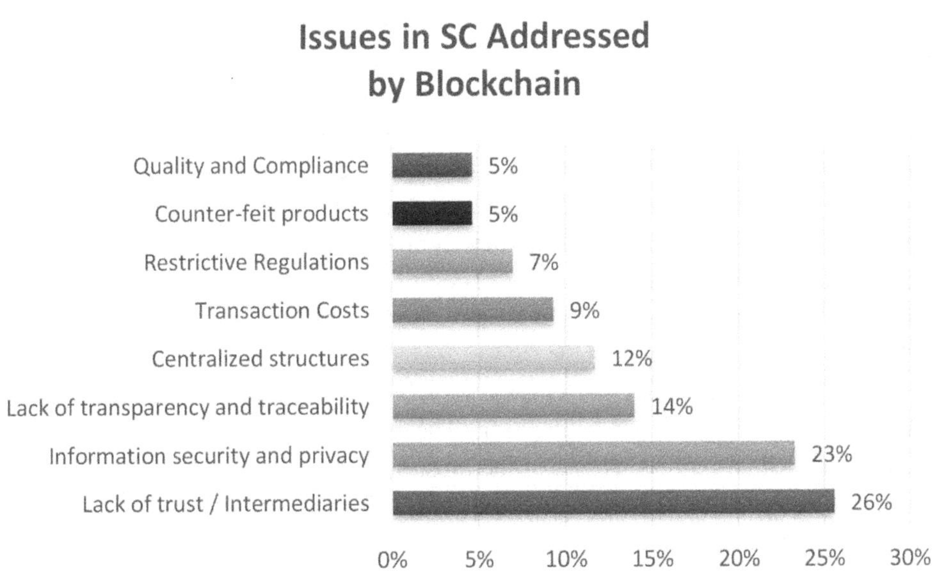

in the blockchain, the data is fully encrypted; cryptography mechanisms involve mathematic models which are highly secure. In addition, public and private keys are used to make sure that data flows only between parties with granted privileges.

The next set of issues identified such as lack of transparency and traceability, centralized structures and transaction costs are also included in 35% of reviewed studies. The growing demands by final consumers as well as the increasingly stringent requirements by regulatory bodies, as happens in the pharmaceutical and food industries demands that organizations know precisely the origin of the products as well as the fulfillment of specific measures such as temperature and humidity. In this sense, blockchain allows digital tracking in each phase of the value chain, enabling real traceability from end to end. As a result, the parties involved can access information in real time and make decisions based on accurate information, resulting in greater transparency and trust among all parties. Moreover, centralized structures are traditionally linked to the supply chain. Organizations with the higher power in the industry exercise greater control over the flow of data, limiting their access because they fear to share information due to factors such as unfair competition. Likewise, particular interests and ambitions are a drag on efficiency in the chain. In blockchain-based technologies, a key component lies precisely in its model distributed and sustained among all the participants in the network. The information is shared openly. It is not under the domain of any platform or entity. In terms of costs, greater complexity across the industries implies higher transaction costs, a large amount of intermediation existing in the supply chain cause a negative impact on this matter. With blockchain, the need for specialized third parties to guarantee processes can be reduced or eliminated. Besides, it offers the possibility of automating study-based processes and ensuring compliance with business rules through intelligent contracts. Moreover, blockchain may help to decrease process time and paperwork, contributing to the whole supply chain to yield better results.

The restrictive regulations, counterfeit products, and quality and compliance accounts for 17% of reviewed studies. Blockchain can act as a facilitator in terms of quality and compliance with specific requirements in regulatory matters, allowing for proper traceability and provenance from raw materials to final product. Furthermore, it is facilitating the detection of contaminated or falsified materials, which can reduce the millions of losses every year to the economy. However, norms and regulations may involve multiple jurisdictions beyond the borders leading to greater complexity and restrictions in the supply chain.

Figure 3, based on Table 1, presents the main fields where blockchain technologies may have a more significant application. The graph shows that the optimization of the supply chain represents one of the most prominent areas where blockchain-based solutions can be applied, appearing in 50% of the studied cases. In general terms, the decentralized blockchain model eliminates the need for third parties to

Figure 3. Application processes blockchain technologies

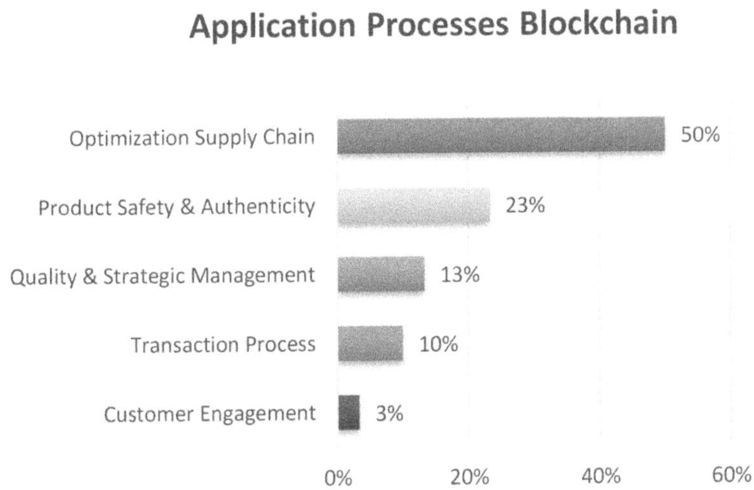

guarantee trusted operations. In addition, the system ensures transparency and end-to-end traceability, fundamental characteristics for the evolution of the management of the supply chain. The following item indicates that in 23% of the cases studied, blockchain technologies can be applied in processes that require the preservation of security and authenticity of goods and services in the supply chain. This emerging technology could bring significant improvements in processes where safety factors are highly valued in the detection and mitigation of fraudulent practices such as illegal mining, contraband, piracy, forgery or manipulation of documents.

Quality and strategic management is another area of greater application which accounts for 13% of studied cases. Blockchain technologies can offer vital support for organizations in processes of planning and control of the movement of materials and finished goods. It also facilitates the integration of data, monitoring of critical indicators and appropriate management of decentralized risks. Besides, transaction processes account for 10% of investigated cases. Blockchain technology has features that allow for quick and efficient execution of transactions such as smart contracts to ensure compliance with business rules. Furthermore, human errors can be mitigated by the automation of processes through blockchain. Finally, customer engagement accounts for 3% of the studied cases. For example, identity management systems based on blockchain could allow consumers to determine with whom they share their data and negotiate a payment for it.

The areas of application and benefits that blockchain technologies promise are innumerable and apart from bitcoin. Many other areas appear quite appropriate for blockchain implementation. The graph shows that blockchain applications in supply chain management appear in first place in 52% of the cases studied. Blockchain

Figure 4. Application areas blockchain technologies

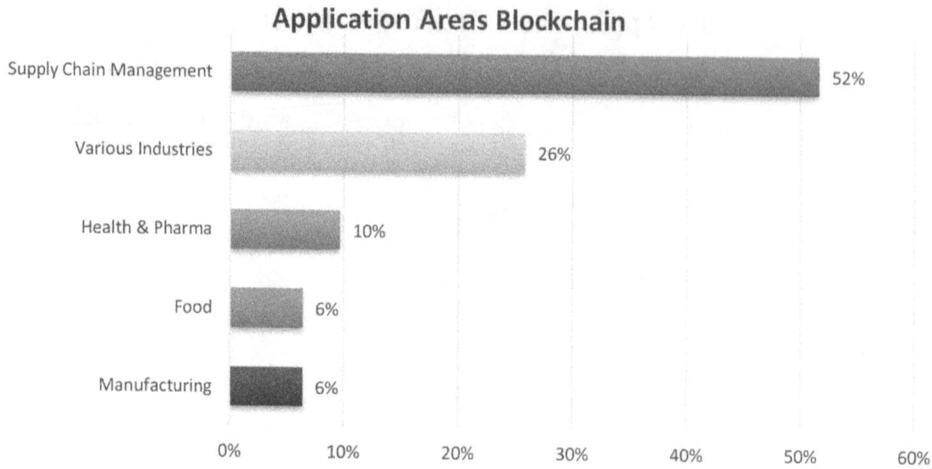

arises as a technology that can revolutionize this industry for solving the problem of distrust that causes greater inefficiencies in the supply chain. Moreover, with 26% of case studies pointed out different industries are suitable to implement blockchain-based solutions such as the consumer industry, aviation, medicine, and governmental bodies.

Health and pharma industries appear in 10% of the studied cases. The main focus of these applications is the detection of counterfeit raw materials and medicines that affect this industry as well as compliance with regulations and standards. In the medical area, the unification of medical records throughout the entire value chain implies great benefits in favor of the health and well-being of patients.

Food industry and manufacturing appear in 12% of the studied cases. In the food industry, blockchain technologies promise to solve the problem of traceability from end to end in the chain by monitoring at each phase the raw materials or products in the process until their final delivery as finished products. The manufacturing industry depends on a large extent on raw materials and products manufactured around the world which makes its operations critical and complex. In addition, serious problems such as counterfeit products, work with illegal labour and environmental damage among other aspects require this industry to exhibit best practices. The blockchain is a technology that can generate transparency and greater visibility helping organizations to discover new insights about their operations.

DISCUSSION

In this chapter, we address the main issues that have affected the supply chain management. The globalization of markets has caused drastic changes in the way organizations trade. Given this scenario, the role of the supply chain has become critical due to the complexity involved in its operations. The reviewed studies give us a clear sample of the current landscape of supply chain, issues which inhibit their development and limit their efficiency and issues that spread through a large part of the industries. These issues range from the lack of trust between stakeholders and the high dependence on intermediaries to ensure quality in the processes. Moreover, the security and privacy of the information generated great concern in the supply chain. The information hosted or controlled by centralized entities makes it more vulnerable to several risks such as improper manipulation, unfair competition, information deviation, and cyber-attacks. Essential properties such as immutability, data integrity and non-repudiation cannot be guaranteed under this scheme. Likewise, the lack of transparency causes adverse effects in the supply chain; the parties involved do not have conditions to track products through their life cycle. Therefore, they face high risks due to non-compliance in regulatory matters as well as the circulation of counterfeit goods that drain the economy every year. Hence, the blockchain-based solution technology offers promising features that would help to deal with these issues to a large extent. Thus, this chapter aims to merge the needs of the supply chain with the potential benefits that blockchain-based solutions could bring into the business.

The lack of trust between stakeholders and the high intermediation appears to be significant in our analysis. The lack of trust lies to a large extent in that the data through the chain does not preserve its integrity, being exposed to tampering or fraud. Moreover, many intermediaries participate in the life cycle of the goods and services, but there is no consensus about sharing quality information. In addition, in a greater intermediation environment, efficiency is significantly reduced, transaction costs increase and risks in terms of information security increase as well. The fundamental principles underlying blockchain technologies are fully capable of generating trust in the supply chain. In a distributed database such as Blockchain, every user who has access to the same information can verify transactions in real time without involving third parties. This may result in considerable improvement in the supply chain in terms of speed, efficiency, and cost and promote trusted relationships between the parties. Recently, some pilot projects have been launched to explore the benefits of blockchain. For instance, given that world trade circulates largely through maritime transport, it was verified that for the shipment of a single container with food from East Africa to Europe about 30 people participated in the process at different stages and close to 200 interactions were carried out between these people (Portofantwerp.

com, 2018). This excessive amount of people and interactions involved increase costs and generate delays that negatively impact supply chain. IBM and Maersk set up an alliance with the purpose of implementing a fully digitized process based on Blockchain technologies whose primary goal was to eliminate the massive amount of paperwork required for a single container. It is known that loading a container is an activity usually takes a short time. However, the same container can be retained for several days because of a missing study, a signature or some pending process by a third party. In addition, the system is plagued by fraud. For example, the bill of lading is subject to manipulation and facilitating illegal practices such as counterfeit products (Groenfeldt, 2018).

However, some researchers claim that automating processes should not be the only option available when considering a blockchain-based solution [9]. It is argued that organizations are not considering the real efficiencies of blockchain-based applications. It would seem that the given approach is only in function of immutability and digitalization of paperwork without considering profound changes at the process level such as re-engineering for improvement processes. Moreover, it is argued that there are already other technologies in use that do just what is proposed as the core function of the blockchain. Considering that blockchain is still in an initial phase, it would be essential to research its real impacts in terms of cost reduction and delivery speed compared to existing technologies (O'Leary, 2008).

Issues regarding information security and privacy can cause enormous concerns for supply chain participants. The results present us with a notion of the atmosphere of distrust that is faced; the parties do not want to share information for fear of compromising their competitive advantages and exposing confidential information to risks such as cheating and unfair competition. Besides, centralized information, which is commonly disseminated in the supply chain, also increases security and privacy risks. Furthermore, when the information resides on a specific side of the chain, for example, a supplier, transporter or third party that audits the process, this may lead to unwanted practices such as tampering, deviation or negligence, attempting against key principles such as the immutability and integrity of the information. Moreover, increasing the vulnerability of risks such as cyber-attacks because of the custody of the information is under the control of a single party.

Indeed, the security conditions may vary according to the blockchain model being used, the requirements of its participants as well as the assets that they intend to trade (Kshetri, 2017). There are two main types of blockchain, *permissionless* and *permissioned*. While in a *permissionless* blockchain anyone can participate, for example, Bitcoin. On the other hand, *permissioned* blockchain access must be granted by any authority according to the user's profile. Despite its initial popularity, blockchain permission less has several limitations in terms of security, privacy, and speed among other aspects. Whereas, *permissioned* blockchain offers a higher level of security and confidence since the number of participants limits its network. (Bussmann, 2017)

Lack of transparency and traceability appear as another critical issue within the supply chain that can be solved through Blockchain. This technology can help address three main concerns that organizations need to know to guarantee the integrity of ownership of goods along the supply chain. They need to understand how transactions are affected in terms of changes in their content, the alteration of their normal flow or the unjustified influence in the order that events should follow. The blockchain is a system designed to facilitate transparency and allow for access and monitoring of goods or services throughout their life cycle. Therefore, these characteristics represent a considerable advantage for business. For instance, companies being able to detect a problem and reorganize or make decisions about events that are happening in real time are something that would radically change the operation in supplies. Besides, with greater transparency, companies can better decide on the management of their resources, optimize their transport systems, physical inventories and distribution plants.

Blockchain's current generation is not yet ready to serve as an efficient means where buyers and sellers carry out their transactions (Earls, 2016). Others argue that there is not yet full knowledge about technology which is understandable because it is still in a nascent phase. Similarly, others point out that it is not easy to adopt new technology such as blockchain in a scenario as complex and established as a supply chain. Likewise, several stages should be observed involving bodies at different levels to understand the complexity and risks inherent in the implementation of this technology in the industry (Earls, 2016). Transaction costs, restrictive regulations, counterfeit products, and quality and compliance are also issuing that affect the supply chain to some extent. As discussed before, high intermediation affects transactional costs, but with Blockchain, these costs can be significantly reduced. In addition, transactions can occur directly between participants in the network in a peer-to-peer way in a matter of seconds rather than many processes that currently take time and generate high costs. For example, this approach could be adopted for the preparation and sending of invoices, the release of payments between the buyer(s) accounts to its supplier(s) and shipment documentation. For instance, IBM and Maersk carried out a follow-up test to a container dispatched from the Mombasa port in Africa to the Netherlands, the cost of this shipment was estimated at $ 2000 and the paperwork inherent to the process around 15% of the cargo. The outcome was satisfactory whereas before this, in the traditional scenario, the cost of paperwork could be equivalent to the cost of the shipment (Groenfeldt, 2017). Shippers should see a remarkable decrease in the current expense of exchanging paper bills by utilizing the blockchain adaptation of the archive.

Compliance with standards and regulations in some industries such as the pharmaceutical or food industries requires excellent support from the supply chain. Determining the origin of products is a growing concern of control bodies and final consumers as well. The blockchain is designed to provide high transparency and means to track products that require special conditions helping to comply in large part with market regulations. However, different regulations in different regions comprise a wide variety of laws. Therefore, the integration of blockchain in traditional structures will suppose an apathetic effort and above all an alignment at large-scale through all industries that will lead to the identification of best practices and the creation of foundations for this Nascent technology (Casey & Wong, 2017).

Although Blockchain has features that suggest it is a technology capable of transforming the supply chain, various obstacles will have to be overcome in order to achieve the predicted benefits. First, the implementation of solutions based on blockchain supposes tremendous efforts in multiples fields. Supply chains are complex so introducing changes as radical as those promoted by blockchain will not be easy to implement in areas which have been subjected to large investments and have been previously established (Earls, 2016). Second, Percy Venegas said that legal expertise in this field is a must. Likewise, organizations must be aware of the financial risks and uncertainties when dealing with a complex variety of participants within the supply chain (Earls, 2016). Third, limitations in technological order due to the gap that exists across different regions should be overcome. Large organizations such as IBM currently lead blockchain technologies. Hence, its large-scale dissemination is still very far away (Kshetri, 2017). Fourth, Blockchain will be disseminated among public models as initially conceived in finance but will also be widely accepted by smaller groups of organizations who seek to protect their market niches and profits. Therefore, the interoperability that allows for the connection between public and private systems implies an arduous task in the creation of standards and agreements across the industries to permit transmission of significant worth and information, between various blockchain networks (Kersten et al., 2017).

The existing security solutions are not fully adapted in Blockchain; hence, exploring different techniques for security and protection solutions could be an exciting path to study and explore the future adaptability of Blockchain in many application areas (Pham et al., 2010; Pham et al., 2011; Pham et al., 2010; Saini et al., n.d.).

Considering blockchain is only an emerging technology, and there is still a long way to go in thinking about appropriate solutions in all types of industry especially supply chain management. It could be argued that there are still significant developments in industries such as health, aviation, pharmaceutical, and food. More case studies should be carried out to test the benefits and real scopes of this innovative technology.

CONCLUSION

This chapter presents a perspective on how blockchain-based solutions can promote broad spectrum improvements of the supply chain by considering various factors such as costs, greater efficiency in processes, as well as security and transparency in the operations. Undoubtedly the most significant contribution of this technology lies in its ability to ensure trust in transactions between unknown parties, leaving aside validation by third parties. From our observations from the peer-reviewed articles, we found that several issues drag inefficiency into the supply chain. A quarter of them indicates a lack of trust and intermediation as the primary cause. Security and privacy issues are also a big challenge for the supply chain. From the reviewed articles it was found that in 23% of cases security issues indicated sharing information between stakeholders inhibits transparency in operations. Fundamental principles support Blockchain's system. Once transactions are recorded, they become transparent and auditable for everyone (this principle appears as the most important with 22%). Moreover, distributed and sustainable principle account for 17% of reviewed articles. No single party governs data instead of that they are distributed between users along the whole network. In addition, immutability was found as a critical principle of blockchain with 16% among the reviewed articles. Data immutability ensures that integration and non-repudiation properties are fulfilled. Tracking and tracing with 11% are found among the fundamental principles of the blockchain. Every time a raw material or product changes hands, the transaction could be documented, creating a full record end-to-end in the supply chain. Among the non-financial industries, supply chain management appears with 52% as one of the areas with the greatest potential for the implementation of blockchain technologies. In addition, 50% of the reviewed articles indicate the optimization of the supply chain processes as an optimum scenario where blockchain can generate an authentic transformation. This work is a suitable element of judgment to introduce organizations into new scenarios that blockchain technologies bring with them. Understanding the fundamentals underlying this technology, knowing its scopes, as well as limitations, constitute a significant contribution in this matter.

ACKNOWLEDGMENT

J.A. and M.N.H. conceived the study idea and developed the analysis plan. J.A. analyzed the data and wrote the initial chapter. M.N.H. helped preparing the figures and tables, and in finalizing the manuscript. All authors read the manuscript.

REFERENCES

Bocek. (2017). Blockchains everywhere - a use-case of blockchains in the pharma supply-chain. *IFIP/IEEE Symposium on Integrated Network and Service Management (IM)*. 10.23919/INM.2017.7987376

Bussmann, O. (2017). *BankThink A public or private blockchain? New Ethereum project could mean both.* Available: https://www.americanbanker.com/opinion/a-public-or-private-blockchain-new-ethereum-project-could-mean-both

Casey, M. J., & Wong, P. (2017). *Global Supply Chains Are About to Get Better, Thanks to Blockchain*. Available: https://hbr.org/2017/03/global-supply-chains-are-about-to-get-better-thanks-to-blockchain.

Chen, Shi, Ren, Yan, & Shi. (2017). A Blockchain-based Supply Chain Quality Management Framework. *The Fourteenth IEEE International Conference on e-Business Engineering*.

Earls, A. R. (2016). *Blockchain not a panacea for supply chain traceability, transparency*. Available: https://searcherp.techtarget.com/feature/Blockchain-not-a-panacea-for-supply-chain-traceability-transparency

Edwards, N. (2017). Blockchain meets the supply chain [online]. *MHD Supply Chain Solutions*, *47*(4), 48–50.

Francisco, K., & Swanson, D. (2018). The Supply Chain Has No Clothes: Technology Adoption of Blockchain for Supply Chain Transparency. *Logistics*, *2*(2), 1–13.

Groenfeldt, T. (2017). *IBM And Maersk Apply Blockchain To Container Shipping*. Available: https://www.forbes.com/sites/tomgroenfeldt/2017/03/05/ibm-and-maersk-apply-blockchain-to-container-shipping/#47cdedb53f05

I. C. &. Lin. (2017). A Survey of Blockchain Security Issues and Challenges. *International Journal of Network Security*, *19* (5), 653-659.

IBM Institute for Business Value. (2017). *Fast forward Rethinking enterprises, ecosystems and economies with blockchains*. IBM.

Khadraoui, A. I. D. (2018). The Security and Traceability of Shared Information in the Process of Transportation of Dangerous Goods. *2018 9th IFIP International Conference on New Technologies, Mobility and Security (NTMS)*.

Korpela & Hallikas. (2017). Digital supply chain transformation toward blockchain integration. *50th Hawaii international conference on system sciences*.

Kshetri, N. (2017). 1 Blockchain's roles in meeting key supply chain management objectives. *International Journal of Information Management*, *39*, 80–89. doi:10.1016/j.ijinfomgt.2017.12.005

Kshetri, N. (2017). Blockchain's roles in strengthening cybersecurity and protecting privacy. *Telecommunications Policy*, *41*(10), 1027–1038. doi:10.1016/j.telpol.2017.09.003

Law, A. (2017). *Smart Contracts and their Application in Supply Chain Management*. Cambridge, MA: Massachusetts Institute of Technology.

Liu, W., Zhu, S., Mundie, T., & Krieger, U. (2017). Advanced Block-Chain Architecture for e-Health Systems. *19th International Conference on E-health Networking, Application & Services (HealthCom): The 2nd IEEE International Workshop on Emerging Technologies for Pervasive Healthcare and Applications (ETPHA 2017)*. 10.1109/HealthCom.2017.8210847

Lo. (2017). Evaluating Suitability of Applying Blockchain. *22nd International Conference on Engineering of Complex Computer Systems (ICECCS)*.

M. &. Iansiti. (2017). The truth about blockchain. *Harvard Business Review*, *95*(1), 118–127.

N. &. Hackius. (2017). Blockchain in logistics and supply chain: trick or treat? *Proceedings of the Hamburg International Conference of Logistics (HICL)*.

Nakasumi, M. (2017). Information Sharing for Supply Chain Management based on Block Chain Technology. *IEEE 19th Conference on Business Informatics (CBI)*. 10.1109/CBI.2017.56

O'Leary, D. E. (2008). Supporting decisions in real-time enterprises: Autonomic supply chain systems. *Information Systems and e-Business Management*, *6*(3), 239–255. doi:10.100710257-008-0086-0

O'Leary, D. E. (2017). Configuring blockchain architectures for transaction information in blockchain consortiums: The case of accounting and supply chain systems. *Intell Sys Acc Fin Mgmt*, *24*(4), 138–147. doi:10.1002/isaf.1417

Pham, Halgamuge, Syed, & Mendis. (2010). Optimizing Windows Security Features to Block Malware and Hack Tools on USB Storage Devices. *The 28th PIERS*, 350-355.

Pham, D. V., Syed, A., & Halgamuge, M. N. (2011, February). Universal Serial Bus Based Software Attacks and Protection Solutions. *Digital Investigation*, *7*(3), 172–184. doi:10.1016/j.diin.2011.02.001

Pilkington, J. H. L. M. (2017). How the Blockchain Revolution Will Reshape the Consumer Electronics Industry [Future Directions]. *IEEE Consumer Electronics Magazine*, *6*(3), 19–23. doi:10.1109/MCE.2017.2684916

Portofantwerp.com. (2017). *Antwerp start-up T-Mining develops Blockchain solution for safe, efficient container release.* Available: http://www.portofantwerp.com/en/news/antwerp-start-t-mining-develops-blockchain-solution-safe-efficient-container-release

G. K. Saini, M. N. Halgamuge, P. Sharma, and J. Purkis, "A Review on Cyberattacks: Security Threats and Solution Techniques for Different Applications", Secure Cyber-Physical Systems for Smart Cities, IGI Global.

Value, I. I. B. (2018). *Blockchain reinvents the consumer experience.* Somers, NY: IBM.

Xu, L., Chen, L., Gao, Z., Lu, Y., & Shi, W. (2017). CoC: Secure Supply Chain Management System Based on Public Ledger. *26th International Conference on Computer Communication and Networks (ICCCN).* 10.1109/ICCCN.2017.8038514

ADDITIONAL READING

Britchenko, I., Cherniavska, T., & Cherniavskyi, B. (2018). Blockchain technology into the. ResearchGate, 307-317.

Efanov, D., & Roschin, P. (2018). The All-Pervasiveness of the Blockchain Technology. *Procedia Computer Science*, *123*, 116–121. doi:10.1016/j.procs.2018.01.019

Ribitzky, R., Clair, J. S., Houlding, D., McFarlane, C., Ahier, B., Gould, M., . . . Clauson, K. (2018). Pragmatic, Interdisciplinary Perspectives on Blockchain and Distributed Ledger Technology: Paving the Future for Healthcare. Blockchain in Healthcare Today, 1-15.

Sill, A., & Sporny, M. &. (2017). Blockchain Standards for Compliance and Trust. IEEE cloud computing, 84-90.

Turk, Z. a. (2017). Potentials of Blockchain Technology for Construction Management. Creative Construction Conference 2017, CCC 2017, 19-22 June 2017 (pp. 638-645). Primosten, Croatia: Elsevier Ltd. 10.1016/j.proeng.2017.08.052

Xu, X., Weber, I., Staples, M., Zhu, L., Bosch, J., Bass, L., & Pautasso, C. &. (2017). A Taxonomy of Blockchain-Based Systems for Architecture Design. 2017 IEEE International Conference on Software Architecture (pp. 243-252). IEEE computer society. 10.1109/ICSA.2017.33

KEY TERMS AND DEFINITIONS

Blockchain: Is made up of a chain of blocks. Each block contains data of transaction records and the information related with each transaction.

Distributed: The ledger is shared and updated every time that a new transaction occurs and replicated among nodes in the network in real time.

Immutable: Records processed in blockchain are linked with their predecessors avoiding their modification or elimination. The manipulation of the information is unlikely since the entire chain should be altered which implies enormous computing power.

Smart Contracts: Digital agreements between transacting parties written in computer code and deployed to the blockchain.

Supply Chain Management (SCM): Is the broad range of activities required to plan, control and execute a product flow from its origin until its distribution to the final customer.

Traceability: Is the ability to trace any product at any time during its life cycle.

Transparency and Auditable: Property that permits to visualize key information about the origin of goods in the supply chain, increasing trust between stakeholders.

Chapter 2
Factors Influencing Blockchain Diffusion in the Supply Chain:
An Empirical Investigation

Samuel Fosso Wamba
https://orcid.org/0000-0002-1073-058X
Toulouse Business School, France

Maciel M. Queiroz
Paulista University, Brazil

ABSTRACT

The literature concerning supply chain management (SCM) and blockchain integration is scarce, and organizations and decision makers still have a limited knowledge about blockchain solution diffusion in SCM. Under these circumstances, studies aimed to improving the understanding of the main blockchain diffusion factors are fundamental. This chapter is a contribution to bridging such gaps, as it aims to unlock the driving factors of blockchain diffusion in the SCM environment. Based on the emerging literature on blockchain, supply chain management, expectancy theory, and diffusion of innovations, a model was developed and validated, considering data from India's supply chain management professionals. The data was analyzed using partial least square structural equation modeling (PLS-SEM). The results indicated that IT deployment capability, compatibility, and trading partner pressure are factors that affect blockchain diffusion significantly. Also, the results bring essential managerial and theoretical implications regarding the blockchain diffusion in the SCM.

DOI: 10.4018/978-1-5225-9078-1.ch002

INTRODUCTION

The Blockchain Technology is a highly disruptive cutting-edge technology that is already bringing tremendous changes in a wide array of business models (Fosso-Wamba & Queiroz, 2018). Blockchain typically consists of a distributed ledger, in which the transactions are organized in blocks and linked to each other in a chain. That is, it operates in a peer-to-peer network, where the transactions are validated and recorded by consensus (Y. Chen, 2018). As a formal definition, Blockchain refers to:

A fully distributed system for cryptographically capturing and storing a consistent, immutable, linear event log of transactions between networked actors (Risius & Spohrer, 2017, p.386).

Originally, Blockchain technology emerged as a technology to perform cryptocurrency transactions in the financial market (Nakamoto, 2008; Prybila, Schulte, Hochreiner, & Weber, 2017). In this landscape, Blockchain applications have reached a relative maturity in the financial market, however it is only recently that they have aroused the interest of other areas, such as Supply Chain Management (SCM) field (Fosso Wamba *et al.*, 2018; Kamble, Gunasekaran, & Arha, 2018).

Thus, the blockchain concept can be understood as being associated with transactions disintermediation, that is, without a central authority to validate and offer transactions credibility. This feature implies some impacts on SCM, involving aspects such as member relationships, collaboration, trust and change in the role-based operations model for cloud agility, among other consequences.

Moreover, Blockchain is a buzzword and has aroused the interest of several scholars and practitioners. In this outlook, it has been studied in many contexts (supply chain, IoT, product traceability, healthcare systems, cybersecurity, among others). Specifically in the area of SCM, Blockchain technology can improve the traceability performance (Lu & Xu, 2017) while generating closer and trustworthy relationships (Aste, Tasca, & Di Matteo, 2017) not only between organizations and their suppliers but also through the entire SCM.

In addition, the steady emergence of blockchain technology as the next game changer in the supply chain, has positioned them among the ten top technology trends that are predicted to deeply transform business operations across industries. However, very few empirical studies have been conducted to assess the real potential of these technologies in transforming supply chain processes. A recent review study on bitcoin, blockchain, and Fintech in the supply chain found that only 5% of identified articles used a survey method approach (Fosso Wamba et al., 2018). Despite the numerous potential benefits of Blockchain, its diffusion is still at the first stages, and all its potential remains unclear. The challenges about how managers can ensure that Blockchain adds value to their organizations, and how the diffusion's enablers look like in SCM, remains unanswered.

Therefore, this study is an initial attempt to fill this knowledge gap in the literature. It seeks to draw on the emerging literature on blockchain (Fosso Wamba *et al.*, 2018; Kamble *et al.*, 2018), the diffusion of innovations (Rogers, 2003), expectancy theory (Vroom, 1964) and SCM (Choi & Dooley, 2009; LeMay, Helms, Kimball, & McMahon, 2017) to investigate the driving factors for Blockchain diffusion. The research question that guides this study is: What are the enablers of Blockchain diffusion in the Supply Chain? In order to answer this question, the aforementioned literature was explored to develop a research model. Thus, a conceptual model was validated using PLS-SEM approach (Akter, Fosso Wamba, & Dewan, 2017) with data collected from India's Supply chain professionals.

Regarding the main contributions, this chapter will shed more light on the literature on blockchain-enabled SCM, with a model that can generate valuable insights to advance this research stream. In the managerial perspective, this study will bring robust insights to practitioners and all other stakeholders of any disruptive digital business models, especially in the SCM context. In addition, a brief description of the chapter structure is provided herein. In the next section, a background is presented, focusing on supply chain management, Blockchain, expectancy theory and diffusion of innovations. Then, the hypotheses are formulated and a research model pointed out, followed by the methodology, the results and an analysis of findings. A discussion of results is then provided, together with managerial and theoretical implications. Finally, a conclusion, with some identified research limitations, ends the study.

BACKGROUND

This section focuses on the blockchain emerging literature, the supply chain management, expectancy theory, and the diffusion of innovations to support the development of the study's hypotheses.

Supply Chain Management: From Traditional Relationship to Digital Age Disruptions

Traditionally, the supply chain management (SCM) field is a complex network (Carter, Rogers, & Choi, 2015; Choi, Dooley, & Rungtusanatham, 2001) involving manufacturers, suppliers, retailers, wholesalers, transportations, customers, among others. The literature on the definition of supply chain management is prolific, as the concept is being defined in various ways. However, this study uses the following definition by Stock and Boyer (2009), which refers to SCM as:

The management of a network of relationships within a firm and between interdependent organizations and business units consisting of material suppliers, purchasing, production facilities, logistics, marketing, and related systems that facilitate the forward and reverse flow of materials, services, finances and information from the original producer to final customer with the benefits of adding value, maximizing profitability through efficiencies, and achieving customer satisfaction (Stock and Boyer 2009, p.706).

Disruptions brought about by the digital age have rendered the relationship between members increasingly complex. For instance, customers today have more information about the services and products, and can make a choice through a complex process with various characteristics. In this regard, the emerging blockchain technology can bring significant improvements to the supply chain management field. Blockchain's key features include its ability to enhance the transparency and trust (Kshetri, 2018) between supply chain members, thereby contributing to adding value at all levels in the network.

Unlocking the Potential of Blockchain as a Disruptive Cutting-Edge Technology

As presented previously, the roots of blockchain technology are in the financial market, more specifically in the cryptocurrency (Nakamoto, 2008; Oh & Shong, 2017; Prybila et al., 2017). However, other industries have been striving to adopt this technology and take advantage of its outstanding properties. The supply chain management field is one of them, as considerable efforts are being put in to add value to various business models owing to blockchain applications. For instance, blockchain helped to better trace the provenance of inputs (Biswas, Muthukkumarasamy, & Tan, 2017) and products (Kim & Laskowski, 2018), to ensure supply chains security and resilience (Min, 2018), and to enhance the exchange of knowledge in manufacturing networks (Li et al., 2018), among others. Unfortunately, the extant literature on blockchain that has relied on an empirical approach is scarce (Fosso Wamba *et al.*, 2018; Kamble *et al.*, 2018).

Blockchain applications in the supply chain management are supposed to bring in features such as smart contracts, digital identity, and cloud storage, etc., thus creating unprecedented disruptions that will transform relationships through the network. Such properties, coupled with the disintermediation and decentralisation of transactions, have been already changing the relationship between members across supply chains. Aspects that should be positively affected include security in product transportation at all levels (Kshetri, 2018), the trust and fraud prevention (R. Y. Chen, 2018), the traceability process to mitigate the adulteration (Biswas et al., 2017), improved clinical trials in healthcare settings (Benchoufi, Porcher, & Ravaud, 2017), among others.

In the supply chain context, it is clear that blockchain can generate benefits to all members, one of them being the optimization of products and services and a better quality of these. Moreover, blockchain can help the members of the supply chain to improve their ways and means of exchanging data, and to provide more transparency and reliability to their relationships with processes. Again, such benefits shall be reaped only if blockchain diffusion in the supply chain management is effective. Furthermore, the pace of diffusion process (slow, fast, very fast) will possibly impact the supply chain, so much so that SCMs with low diffusion process will be disadvantaged as compared to others.

Expectancy Theory: The Effect on Workers Productivity and Technology Diffusion

The expectancy theory (Vroom, 1964) aims to explain the work and motivation relationship in the behavioral choice process. It is the motivational variable that acts as an influencer of an employee's behavioral choice. Regarding this theory, the behavioral choice takes into account the valence and expectancy. The valence refers to the efforts that an employee puts in with a view to achieving the anticipated outcomes. This is for example the case for the overall attractiveness (Chou & Pearson, 2012). Expectancy is associated with a likelihood estimation that an employee has to apply in order to achieve the expected outcomes, an outcome refering to anything that an employee hopes to achieve (Mitchell & Beach, 1977). As for the expectancy theory core, it is related to the explanation about the behavioral choice, faced by workers' alternatives, in which the person will elect the alternative conducted by the most positive, or the less negative motivational strength. In other words, there is a maximization criterion that governs the choices (Tubbs, Boehne, & Dahl, 1993).

Additionally, the extant literature highlights the expectancy theory in job satisfaction (Graen, 1969), job performance (Hackman & Porter, 1968), and in goal-setting context (Tubbs et al., 1993). Thus, it is clear that the expectancy theory significantly affects the organizations' performance (Vroom 1964; Mitchell and Beach 1977). This theory has been used in several supply chain contexts, including the supplier development activities adoption (L. Chen, Ellis, & Suresh, 2016), the information technology job satisfaction, the customer experience satisfaction (Fawcett, Fawcett, Cooper, & Daynes, 2014), and the contracted provider motivation (Lambright, 2010). Therefore, as the supply chain field is faced with increasing complexities, mainly by the digital age disruptions, expectancy theory is a suitable approach to provide a better understanding of the motivation factors that affect the blockchain diffusion in supply chains.

Moreover, the organization's motivation for some benefits (e.g., technology partners) implies the need to develop closer relationships; however, if an expectation for a superior benefit arises, the relationship can be replaced by another with a perceived superior benefit (Golicic & Mentzer, 2005). Also, both, internal and external culture can bring significant impact on employee expectations (McAfee, Glassman, & Honeycutt, 2002). Considering a cutting-edge technology like blockchain, the literature on which is still in their initial stages, the expectancy theory is well fit to help understand the organization's diffusion behavior, as the intra-organizational context and even the entire supply chain network can take advantage of the many benefits of blockchain. One of these is the blockchain ability to improve on supply chain integration and mitigate the impact of complexity (Turkulainen, Roh, Whipple, & Swink, 2017), while triggering a high level of trust and commitment. In doing so, the job performance of workers is significantly improved in SCM (Kamble et al., 2018; Queiroz & Fosso Wamba, 2019), together with their expectations about what blockchain will bring to their activities in terms of improvements and efficiency.

Diffusion of Innovations: What About Blockchain Diffusion in the Supply Chain Management?

The diffusion of innovations is related to how new models (ideas/technologies) propagate through the society (Rogers, 2003). As a formal rendering, this concept is defined as the *"process by which an innovation is communicated through certain channels over time among the members of a social system"* (Rogers, 1983, p.5). The diffusion can be accelerated in a supply chain if a central user disseminates technology performance in order to perceive its advantage (Hung, Tsai, & Chuang, 2014). That is, the performance of a new technology should be reported at this level not only by the adopter and user located in the edge of the network, but also by a strategic user located at the center of the network and with a maximum of connections to spread the innovation.

The literature on the diffusion of innovation describes an innovation according to five characteristics: relative advantage, compatibility, complexity, trialability, and observability (Rogers, 1983). It means that these characteristics influence the innovation adoption rate, and consequently, the diffusion process through the supply chain management. Therefore, considering the embryonic development of Blockchain and of the related literature, especially in the supply chain management context, this study identified and integrated three fundamentals constructs (trading partner pressure, It deployment capability, and compatibility).

CONCEPTUAL MODEL AND HYPOTHESES

IT Deployment Capability

IT deployment capability refers to "*the firm's ability to mobilize and deploy IT-based resources (i.e., computers, networks, databases, and technological platforms*" (Lin, 2017, p.702). In this context, IT deployment capability can be a critical resource for integrating blockchain with SCM. Consequently, it has a direct association with and influence on blockchain diffusion. Therefore, this study proposes the following hypothesis:

Hypothesis One: *IT deployment capability has a significant positive effect on blockchain diffusion.*

Compatibility

The system's compatibility with organizations' current systems is critical from one side to another, but also for adoption and diffusion. Compatibility refers to the degree to which the systems are consistent with organizations' work practices or preferences (Hung et al., 2014). In this vein, blockchain compatibility impacts the current systems, not only within an individual organization but also in the entire SCM setting. This leads to hypothesize that:

Hypothesis Two: *Compatibility has a significant positive effect on blockchain diffusion.*

Trading Partner Pressure

A trading partner relationship refers to a business relationship that can involve two or more organizations, or organizations and customers. The trading partner integration is essential to supporting the development of collaborative solutions. (Y. M. Wang, Wang, & Yang, 2010). However, there is no doubt that the trading partner exerts influence not only on technology adoption but also on the diffusion of such technology. Hence, this study proposes the following hypothesis:

Hypothesis Three: *Pressures from trading partners have a significant positive effect on blockchain diffusion.*

Figure 1. Research model

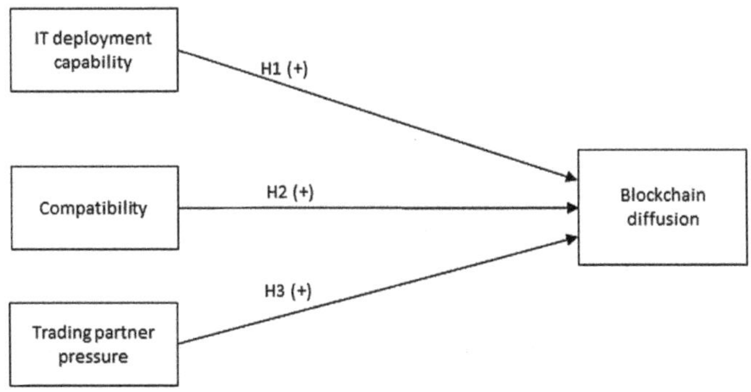

METHODOLOGY

This study used a survey method approach to collect data (Richey, Daugherty, & Roath, 2007; Schoenherr, Griffith, & Chandra, 2014), coupled with items from the extant literature that were adapted to fit our research context. A 7-point Likert scale (ranging from "strongly disagree" to "strongly agree") was then used to measure all constructs. The survey was administered by a leading market research firm, who collected data from India-based supply chain professionals operating in different industries. Specifically, the sample obtained 344 valid responses.

Nonresponse and Common Method Bias

This study performed a nonresponse bias to find out if there is any difference between respondents (early and late respondents), that is, a comparison between respondents that returned the questionnaire in the last week and respondents that returned in the first weeks (Armstrong & Overton, 1977). A t-test was employed, and no statistical differences were found between the respondents. In addition, when a survey uses a single respondent, a common method bias can inflate the relationship between independent and dependent variables (Podsakoff & Organ, 1986). To ensure that common method bias does not affect the model, a Harman's one-factor test was conducted (Y. Y. Wang, Wang, & Lin, 2018; Zelbst, Green, Sower, & Reyes, 2012). Thus, common method bias was not a problem to the model.

Results and Analysis

The analysis of the model was performed by means of partial least-squares structural equation modeling (PLS-SEM) approach (Hair Jr., Hult, Ringle, & Sarstedt, 2017;

Ringle, Christian M., Wende, Sven, & Becker, 2015). PLS-SEM is a robust approach to analyze complex and simple models (Hair Jr. et al., 2017), and it has been increasingly used in supply chain management and business logistics literature (Golicic et al., 2012; Grawe et al., 2015) as a reliable tool, notably because it supports small sample sizes, nonnormal data, and reflective and formative constructs (Hair Jr. et al., 2017; Hair Jr. et al., 2014). Moreover, PLS-SEM displays high levels of statistical power to perform large numbers of variables, and complex models (Akter et al., 2017). We, therefore, used SmartPLS 3.0 (Ringle, Christian M., Wende, Sven, & Becker, 2015) to test all the hypotheses proposed and the relationship between the constructs.

Table 1 reports the main descriptive statistics for the sample. The male respondents formed the majority, achieving 70.1 percent. In addition, the age of the participants was concentrated in the range 26-41. More specifically, the participants' ages were in the range 26-33, accounting for 51.2 percent, followed by the 34-41 bracket, with 28.8 percent. With regard to the highest education level, postgraduate and bachelor's degrees were the most popular, with 52.6 and 34.0 percent respectively. The rate of respondents' experience in their current organization was interesting. For example, almost 80 percent of respondents had already spent between 2 and 10 years in their recruiting organization. Those in the interval 2-5 years accounted for 40.1 percent of the respondents, followed by those of the bracket 6-10 years of experience, with 38.1 percent. However, the interval of 11-15 years was also expressive, achieving 14.2 percent.

Still in the area of experience, the respondents' experience in blockchain-related projects was interesting. For instance, more than 80 percent had a maximum of five years of general experience, with only 27.3 percent having less than one year of experience in blockchain-related projects, and 56.4 percent having between 2 and 5 years of experience in the same technology. As for the recruiting industries of respondents, they included a wide array of traditional segments such as manufacturing with 35.2 percent, information, and communication (14.5 percent), Administrative and support service activities (7.3 percent), Transportation and storage (4.4 percent), among others.

Measurement Model

The model reliability was assessed by means of loading factors, constructs reliability integrating Cronbach's alpha, composite reliability, and average variance extracted. Table 2 presents the loading factors. All values obtained were higher than 0.7, and were therefore consistent with other results found out in the literature (Hair Jr. et al., 2017). In Table 3, the constructs reliability measures are highlighted, and all values outperformed the recommended literature threshold. Thus, the model was supported and validated. All Cronbach's alpha values and composite reliability were higher

Table 1. Descriptive statistics

	n	Percentage
Gender		
Male	241	70.1
Female	103	29.9
Age		
18-25 years old	34	9.9
26-33	176	51.2
34-41	99	28.8
42-49	22	6.4
50+	13	3.8
Educational Level		
No formal education	0	0.0
Primary	0	0.0
Secondary	5	1.5
Diploma/polytechnic	41	11.9
Bachelor degree	117	34.0
Postgraduate degree (Master/Ph.D.)	181	52.6
Number of Years Working in the Organization		
Less than one year	13	3.8
2-5 years	138	40.1
6-10 years	131	38.1
11-15 years	49	14.2
16-20 years	8	2.3
Over 20 years	5	1.5
Number of Years of Experience in Blockchain-Related Projects		
Less than one year	94	27.3
2-5 years	194	56.4
6-10 years	38	11.0
11-15 years	15	4.4
16-20 years	1	0.3
Over 20 years	2	0.6

continued on following page

Table 1. Continued

	n	Percentage
Industry		
Accommodation and food service activities	5	1.5
Administrative and support service activities	25	7.3
Agriculture, forestry and fishing	4	1.2
Arts, entertainment and recreation	5	1.5
Construction	15	4.4
Education	25	7.3
Electricity, gas, steam and air conditioning supply	12	3.5
Financial and insurance activities	22	6.4
Human health and social work activities	11	3.2
Information and communication	50	14.5
Manufacturing	121	35.2
Mining and quarrying	4	1.2
Professional, scientific and technical activities	18	5.2
Public administration and defence; compulsory social security	0	0.0
Real estate activities	4	1.2
Transportation and storage	15	4.4
Water supply; sewerage, waste management	1	0.3
Wholesale and retail trade; repair of motor vehicles and motorcycles	5	1.5
Other service activities	2	0.6

than the 0.70 threshold, while the average variance extracted outperformed the 0.50 threshold indicated in the relevant literature. Lastly, the discriminant validity (Table 4) was employed in order to ensure the quality of constructs (Fornell & Lacker, 1981). In this regard, all values obtained in the discriminant validity were in line with the standards set forth in the literature (Fornell & Lacker, 1981).

Structural Model

As highlighted previously, this study employed the software SmartPLS (Hair Jr. et al., 2017; Ringle, Christian M., Wende, Sven, & Becker, 2015) to analyze the relationship of the constructs' model. In this outlook, Table 5 reports the main statistics for the path coefficients. The analysis included beta values, standard deviation, t-statistics, and p-values.

Table 2. Outer loadings

Items	COMP	DIFUS	ITDE	TPPS
COMP1	0.867			
COMP2	0.864			
COMP3	0.859			
DIFUS1		0.790		
DIFUS2		0.797		
DIFUS3		0.766		
DIFUS4		0.801		
DIFUS5		0.790		
DIFUS6		0.759		
DIFUS7		0.739		
DIFUS8		0.754		
DIFUS9		0.760		
ITDE1			0.797	
ITDE2			0.863	
ITDE3			0.839	
ITDE4			0.802	
TPPS1				0.879
TPPS2				0.872
TPPS3				0.822

Note: COMP = Compatibility; DIFUS = Blockchain Diffusion; ITDE = IT deployment capability; TPPS = Trading partner pressure.

Table 3. Constructs reliability

Construct	Cronbach's Alpha	Composite Reliability	Average Variance Extracted (AVE)
COMP	0.830	0.898	0.746
DIFUS	0.916	0.930	0.598
ITDE	0.844	0.895	0.682
TPPS	0.821	0.893	0.736

Table 4. Discriminant validity

Construct	COMP	DIFUS	ITDE	TPPS
COMP	0.864			
DIFUS	0.552	0.773		
ITDE	0.67	0.537	0.826	
TPPS	0.651	0.604	0.607	0.858

As a reminder, that the various assumptions underpinning our research were stated as follows: *IT deployment capability has a significant positive effect on blockchain diffusion* (Hypothesis One); *Compatibility has a significant positive effect on blockchain diffusion* (Hypothesis Two); *and Pressures from trading partners have a significant positive effect on blockchain diffusion* (Hypothesis Three). The results obtained supported the first hypothesis ($\beta = 0.193$, $t = 2.746$, $p = 0.006$), the second hypothesis ($\beta = 0.185$, $t = 2.419$, $p = 0.016$), and the third hypothesis ($\beta = 0.369$, $t = 6.887$, $p = 0.000$). So it is clear that all were the three hypotheses were supported, ascertaining the fact that IT deployment capability, Compatibility, and Pressures from trading partners exert a significant effect on Blockchain Diffusion. Table 6 synthesizes the proposed hypotheses. In the next sections, the main implications of these results will be discussed with the objective to provide scholars and practitioners interested in this hot topic with useful insights into the issue, from both the academia and industry perspectives.

Furthermore, Table 7 reports the results for the variance explained by the model. The R^2 values in PLS-SEM ascertain the robustness of the structural model (Al-Busaidi & Olfman, 2017). The model accounted for the 42.1 percent of the variance in the blockchain diffusion, which indicates that the independent variables are good predictors of the blockchain diffusion.

Table 5. Statistics for path coefficients

Path	Beta	Standard Deviation (SD)	*t*-Statistics	*p*-Values
ITDE -> DIFUS	0.193	0.069	2.746	0.006
COMP -> DIFUS	0.185	0.076	2.419	0.016
TPPS -> DIFUS	0.369	0.054	6.887	0.000

Table 6. Summary of the Results analysis of the proposed hypotheses

Hypotheses	Conclusion
Hypothesis One: IT deployment capability has a significant positive effect on blockchain diffusion	Supported
Hypothesis Two: Compatibility has a significant positive effect on blockchain diffusion	Supported
Hypothesis Three: Pressures from trading partners have a significant positive effect on blockchain diffusion	Supported

Table 7. Results of dependent constructs

Construct	R Square	R Square Adjusted
DIFUS	0.426	0.421

DISCUSSION

Managerial Implications

This work brings valuable insights into in a better understanding of both supply chain management and blockchain. Specifically, blockchain as applied to the area of supply chain management is still to be well developed within most organizations, as their low adoption intention is nurtured by the fact that the technology remains at its early development stages. In this context, recent empirical studies (Kamble et al., 2018; Queiroz & Fosso Wamba, 2019) that came up to shed more light on blockchain adoption behaviours in SCM were particularly welcome. Obviously, there are a set of organizations that need to further their awareness about the many advantages of blockchain. But there are also those already in the process of adopting blockchain, and even at the diffusion stage already. Overall, supply chain managers and practitioners need to reorganize their current business model operations and try to better understand how disrupted supply chain operations may be, for instance.

This study found that IT deployment capability, compatibility, and trading partner pressure affect blockchain diffusion across supply chains. This implies that managers and organanizations need to consider not only the effects of IT infrastructure, but also the required skills to facilitate blockchain implementation across the supply chains of their respective organizations. Apart from acquiring cutting-edge technologies and IT resources, they should develop IT resource mobilization strategies in order to maximize their organizations' potential. For instance, maximizing the the diffusion, adoption and use of blockchain between organization's partners depends on IT resource mobilization. As a result, unsatisfactory IT deployment capability inevitably undermines blockchain diffusion in the supply chain and makes all supply chain members of involved become less competitive than their counterparts who evolve in a network where blockchain diffusion is well articulated.

As for the effect of compatibility, managers are expected to master not only their organizations' current systems, but also the blockchain integration process at organizational and inter-organizational levels. Moreover, since compatibility can repel blockchain diffusion, IT professionals and managers are expected to deeply analyze how best blockchain can be integrated with other systems and even with key supply chain members.

For it has been demonstrated that trading partner pressure had a strong influence on blockchain implementation in the SCM context, any pressure from trading partners has to be identified and managed. Once of such sources of pressure is the bargaining power of partners, which can facilitate blockchain diffusion. Failure to do this would give rise to a number of side effects (e.g., inefficiencies, costs) across the supply chain.

Theoretical Implications

One of the theoretical implication is related to how our research model was developed and validated to support a better understanding of essential behaviors toward blockchain diffusion in the supply chain management. The proposed model can well be expanded and adapted by considering moderating variables such as the organization size, the experience of the workers with cutting-edge technologies, etc., and by adding new constructs, depending on supply chain complexity, and supply chain members' behaviors and the relationships.

The path coefficients for IT deployment capability and compatibility were 0.193 and 0.185, respectively, thus confirming that the organization infrastructure exerts a high influence on blockchain diffusion in the supply chain management. Therefore, more studies are needed to investigate the role of organizations' capability in blockchain diffusion through the supply chain network. In this perspective, the Resource Basic Theory (Barney, 1991; Wernerfelt, 1984) and Dynamic capabilities (Teece, Pisano, & Shuen, 1997) can support a deeper understanding of behaviors around IT deployment capability and compatibility, considering blockchain diffusion.

The path coefficient of trading partner pressure on the diffusion (0.369) confirmed that it influences significantly blockchain diffusion. It should be noted that not only pressure from suppliers is considered in blockchain diffusion, as the digital age has brought about critical disruptions in the roles and relationships of customers across the entire supply chain. As a result, pressure from customers cannot be neglected.

With regard to pressure from customers in an environment characterizaed by blockchain diffusion, investigations need to be furthered. And so goes with the relationship between blockchain and supply chain. Moreover, an emerging economy approach is being reported for studying behaviors in blockchain diffusion. Yet, such literature may be extended, notably by comparing the results of this study not only with those of reseach conducted in emerging economies, but also with those from developed countries, so as to report similarities and differences in blockchain diffusion when managing supply chains. Furthermore, it would be interesting to provide an analysis differences in terms of blockchain diffusion by industry type.

Limitations

We developed a model based on only four constructs, even though the variance of the model (42.1 percent) showed a good relationship between the independent variables, which appear as predictors of blockchain diffusion. So, there is a need to adapt and/or extend the model for other supply chain environments. In addition, it is difficult to generalize the results obtained here as research was conducted in only one country. But at the same time, scholars, researchers and practitioners are hereby provided with an opportunity to develop more empirical studies in order to establish comparisons and support generalizations.

CONCLUSION

This study's purpose was to investigate and analyze blockchain diffusion in the field of supply chain management. The research model was validated and tested using PLS-SEM approach (Hair Jr. et al., 2017; Ringle, Christian M., Wende, Sven, & Becker, 2015). The survey was applied by a professional in supply chain management from an emerging representative economy (India). The reliability of the research model (measurement model) was assessed by employing the Cronbach's alpha, composite reliability, and average variance extracted (AVE). All values were consistent with the thresholds established in the literature. Also, a structural model (path coefficients) showed that all hypotheses were supported. The model explained 42.1 percent of the variance in blockchain diffusion. In other words, all constructs were proved to be good predictors of blockchain diffusion in the SCM context. The results provide managers, scholars, and practitioners with strong managerial and theoretical implications, especially for those interested in improving their understanding of blockchain, its benefits, and complexities, within the supply chain environment. Also, this study contributes to a more deeper understanding of the blockchain integration with supply chain, as well of blockchain diffusion with a model that can be applied easily in other countries.

REFERENCES

Akter, S., Fosso Wamba, S., & Dewan, S. (2017). Why PLS-SEM is suitable for complex modelling? An empirical illustration in big data analytics quality. *Production Planning and Control*, *28*(11–12), 1011–1021. doi:10.1080/09537287.2016.1267411

Al-Busaidi, K. A., & Olfman, L. (2017). Knowledge sharing through inter-organizational knowledge sharing systems. *VINE Journal of Information and Knowledge Management Systems*, *47*(1), 110–136. doi:10.1108/VJIKMS-05-2016-0019

Armstrong, J. S., & Overton, T. S. (1977). Estimating Nonresponse Bias in Mail Surveys. *JMR, Journal of Marketing Research*, *14*(3), 396–402. doi:10.1177/002224377701400320

Aste, T., Tasca, P., & Di Matteo, T. (2017). Blockchain Technologies: The Foreseeable Impact on Society and Industry. *Computer*, *50*(9), 18–28. doi:10.1109/MC.2017.3571064

Autry, C. W., Williams, B. D., & Golicic, S. (2014). Relational and process multiplexity in vertical supply chain triads: An exploration in the U.S. restaurant industry. *Journal of Business Logistics*, *35*(1), 52–70. doi:10.1111/jbl.12034

Barney, J. B. (1991). Firm Resources and Sustained Competitive Advantage. *Journal of Management*, *17*(1), 99–120. doi:10.1177/014920639101700108

Benchoufi, M., Porcher, R., & Ravaud, P. (2017). Blockchain protocols in clinical trials: Transparency and traceability of consent. *F1000 Research*, *6*, 66. doi:10.12688/f1000research.10531.1 PMID:29167732

Biswas, K., Muthukkumarasamy, V., & Tan, W. L. (2017). Blockchain Based Wine Supply Chain Traceability System. *Future Technologies Conference*.

Carter, C. R., Rogers, D. S., & Choi, T. Y. (2015). Toward the theory of the supply chain. *The Journal of Supply Chain Management, 51*(2), 89–97. doi:10.1111/jscm.12073

Chen, L., Ellis, S. C., & Suresh, N. (2016). A supplier development adoption framework using expectancy theory. *International Journal of Operations & Production Management, 36*(5), 592–615. doi:10.1108/IJOPM-09-2013-0413

Chen, R. Y. (2018). A traceability chain algorithm for artificial neural networks using T–S fuzzy cognitive maps in blockchain. *Future Generation Computer Systems, 80*, 198–210. doi:10.1016/j.future.2017.09.077

Chen, Y. (2018). Blockchain tokens and the potential democratization of entrepreneurship and innovation. *Business Horizons, 61*(4), 567–575. doi:10.1016/j.bushor.2018.03.006

Choi, T. Y., & Dooley, K. J. (2009). Supply networks: Theories and models. *The Journal of Supply Chain Management, 45*(3), 25–26. doi:10.1111/j.1745-493X.2009.03168.x

Choi, T. Y., Dooley, K. J., & Rungtusanatham, M. (2001). Supply networks and complex adaptive systems: Control versus emergence. *Journal of Operations Management, 19*(3), 351–366. doi:10.1016/S0272-6963(00)00068-1

Chou, S. Y., & Pearson, J. M. (2012). Organizational citizenship behaviour in IT professionals: An expectancy theory approach. *Management Research Review, 35*(12), 1170–1186. doi:10.1108/01409171211281282

Fawcett, A. M., Fawcett, S. E., Cooper, M. B., & Daynes, K. S. (2014). *Moments of angst: A critical incident approach to designing customer-experience value systems* (Vol. 21). Benchmarking. doi:10.1108/BIJ-09-2012-0059

Fornell, C., & Lacker, D. F. (1981). Evaluating structural equation models with unobservable variables and measurement error. *JMR, Journal of Marketing Research, 18*(1), 39–50. doi:10.1177/002224378101800104

Fosso Wamba, S., Kamdjoug, K., Robert, J., Bawack, R., & Keogh, J. (2018, June). Bitcoin, Blockchain, and FinTech: A Systematic Review and Case Studies in the Supply Chain Bitcoin, Blockchain, and FinTech : A Systematic Review and Case Studies in the Supply Chain. *Production Planning and Control.*

Fosso-Wamba, S., & Queiroz, M. M. (2018). *Call for Papers for the Theme special section: Blockchain in the Operations and Supply Chain Management.* Retrieved from https://www.journals.elsevier.com/international-journal-of-information-management/call-for-papers/blockchain-in-the-operations-and-supply

Golicic, S. L., & Mentzer, J. T. (2005). Exploring the drivers of interorganizational relationship magnitude. *Journal of Business Logistics*, *26*(2), 47–71. doi:10.1002/j.2158-1592.2005.tb00205.x

Graen, G. (1969). Instrumentality theory of work motivation: Some experimental results and suggested modifications. *The Journal of Applied Psychology*, *53*(2), 1–25. doi:10.1037/h0027100 PMID:5797825

Grawe, S. J., Daugherty, P. J., & Ralston, P. M. (2015). Enhancing dyadic performance through boundary spanners and innovation: An assessment of service provider-customer relationships. *Journal of Business Logistics*, *36*(1), 88–101. doi:10.1111/jbl.12077

Hackman, J. R., & Porter, L. W. (1968). Expectancy theory predictions of work effectiveness. *Organizational Behavior and Human Performance*, *3*(4), 417–426. doi:10.1016/0030-5073(68)90018-4

Hair, J. F. Jr, Hult, G. T. M., Ringle, C. M., & Sarstedt, M. (2017). *A Primer on Partial Least Squares Structural Equation Modeling. Long Range Planning* (2nd ed.; Vol. 46). Sage Publications. doi:10.1016/j.lrp.2013.01.002

Hair, J. F., Sarstedt, M., Hopkins, L., & Kuppelwieser, V. G. (2014). Partial least squares structural equation modeling (PLS-SEM): An emerging tool in business research. *European Business Review*, *26*. doi:10.1108/EBR-10-2013-0128

Hung, S.-Y., Tsai, J. C.-A., & Chuang, C.-C. (2014). Investigating primary health care nurses' intention to use information technology: An empirical study in Taiwan. *Decision Support Systems*, *57*, 331–342. doi:10.1016/j.dss.2013.09.016

Kamble, S., Gunasekaran, A., & Arha, H. (2018). Understanding the Blockchain technology adoption in supply chains-Indian context. *International Journal of Production Research*, *0*(0), 1–25. doi:10.1080/00207543.2018.1518610

Kim, H. M., & Laskowski, M. (2018). Toward an ontology-driven blockchain design for supply-chain provenance. *Intelligent Systems in Accounting, Finance & Management*, *25*(1), 18–27. doi:10.1002/isaf.1424

Kshetri, N. (2018). 1 Blockchain's roles in meeting key supply chain management objectives. *International Journal of Information Management, 39*, 80–89. doi:10.1016/j.ijinfomgt.2017.12.005

Lambright, K. T. (2010). An update of a classic: Applying expectancy theory to understand contracted provider motivation. *Administration & Society*, *42*(4), 375–403. doi:10.1177/0095399710362714

LeMay, S., Helms, M. M., Kimball, B., & McMahon, D. (2017). Supply chain management: The elusive concept and definition. *International Journal of Logistics Management, 28*(4), 1425–1453. doi:10.1108/IJLM-10-2016-0232

Li, Z., Wang, W. M., Liu, G., Liu, L., He, J., & Huang, G. Q. (2018). Toward open manufacturing. *Industrial Management & Data Systems, 118*(1), 303–320. doi:10.1108/IMDS-04-2017-0142

Lin, H.-F. (2017). Antecedents and consequences of electronic supply chain management diffusion: The moderating effect of knowledge sharing. *International Journal of Logistics Management, 28*(2), 699–718. doi:10.1108/IJLM-01-2016-0023

Lu, Q., & Xu, X. (2017). Adaptable Blockchain-Based Systems: A Case Study for Product Traceability. *IEEE Software, 34*(6), 21–27. doi:10.1109/MS.2017.4121227

McAfee, R. B., Glassman, M., & Honeycutt, E. D. Jr. (2002). the Effects of Culture and Human Resource Management Policies. *Journal of Business Logistics, 23*(1), 1–18. doi:10.1002/j.2158-1592.2002.tb00013.x

Min, H. (2018). Blockchain technology for enhancing supply chain resilience. *Business Horizons*. doi:10.1016/j.bushor.2018.08.012

Mitchell, T. R., & Beach, L. R. (1977). Expectancy theory, decision theory, and occupational preference and choice. In *Human Judgment and Decision Processes in Applied Settings* (pp. 203–226). Elsevier. doi:10.1016/B978-0-12-397240-8.50016-1

Nakamoto, S. (2008). *Bitcoin: A Peer-to-Peer Electronic Cash System*. Academic Press. doi:10.100710838-008-9062-0

Oh, J., & Shong, I. (2017). A case study on business model innovations using Blockchain: Focusing on financial institutions. *Asia Pacific Journal of Innovation and Entrepreneurship, 11*(3), 335–344. doi:10.1108/APJIE-12-2017-038

Podsakoff, P. M., & Organ, D. W. (1986). Self-Reports in Organizational Research: Problems and Prospects. *Journal of Management, 12*(4), 531–544. doi:10.1177/014920638601200408

Prybila, C., Schulte, S., Hochreiner, C., & Weber, I. (2017). Runtime verification for business processes utilizing the Bitcoin blockchain. *Future Generation Computer Systems*. doi:10.1016/j.future.2017.08.024

Queiroz, M. M., & Fosso Wamba, S. (2019). Blockchain adoption challenges in supply chain: An empirical investigation of the main drivers in India and the USA. *International Journal of Information Management, 46*, 70–82. doi:10.1016/j.ijinfomgt.2018.11.021

Richey, R. G., Daugherty, P. J., & Roath, A. S. (2007). Firm Technological Readiness and Complementarity: Capabilities Impacting Logistics Service Competency and Performance. *Journal of Business Logistics*, *28*(1), 195–228. doi:10.1002/j.2158-1592.2007.tb00237.x

Ringle, C. M., Wende, S., & Becker, J.-M. (2015). *SmartPLS 3*. Bönningstedt: SmartPLS.

Risius, M., & Spohrer, K. (2017). A Blockchain Research Framework: What We (don't) Know, Where We Go from Here, and How We Will Get There. *Business & Information Systems Engineering*, *59*(6), 385–409. doi:10.100712599-017-0506-0

Rogers, E. (1983). Diffusion of Innovations (3rd ed.). New York: Free Press.

Rogers, E. (2003). Diffusion of Innovations (5th ed.). New York: Free Press.

Schoenherr, T., Griffith, D. A., & Chandra, A. (2014). Knowledge management in supply chains: The role of explicit and tacit knowledge. *Journal of Business Logistics*, *35*(2), 121–135. doi:10.1111/jbl.12042

Stock, J. R., & Boyer, S. L. (2009). Developing a consensus definition of supply chain management: A qualitative study. *International Journal of Physical Distribution & Logistics Management*, *39*(8), 690–711. doi:10.1108/09600030910996323

Teece, D. J., Pisano, G., & Shuen, A. (1997). Dynamic capabilities and strategic management. *Strategic Management Journal*, *18*(7), 509–533. doi:10.1002/(SICI)1097-0266(199708)18:7<509::AID-SMJ882>3.0.CO;2-Z

Tubbs, M. E., Boehne, D. M., & Dahl, J. G. (1993). Expectancy, Valence, and Motivational Force Functions in Goal-Setting Research: An Empirical Test. *The Journal of Applied Psychology*, *78*(3), 361–373. doi:10.1037/0021-9010.78.3.361

Turkulainen, V., Roh, J., Whipple, J. M., & Swink, M. (2017). Managing Internal Supply Chain Integration: Integration Mechanisms and Requirements. *Journal of Business Logistics*, *38*(4), 290–309. doi:10.1111/jbl.12165

Vroom. (1964). *Work and motivation*. New York: Wiley.

Wang, Y. M., Wang, Y. S., & Yang, Y. F. (2010). Understanding the determinants of RFID adoption in the manufacturing industry. *Technological Forecasting and Social Change*, *77*(5), 803–815. doi:10.1016/j.techfore.2010.03.006

Wang, Y. Y., Wang, Y. S., & Lin, T. C. (2018). Developing and validating a technology upgrade model. *International Journal of Information Management*, *38*(1), 7–26. doi:10.1016/j.ijinfomgt.2017.07.003

Wernerfelt, B. (1984). A resource-based view of the firm. *Strategic Management Journal*, *5*(2), 171–180. doi:10.1002mj.4250050207

Zelbst, P. J., Green, K. W., Sower, V. E., & Reyes, P. M. (2012). Impact of RFID on manufacturing effectiveness and efficiency. *International Journal of Operations & Production Management*, *32*(3), 329–350. doi:10.1108/01443571211212600

APPENDIX

Table 8.

Variable	Id	Indicators	Adapted From
IT Deployment Capability		Your firm currently has …	(Lin, 2017)
	ITDE1	Adequate IT infrastructure to facilitate blockchain enabled-supply chain implementation	
	ITDE2	Robust IT plan to facilitate Blockchain enabled SC applications implementation	
	ITDE3	Strong IT development capability to facilitate Blockchain enabled SC applications implementation	
	ITDE4	The skills to integrate the various IT components to facilitate e-SCM implementation	
Compatibility	COMP1	Using the blockchain will fit well with the way I work	(Hung et al., 2014)
	COMP2	Using the blockchain will fit into my work style	
	COMP3	The setup of the blockchain will be compatible with the way I work	
Trading Partner Pressure	TPPS1	The major trading partners of my company encouraged implementation of blockchain technology	(Y. M. Wang et al., 2010)
	TPPS2	The major trading partners of my company recommended implementation of blockchain technology	
	TPPS3	The major trading partners of my company requested implementation of blockchain technology	
Blockchain Diffusion		Your firm implements the Blockchain enabled SC in …	(Lin, 2017)
	DIFUS1	Supporting accounting management	
	DIFUS2	Supporting product and service delivery management	
	DIFUS3	Supporting warehousing and inventory management	
	DIFUS4	Supporting productions and operations management	
	DIFUS5	Facilitating purchase ordering and fulfillment management among supply chain partners	
	DIFUS6	Facilitating electronic data interchange among supply chain partners	
	DIFUS7	Facilitating product authentication	
	DIFUS8	Facilitating product tracking and tracing	
	DIFUS9	Facilitating immediate supply chain information sharing among trading partners	

Chapter 3
An Exploratory Study on Blockchain Application in a Food Processing Supply Chain to Reduce Waste

Emily Anne Carey
Samsung Electronics, UK

Nachiappan Subramanian
Independent Researcher, UK

ABSTRACT

This chapter aims to explore the feasibility of using blockchain in the beef supply chain to reduce waste. A mono-method, qualitative, inductive, single case study approach was taken on a cross-sectional scale from June 2018 to August 2018, with two individuals interviewed: a beef and a blockchain expert. The case study also involved observations, a field visit, and other secondary source data. Beef is a high demand, valuable food product with a limited shelf life. By using blockchain in conjunction with RFID and sensor technologies, farming and processing stages in the beef supply chain can be streamlined. Firstly, using the technology to monitor the animals on the farm and during transportation can reduce the amount of water and energy wasted. Secondly, blockchain can be used to establish exactly when and where the meat is cut and packaged, improving the accuracy of information between supply chain entities, resulting in improved inventory management, specifically more accurate delivery times and lengthened product shelf lives.

DOI: 10.4018/978-1-5225-9078-1.ch003

INTRODUCTION

Beef Supply Chain and Waste

The British meat market consists of poultry, beef, pork, lamb and others. Beef is a high value, high demand food product, with global popularity and its supply chain is complex. Cattle are raised on beef farms from 3 months to 30 months dependent on breed and market demand. When they reach the required age, they are sent to the abattoir and processor, where they are butchered, boned, and processed into different beef products, such as mince, steak, burger, joint, stir fry, etc. then the processed products are packaged and labelled (Mishra & Singh, 2016). There is a preferential bias in developed countries like the UK towards the more expensive cuts of beef such as the hindquarters, for example sirloin and rump steaks (Traill, 1997). This carcass imbalance causes an oversupply of the lower-value forequarter cuts, which is a problem for beef processors. This issue is especially prominent because of the foodservice industry, where high-value cuts of meat are a common menu component. Food suppliers tend to overproduce to be able to cope with demand at short-notice, to avoid the possibility of being delisted from large retailers (Parfitt et al., 2010).

WRAP (2007) estimates that each household throws away edible food worth £4.80 to £7.70 each week, which adds up to £15,000 to £24,000 in a lifetime. Whilst the UK Food Standards Agency has published a figure, estimating that food fraud, such as the Salmonella Peanut Butter Outbreak in 2009 and the Horsemeat Scandal of 2013, costs families £1.17 billion a year (National Food Crime Unit, 2017). Moreover, food waste that is sent to landfill that produces methane and carbon dioxide in its decomposition, which contributes to global warming. Also contributing to the negative environmental impact of food waste, are natural resource depletion and embedded carbon from previous life cycles of food before it becomes waste. There is also the ethical note in terms of food waste, which is the lack of balance between food waste and food poverty. The World Health Organisation (WHO) estimates 420,000 people die and a tenth of people fall ill each year after eating contaminated food (National Food Crime Unit, 2017). The two most advantageous options for food waste reductions according to Papargyropoulou et al., (2014) are food surplus prevention and reduction of avoidable food waste, as waste is largely from by-products and unsold prepared food products.

Blockchain Technology

Blockchain is a decentralised, distributed ledger and a single database where every member with access has an identical copy, allowing every database entry to be shared. Continual crosschecking also ensures the integrity of existing entries. One party does not own it and records made in the database are permanent and unalterable (Abeyratne & Monfared, 2016).

Blockchain technology is known to improve visibility, productivity and security for businesses. A blockchain ledger acts as a single location where all members of the business supply chain record their actions. It simplifies record keeping, it is updated in real-time and it is readily accessible by all parties with permission. As a distributed ledger, blockchain allows many kinds of business transactions to be decentralised, preventing any one party from exclusively owning all the information within the supply chain. This enables reductions in cost and complexity, facilitates faster transactions and disintermediates the supply chain. Businesses carry legal liability for their services and products and are therefore, with blockchain, more interdependent than ever before (Johnson, 2018).

Blockchain is already popular in financial business, but only since 2016 has blockchain had the ability to work in a supply chain function. The technology is increasing in popularity and IBM currently has 400 companies trailing their blockchain service including Unilever, Nestlé, Tyson Foods and on the largest scale, Walmart (Supply Management, 2017a). However, doubts about the operational feasibility of blockchain have been highlighted in Supply Management (2017b) such as blockchain is decentralised and there is uncertainty about who will pay for it.

Potential to Reduce Supply Chain Waste Using Blockchain Technology

Blockchain can record every interaction a product has from the farm to the shop with cryptography, making the supply chain fully transparent (Creasy, 2017). Faulty products can individually be identified using blockchain and removed from the shelves without the need for large-scale costly recalls (Steiner et al., 2015). In addition, consumers receive transparency and security with the knowledge that they know exactly what they are buying and where it came from, improving customer loyalty. Food safety and food quality consumer information asymmetry has been a problem for years. Traceability can ultimately be cost saving, reduce risk and strengthen liability incentives for food safety (Hobbs, 2003).

Common problems that cause waste in food supply chains include: food fraud, illegal production, foodborne illness and food recalls. Blockchain technology can be used to resolve these problems by providing transparency, efficiency and food safety. Brigid McDermott, the Vice President of blockchain business development at IBM shares how currently, identifying the precise point of contamination that has caused a food scare and it can take weeks without blockchain (Creasy, 2017). Blockchain implementation to a food supply chain digitises tracking and storage of all information at every stage of the supply chain, avoiding food fraud and increasing consumer confidence. Blockchain technology improves how food is tracked, transported and sold, eliminating inaccuracies caused by paper tracking and manual inspections. Retailers can better manage shelf life and distribution processes can be streamlined, improving competition between online retailers like Amazon. Additionally, costs can be cut and food waste and foodborne illness can be reduced if few batches of contaminated food product can be recalled as opposed to recalling the entire supply of food product from every retailer (Koonce, 2017).

Academic Rationale

The food supply chain comprises of countless players that are functionally and geographically diverse, with many of these entities largely unaware of each other. The fragmented structure of current food supply chains inhibits the flow of information up and down the supply chain. These long-standing structural issues in food supply chains pose challenges that lend themselves to blockchain-based supply chain solutions. Despite the obstacles of implementing blockchain, supply chains in the food industry represent one of the most important applications of the technology. Given the potential benefits and promise offered by blockchain-based applications, the question is not whether it will be implemented, but when (Material Handling and Logistics (MHL News), 2017). However, this statement is very presumptuous that blockchain will be successful and widespread. Therefore, it is important that the feasibility of blockchain application is properly assessed.

Creasy (2017), looks at the potential for blockchain in the food industry. As well as preventing food counterfeit and fraud, blockchain can play an important role in quickly addressing food scares and emergency product recalls. It has been found by IBM, that as customers are becoming more digital, they are expecting more information about the origin and journey that their product took to get to the retailer, and blockchain can certify this information. Consumers would be able to receive the entire desired product and supply chain information by scanning an icon (such as a QR code) on the product, enabling packaging space to be freed.

There is a gap in the literature involving the reduction of waste in beef supply chains and also little research about reducing waste using blockchain. The major contributions to the research area that this study will provide, will be a full assessment into the feasibility of using blockchain technology in a beef supply chain and in addition, if blockchain can be used to reduce waste in a beef supply chain.

The primary research question in this study is "can the application of blockchain reduce waste in a beef supply chain" and the following questions will support the investigation, "what kind of waste occurs in a beef supply chain" and "can blockchain technology be applied to beef supply chains". To achieve answers to these questions, the following research objectives are to be investigated; "identify all waste in a beef supply chain" and "identify how blockchain be applied to a beef supply chain".

LITERATURE REVIEW

Beef Supply Chain Waste

A definition found in a review by Parfitt et al., (2010) defines waste in the food supply chain as "wholesome edible material intended for human consumption, arising at any point in the food supply chain that is instead discarded, lost, degraded or consumed by pests". Food loss occurs at the production, post-harvest and processing stages of a food supply chain and refers to the loss of edible food mass throughout the part of the supply chain that leads to edible food, excluding feed and parts of the supply chain not for human consumption. Food waste occurs at the retail and consumption end of the food supply chain and relates to the retailer and customer behaviour (FAO, 2011). For the duration of this study, I will be referring to the food losses and food waste along a food supply chain as the collective term 'food waste'.

Results from interviews conducted by Mena et al., (2011) showed that based on actual waste records, more than 7% of beef products are wasted by consumers. One reason for this relatively high level of waste is due to the short shelf life of the beef products. The tight selling deadline gives beef supply chains little room to manoeuvre, in order to maximise the product use by date for customers. Another cause of consumer waste of beef products is volatile demand related to the weather.

Improving beef product characteristics is important in order to meet consumers' demand for specific attributes, such as fat content, colour, flavour and texture. Different attributes are important to different consumers; colour, leanness, type of cut, price, sell-by-date, country of origin, production methods used (i.e. halal, kosher), and farm of origin (Fearne et al., 2001).

Garrone et al., (2014) reveals that livestock farming involves overproduction at the farming stage as the main source of surplus food waste. However, livestock farming has a low degree of recoverability regarding surplus product, as the products are unable to be consumed immediately. In addition to high management intensity, as products are perishable and require refrigeration during storage and transportation.

FAO (2011) identifies five system boundaries to distinguish different types of food waste for beef, as seen in Table 1.

According to Mishra and Singh (2016) and the maximum amount of waste has shown to be produced at the consumer end of the supply chain. Consumers expect their beef products to be a fresh red colour, of the correct fat content, tender when cooked, not to have a bad smell and not to contain any foreign bodies. Some

Table 1. End-to-end food waste or loss

Type of Food Waste/Loss	Beef Supply Chain Case
Agricultural production	Food losses refer to animal death during breeding
Post-harvest handling and storage	Losses refer to death during transport to slaughter and condemnation at slaughterhouse
Processing	Losses refer to trimming spillage during slaughtering and additional industrial processing
Distribution	Losses and waste in the market system, for example, at wholesalers or supermarkets
Consumption	Losses and waste at the household level

Source: FAO, 2011

reasons for the generation of waste include discolouration before expiry date, lack of tenderness, the presence of extra fat and inefficient trimming procedures in the boning hall of the abattoir. Additionally, the oxidation of beef, the presence of foreign bodies in beef product and inefficient cold chain management can generate waste.

According to 'Improving efficiency, generating better returns & tackling environmental impacts in beef supply chains' (2017), if beef carcases are too fat, yields and returns are adversely affected, and it is estimated that beef carcases that are out of the specification range cost British farmers £12.5 million per year. The beef industry is known for its lack of integration, and improvements to quality standards and cost savings could be made by better communications between retailers and producers, as a result of the two supply chain ends working more closely together 'Improving efficiency, generating better returns & tackling environmental impacts in beef supply chains' (2017).

To improve residual material waste, to benefit both the environment and finances, material edible by humans must be maximised and residual materials that are not fit for human consumption must be minimised. The material waste not fit for human consumption is split into two categories; category one, which includes cattle that are dead on arrival to the slaughterhouse, post-mortem failures, and soiled or medicine-contaminated materials. The second category, category one includes high risk waste material, and material unsuitable for human or animal consumption. Methods to reduce residual material waste include providing sufficient bins in the slaughterhouse to enable better segregation of the different categories of material, exploring alternative waste outlets or engaging with producers to reduce avoidable carcase rejections (Improving efficiency, generating better returns & tackling environmental impacts in beef supply chains, 2017).

Blockchain Application

Blockchain gained prominence in 2009, but researchers and practitioners are still at the beginning of trying to fully understand its potential, especially the technical challenges and limitations of the technology (Fridgen et al., 2017). According to DecisionNext (2018), every stage of the value chain in a food supply chain estimates demand. Information distances could be shortened and visibility increased between different stages of the supply chain by applying blockchain. In addition, streamlining supply chains and increasing accurate information sharing can reduce food waste.

According to Authenticate (2018), some challenges associated with the application of blockchain to a food supply chain include sourcing data, data duplication, reasonable transparency and the implications of the majority of food being disassembled rather than assembled. In regards to data sourcing, additional data capture is viewed as a burden by most food producers, as they see no immediate benefit. Existing ERP systems may not support blockchain entries and making separate entries can cause additional costs as well as receiving little support. Commercially, extra transparency may not be desirable and the right balance must be created between traceability and transparency and protecting individual businesses. Furthermore, the common disassembly of food products means that it becomes harder to trace a product after it has been processed. Therefore, in the meantime, single animal commodities are likely to be the products taken on by blockchain strategies first (Authenticate, 2018).

Blockchain provides an enormous task for management, as tools must be developed to manage the vast quantity of devices and extra data that this technology brings (Haddud et al., 2017). Tian (2016), looks at how blockchain can be combined with RFID and used in a traceability system in an agri-food supply chain. There are benefits of this in tracking and traceability management, enhancing the credibility of food safety information and fighting against food fraud. Some disadvantages mentioned in this study include the high cost of RFID and the current immaturity of blockchain. In addition, a study by Saberi et al., (2018) listed four potential barriers for block chain implementation and they are intra and inter organisational, technical and external barriers.

Moreover, research shows that in order for blockchain to be implemented effectively across the supply chain, IT infrastructures must be in place from end-to-end. Digital profiles must be updated at every stage of the supply chain using technology such as RFID. Blockchain-run 'smart contracts' should also be integrated into the system to improve the security of transactions and give the ability to monitor the progress of a business process (Abeyratne & Monfared, 2016).

Pilot Trials

The food industry is one of the next big targets for blockchain technology. With multiple pilot tests happening in big retailers, and food contamination and fraud outbreaks in the news, it is a good time for blockchain technology to make an impact in the food industry (Food Logistics, 2017).

Last year Walmart, IBM and Tsinghua University piloted blockchain technology in tracing a pork supply chain in China and mango sourcing in the US. The pilot investigated whether blockchain could be a viable alternative to paper tracking and manual inspections. Data such as farm details, batch numbers, factory data, processing data, shipping data and expiry dates were added to the blockchain supply chain data collection and matched to each product. Certificates and testing and auditing documentation were also added to the blockchain. The results from the pilot tests were positive, with a mango being traced within two seconds, compared to the multiple days it would have taken with the previous method. Walmart said using blockchain in this way could improve supply chain efficiency, promote sustainability and reduce food waste. There were some concerns during the pilot about information sharing but the trail ultimately led to the formation of the Blockchain Food Safety Alliance, comprising of IBM, JD.com, Walmart and Tsinghua University. The alliance has a goal to improve food tracking, traceability and safety in China. In addition, the alliance has created a standards-based method of collecting data about food safety, origin and authenticity, using blockchain technology (Supply Management, 2018).

In a pilot trial, JD.com rolled out a blockchain platform for Chinese customers to track their beef orders from overseas (specifically Australia). The customers were able to see where their beef product has been, how it was transported and how the cow was raised. China currently has 292.5 billion active users of blockchain, and the Chinese middle class place an importance of knowing where their beef has come from (Medium, 2018). The high levels of transparency in this beef supply chain and success from the project will surely encourage others in the Chinese market and even around the world to follow suit. The Chinese beef market is especially important to the UK because China has lifted the ban on exports of British beef, with estimates of £250 million to be made from the market in the next five years (GOV.UK, 2018).

To demonstrate the importance of food waste reduction in the UK, in July 2018 the FSA (Food Standards Agency) trialled the first use of blockchain as a regulatory tool to ensure compliance in the food industry. Blockchain technology was used in a cattle slaughterhouse during the pilot, due to the amount of inspections and collation of results that take place in a slaughterhouse. Further blockchain pilots have been planned with cattle and subject to the pilot successes and industry backing, blockchain technology will become a permanent fixture in the food industry (Food Standards Agency, 2018).

Ripe.io is a leading start-up that uses blockchain technology and the Internet of Things to enable data transparency from farm to fork, focussing on agriculture (Ripe.io, 2018). A podcast interview with Raja Ramachandran, the CEO of Ripe.io, revealed that one of the reasons for starting the company was finding out that food spoilage is a $250 billion problem globally and 25% of our water goes to spoiled food. As well as finding out that the top 50 food suppliers make up less than 20% of all sales. Ripe.io want to enable consumers and businesses to improve upon this $4 trillion industry (Wolf, 2017).

In another interview, Raja Ramachandran gives his perspective on blockchain as a method to reduce waste in a food supply chain. He said food waste from the USA is mainly from consumer waste, at the farm level and a little bit from transportation. Contributing factors to the waste include temperature control, bacteria formation, and produce that is discoloured or misshapen. He does not believe that blockchain will be able to solve the problem of food waste and says consumer waste will be the hardest to reduce. Blockchain gives analytics and the ability to identify how to improve systems in order to reduce waste. Therefore, the first solution to waste reduction could be to use blockchain analytics to investigate problems such as temperature mismanagement, or to monitor water usage. Secondly, market spaces should be created to specifically sell 'non-perfect' food products that are still safe and edible. Thirdly, involving scientific invasive technologies in the blockchain can reduce the bacteria formation problem. Finally, improving the time taken to trace the food products will also allow for better detection of food waste problems. However, if these identified problems are isolated, there will only be a partial solution, the problems must be solved altogether. A system of records shared by the industry will be required, and this is where blockchain can provide the level of visibility required and a full longitudinal look at the supply chain (Hammerich, 2018). In addition, blockchain's development is constant and does not fit into a single artefact and specifically further researchers need to investigate the type of blockchain fit for particular food supply chain (Treiblmaier, 2018).

From this literature review it can be seen that there are few studies about waste reduction in the supply chain and specifically in a beef supply chain. There are also few studies that relate blockchain technology to reduction of waste in food supply chains and no studies relating reduction of waste to blockchain technology in a beef supply chain. Therefore, this study paper will attempt to assess the feasibility of waste reduction in a beef supply chain as a result of the application of blockchain. This will be done by explaining what waste is found in a beef supply chain, how blockchain technology can be applied, and then a link between blockchain application and waste reduction in a beef supply chain will be explored.

METHODOLOGY

Philosophy and Approach

This research will be carried out using a qualitative design. The research philosophy associated with a qualitative research design is interpretivism, using a subjective ontological approach. The interpretivist approach is taken so that the researcher can investigate the subjective and socially constructed meanings around the research area (Saunders, Lewis, & Thornhill, 2012). The approach to conducting this research project will be induction; looking at numerous specific events leading to general conclusions or relationships. This approach is best used for the exploratory nature of the research questions as the research topic is immature, and also to produce a richer theoretical perspective in the literature (Saunders, Lewis, & Thornhill, 2012).

Methodological Choice, Strategy and Time Period

This research will be carried out as a mono-method qualitative investigation. The mono-method methodological choice uses a single data collection technique and a qualitative research design (Saunders, Lewis, & Thornhill, 2012). The mono-method approach was chosen due to the nascent nature of the research topic and the lack of current research on the topic. Focussing on a mono-method will easily allow for conclusions to be drawn in a clear and concise manner, on a topic that can sometimes be perceived as complex. Multiple-methods are not required for this particular investigation and along with mixed-methods, would be difficult to adopt given the time and resource constraints.

In the literature, a case study approach was implemented for a study about blockchain integration in the supply chain (Korpela et al., 2017). Case studies can provide multiple perspectives of a situation in a specific context (Järvensivu & Törnroos, 2010) and in this case, different perspectives on the ability of blockchain to reduce waste in beef supply chains. A case study approach is often adopted in exploratory research therefore the research strategy for this study will be based on the use of a singular case study.

The unit of analysis for the case studies will be a supply chain professional from the beef supply chain for a firm-level perspective, and also a professional with expertise in blockchain in food supply chains. Interviewing a beef industry professional will provide evidence to address the research objective to identify all waste in a beef supply chain. Whereas interviewing a blockchain expert will address the research objective about identifying how blockchain can be applied to a beef supply chain, from a supply chain and technology perspective. Yin (2013) emphasize the importance of case study context with the distinctiveness between the context and the phenomenon. A single case study design can be justified due to the nascent technology and new phenomenon of the research topic.

In terms of a time horizon, investigations into the research questions and primary data collection was conducted on a cross-sectional scale, between June 2018 and August 2018. This timescale is more suited to the research question than a longitudinal study, due to the fast-pace of technological development and business implementation processes (Saunders, Lewis, & Thornhill, 2012).

Data Sample

In order to fully understand beef supply chain waste and also properly assess the feasibility of blockchain application, it was necessary to interview more than one partner from different areas of a beef supply chain. A butcher was approached as this individual produced, processed and sold the beef themselves, so could give a deep insight into the beef supply chain on a local level. An employee from blockchain technological company was approached as this individual could give a high level insight into the beef supply chain from a business and large-scale perspective, as well as insight into blockchain application.

Qualitative Procedure

Within an exploratory, qualitative single case study research design, interviews will be the primary mode of data collection. Non-standardised, in-depth individual unstructured interviews will take place, either in person, or on the phone with an inductive approach. The interviews for a qualitative study must be flexible, the questions must be open and an interview guide should not be too structured (Bryman & Bell, 2011).

The informant interviews will be audio-recorded where permission is given, and then transcribed. NVIVO software will not be used in this study as only two interviews will be conducted and the creation of codes and themes will not be necessary for the ability to analyse the data thoroughly. NVIVO data analysis can also fragment and decontextualize the data, therefore it is of the judgement of the researcher to instead transcribe and analyse data directly from the interviews (Bryman & Bell, 2011).

Several measures must be overcome to prevent researcher and interviewee bias in the in-depth interviews. For example, careful consideration of initial question and approach to questioning, appropriateness of time, place and appearance at the interview, the impact of researcher's general behaviour in the interview and ability of researcher to accurately record data (Saunders, Lewis, & Thornhill, 2012).

The individual interviews will be taken from two candidates, firstly, a supply chain professional from the beginning of the beef supply chain, who is a farmer and butcher. The second interview will be conducted with the president and founder of technological company that provides blockchain solutions to food supply chains. It is important to select candidates for interviews from each of the food and technology sectors in order to determine what wastes are found in the beef supply chain, and also if blockchain technology has the potential to be applied to a beef supply chain.

The delphi technique was implemented in some of the literature. This group communication method was used to allow groups of individuals to tackle the complex problem of blockchain integration in supply chain (Korpela et al., 2017). Triangulation was also used in various studies to increase validity and reliability. However, this study investigates using a mono-method technique. Perhaps if interview data is requiring further validation, triangulation using multiple sources of evidence could then be used as a follow-up post-interview.

Data Quality Checks

Interviewing a supply chain professional from the beef supply chain and also interviewing a blockchain professional with an expertise in food supply chains will benefit the study by providing a more balanced argument, with the benefits, negatives and details of potential blockchain solutions and beef supply chain waste. Inevitably, there may be limitations around the level of detail that interviewees are able to or want to share, in regards to their company's supply chain strategies. Due to the popular nature of blockchain research at the current time, many potential interviewees were unable to provide the time for even a short interview. Therefore, in the given timeframe, it was not possible to interview any further professionals from different areas of the beef supply chain.

Constraints experienced during the study may be around company accessibility - there may be no response to the request for an interview. A minor constraint in data acquisition would be if permission were not to be given to audio-record the interviews. This would impact the validity of responses, as physical note taking would have to occur instead. Therefore, to prevent the data becoming inaccurate and to ensure reliability in the data, interview transcripts will be sent back to the interviewees, to check the accuracy of their responses.

To further increase reliability of the data, variables in the interview guides will be clearly laid out to prevent researcher or participant bias. Also, the researcher will try to be objective during the interview, which will increase the reliability of the data. In addition, ethical approval was granted before approaching the individuals for an interview. A consent form was created, and a sample information sheet and interview guide were also sent off and granted approval by the Business School.

FINDINGS

Beef Supply Chain Perspective

The first interview involved interviewing the CEO of Butcher and Grazier in the UK. The firm owns their own farm, where they rear the cattle (and other animals) and then they process and sell the meat. They supply meat at the butcher shop and also to select retailers and restaurants across the county.

Regarding issues around carcass yields or excess fat, CEO does not buy cattle that are too fat and there are strict specifications to which he abides by. Residual material use is low and this is their biggest source of waste. Waste due to appearance standards is only an occasional problem and animal loss on the farm happens only rarely. Other issues regarding the meat quality was a subject that was brushed over, as case farm, processes and sells the meat directly, so the person is in control and sure of the meat quality. There's also low mileage and a fast processing speed as the person's farm is local, therefore waste during transportation is not a significant problem.

Residual material refers to the bones and trimmings amongst other bits, which are discarded when the meat is processed and turned into the final product. These discarded materials are thrown out constantly in the business but are rendered down. Residual material cannot be helped though, so perhaps an alternative disposal method is required, but EU law is selective about the disposal methods used for raw meat. Appearance standards refer to discolouration in this case. Often when the meat does not turn out the way it should, such as discolouration or bruising, it is due to circumstances that are not observed, such as knocks to the cattle or carcasses. However, these are few and far between.

CEO supplies to select retailers and local restaurants as well as selling in his own butcher shop. The products to the retailers are sold in packaging with dates and the meat he sells in his butcher shop aren't packaged or dated in this way. However, CEO says that the packaging and labelling is the only difference between the meat he sells and the meat he supplies retailers and that the meat is cut the exact same way. He says that customers from his butcher shop either buy meat to eat on the same day or else they buy in bulk and freeze it, and he verbally advises his customers over the use by date.

Regarding traceability, each cattle carries an ear tag and is also assigned a 'passport', in which the numbers in these are used to identify the animal and relate to its ancestral history. Asked if he thinks the current systems are good enough for traceability, CEO replies rather defensively that he does not see why there would be a problem with the current system and that there is not a problem on his behalf.

The butcher was reluctant to be interviewed for very long. He was reluctant and dismissal at the idea of change, not seeing any problem with the current systems and processes. He was unsure and slightly defensive when it came to asking about waste. He did not seem to know much or was not willing to share information about the different technology currently used or for potential use, nor have any figures to hand.

Technology Perspective

The second interview was with the president and founder of an emerging technological company, a company specialising in blockchain for food supply chains. The interview began with the founder saying that meat supply chains, including beef, are highly inefficient. He said that the beef supply chain could be improved at the beginning, middle and end of the supply chain, and proceeded to explain his thoughts.

Waste at the end of the supply chain is simply about the ability of consumer facing entities to better manage their inventories. The link between the producer, distributor and consumer facing entity is that they all have their data in silos. There is not a lot of information that comes from the slaughterhouse, into the processor, all the way through. Shelf life and sell by dates are "a load of rubbish", they do not really mean anything. If the retailers have more insight into the timestamps of the processing in the slaughterhouses, then they have more data that can time the delivery, in terms of inventory to the grocery store or other consumer facing entity. For example, more data can enable more accurate deliveries, which can prevent late food deliveries due to a poorly connected supply chain. It is about having more data available, which then gives an easier decision as to when the shelf life runs out. These statements are contrary to Beef supply chain (Interview One), who was adamant that there was no waste in the process. However, founder of technological company was speaking on a higher macro level than CEO of beef supply chain and founder of technological

Table 2. Firm level issues

Waste Type	Causes
Residual materials at the processing stage	The bones and trimmings amongst other bits, which are discarded when the meat is processed and turned into the final product. These discarded materials are thrown out constantly, but this cannot be helped.
Beef products below appearance standards	Discolouration can be caused by knocks to the animals or carcasses. Often the cause of the discolouration is unknown.
Animal loss before slaughter	Not specified.
Damage to beef products during transportation	Poor temperature control during transportation, meat can travel long distances along the supply chain.

company has experience of a lot of different beef supply chains, rather than CEO, who only has experience of his own small-scale beef supply chain.

Data in terms of ownership of the consumer-facing brand, like supermarkets, is shared on a really sensitive basis. Data from the distributor, the farm, and the slaughterhouse does not always get into the hands of the supermarkets or grocery store in time, it might be two or three days late. By collapsing the data, as it is available in relatively real time, the supply chain should see improved inventory management and a better sense of shelf life and waste at the end point.

At the beginning of the supply chain, waste can be measured in terms of how water is used, how energy is used, how feed is used, and how the animals are kept healthy. When we look at waste, it is not just a piece of meat that comes from a cow; it is also all of the periphery things around the creation of that piece of meat that can also be reduced. So, by tagging individual cows, which can be done through RFID tags or other technology, we are now able to make sure that the animals can stay healthy, because there is a better way to make sure that they are kept well and healthy. Making sure that they are up to date with all of their vaccinations, the medical records are kept up to date and they are receiving the correct quantity of feed does this. Hence, a lot of those processes can be streamlined to reduce the amount of water consumption that is also going in. Waste is not just about food that is thrown away from the supermarket shelf, we define waste also basically around sustainability. All of those dimensions that go into raising and growing an animal can be improved and waste can be reduced in the cattle.

In the middle of the supply chain, at the processor and slaughterhouse, there is a void of data. This information is very relevant to understand the real shelf life of a piece of meat. The inventory does not work as it should for an animal that is slaughtered on a Tuesday and shipped out on a Thursday. There is a lot of waste where bits of meat are being thrown off the shelf because they are past their sell by date, when in fact they technically have two more days of shelf life than what is stated on the label. It is important, in order to reduce this waste, to establish exactly when the meat gets cut and packaged, to enable a more accurate and probably longer shelf life. What blockchain can do is to bring all of these data sets together and share more information along the supply chain, which is very difficult to do in a traditional silo of data.

When asked about how accepting he thinks different members of the supply chain will be, Founder responded that technology companies are working on this problem on a global basis. The company has projects in pork, beef, chicken, dairy, fruits, vegetables, all over the world and in many different categories. The right condition for his company to work is where there is a willing collaboration of supply chain partners to achieve something. The company ask the supply chain partners what it is they want to achieve, whether it is ensuring the quality of the meat, exposing to the consumer that the product is grown sustainably, or is it organic. It is currently very hard to quantify and get evidence for these statements, especially 'local'. The people that the company work with all have a common goal that they want to achieve, either exposing quality, traceability, or authenticity, etc.

Technology company have seen great traction from the innovators, but there is going to be older, more traditional food companies that are going to resist this transformation, as is human nature. However, seeing their market share drop, no more questions will be asked and they will have to make the change. For example, if you look at all of the companies that did not get on the e-commerce bandwagon and continued selling different services with just brick and mortar, brands that have been wiped off the face of the earth. It will be a very similar chain of events, if something is not blockchain certified then people are not going to buy it, so this is a very good comparison to make.

Blockchain certification will be the future of everything you buy, not just food. We are already seeing that first hand in the USA. Based on the two interviews and literature review, we listed the causes and mitigation strategies for waste in the food supply chain in Tables 3 and 4.

DISCUSSION

Interview one with CEO of beef supply chain identified several causes of waste, although the identified types of waste were said to be minimal or rare, these are highlighted in Table 5.

Table 3. Supply chain and technology related issues

Waste Type	Causes
Beginning of SC	
Water and energy throughout every step	Lack of monitoring consumption.
Animal loss before slaughter	Poor health.
Middle of SC	
Waste caused by poor inventory Management	Void of data at the processor and slaughterhouse.
End of SC	
Shelf-life waste	Data in silos down the supply chain and lack of data from the slaughterhouse leads to inaccurate shelf life dates.
Waste caused by poor inventory Management	Lag in data retrieval from farm, slaughterhouse and distributor.

Table 4. Mitigation Strategies for end to end supply chain

Waste	Blockchain Mitigation
Beginning of SC	
Water and energy throughout every step	Processes can be streamlined to reduce the amount of water and energy consumption.
Animal loss before slaughter	Ensuring animals stay healthy by tagging individual animals with RFID tags, and through these, keeping vaccinations up to date, medical records up-to-date and consumption of the optimal quantity of feed.
Middle of SC	
Waste caused by poor inventory Management	Establish exactly when the meat gets cut and packaged, to enable a more accurate and probably longer shelf life, by bringing all of these data sets together and sharing more information along the SC.
End of SC	
Shelf-life waste	Processing timestamps in the slaughterhouse will allow for more accurate delivery times to retailers and hence better product shelf life.
Waste caused by poor inventory Management	Collapsing the data as it is available in relatively real-time should improve inventory management and give a better sense of shelf life and waste at the end of the SC.

Table 5. Mitigation strategies (interview one)

Waste	Blockchain Mitigation
Residual materials at the processing stage	N/A
Beef products below appearance standards	An end-to-end blockchain system would provide customers with a detailed view of the product's timeline in an easy to access format, which customers would trust and realise that the product is of equal quality to the others. In addition, detailed data, for example concerning temperature control could indicate why product is discoloured.
Animal loss before slaughter	When the farmer is not sure what caused the loss, blockchain data logs would be able to prove the history and assist with route causing the problem, i.e. potential hereditary diseases or weaknesses to certain infections.
Damage to beef products during transportation	Quickly identify where the meat quality of a particular product might have been tarnished.

The different types of beef waste identified by FAO (2011) were confirmed in the findings in Interview One and Two. However, beef waste due to below-standards appearance was not mentioned in FAO (2016) but was highlighted by CEO of beef supply chain in Interview One. Furthermore, loss and waste in the market system was mentioned in FAO (2016), but poor inventory management and shelf life waste were not specifically identified as root causes of waste. Therefore, this study further advances empirical base in the literature about beef supply chain waste identification.

Founder of technology company in Interview Two confirmed that blockchain can be used to reduce waste in beef supply chains in his explanation of how each waste can be mitigated or significantly reduced using blockchain technology. Echoing Raja Ramachandran in an interview by Hammerich (2018), the Food Standards Agency (2018) and Brigid Mcdermott from IBM (Creasy, 2017).

Tian (2016) looked at how blockchain and RFID could be used in combination in an agri-food supply chain. This study, which focuses on the beef supply chain, supports research by Tian in statements about the combination of blockchain and RFID bringing tracking and traceability management benefits to food supply chains. Furthermore, research by Abeyratne and Monfared (2016) shows IT infrastructure must be in place from end to end in the supply chain for blockchain to be implemented effectively, which supports blockchain mitigation strategies proposed in this study. Mitigation strategies supported by Abeyratne and Monfared include having an end-to-end blockchain infrastructure and streamlining all processes throughout the supply chain using blockchain. These mitigation strategies would help reduce waste from beef below appearance standards, in addition to reducing water and energy waste, amongst other wastes.

Referring to Interview Two, in which Founder talks about acceptance of the application of blockchain to the beef supply chain, there are links to the literature under lean supply chain concepts that can support his statements. Vitasek, et al., (2005), said that cultural change is the biggest recurring obstacle to successfully applying lean supply chain concepts. Cultural change in this instance is referring to resistance from people who will be asked to embrace and implement the change to learn new lean supply chain concepts.

In addition, to tie in to when Founder was speaking about how there must be more data sharing up and down the supply chain to streamline processes, Taylor (2006) supports this point. Taylor (2006) says that the full cooperation of all value chain partners is required for a really efficient value chain. In addition, cooperation can contribute to a stable, dedicated and competitive supply chain, by improving on quality, cost and delivery year on year. Additionally, DecisionNext (2018) supported this point, believing that streamlining supply chains and increasing accurate information sharing can reduce food waste.

However, as demonstrated in Interview One by the reluctance of CEO of Beef supply chain to accept the possibility and need for change from current processes, Taylor (2006) also backs up this behaviour. Taylor (2006) says that such improvements to the red meat sector will require a significant change in attitude for every member of the value chain, to overcome some of the lack of trust and hostility between value chain members. In addition, improvements will require the desertion of many long held business norms, which could be a cause of reluctance to improvement.

Differences between firm level views and supply chain views can be explained through Ignorance Theory. Firm level views can be seen as ignorant, arising from the absence of knowledge; this can be explained when a firm refrains from acquiring knowledge when the cost of education exceeds the potential benefit of what the knowledge would acquire (Roberts, 2012). For example, CEO of beef supply chain could be reluctant to learn about blockchain any further because he does not know enough about it to see its potential and only views the technology as an added cost. Whereas Founder of technology company, has a supply chain perspective and is knowledgeable about the technology, therefore, his views are not tainted by ignorance.

Furthermore, Founder of technology company ends the interview with the statement about blockchain certification being essential for every purchase in the future. This cements his and his company's belief in the blockchain technology and in its successful application in food and beef supply chains.

Weaknesses and Limitations

One weakness of the research is the lack of exploration into other processes and technologies that are currently used in beef supply chains and their ability to reduce waste. We believe, with further research, the extra information could balance our argument and lead to a more rounded conclusion. In addition, if more professionals from different parts of the beef supply chain were interviewed, further detail and insight could have been provided on beef supply chain waste, increasing the validity of the study.

Contribution to Knowledge

The findings in this study provide a good insight into the feasibility of blockchain application to a beef supply chain and the consequential waste reduction. Wastes in the beef supply chain were identified and general knowledge was increased compared to the current literature. An additional contribution to supply chain literature is that blockchain has been shown to have the potential to reduce different waste in a beef supply chain. Now that the feasibility of blockchain application to a beef supply chain has been acknowledged, this study could be used as a platform for further research into waste reduction in the beef supply chain. Eventually all food supply chains will utilise blockchain technology in various forms and to various extents, and food waste will be reduced on a global scale, as a result.

CONCLUSION AND RECOMMENDATIONS

The key takeaway of the chapter is to confirm the feasibility of blockchain technology application in beef supply chain. The affirmation that waste can be reduced in a beef supply chain by the application of blockchain technology. All areas of the beef supply chain can be streamlined by increasing data sharing between the different entities and adopting various blockchain enabled processes. Finally, even though the global direction of food supply chains is to become leaner, more cost effective and more visible and traceable, there is resistance especially from the farming area of the beef supply chain for any change in regards to technology and improving current practices.

Beef supply chain waste was identified, including residual materials at the processing stage, animal loss before slaughter, use of water and energy throughout the supply chain, damage to beef products during transportation, beef products below appearance standards, and waste caused by poor inventory management. Targeted blockchain mitigation strategies were then proposed. Using blockchain and other technologies combined in the farm, transportation and slaughterhouse supply chain stages, animal loss, damaged meat products and water and energy use can be reduced. Exactly where and when the meat is cut and packaged can be determined and communicated between supply chain entities more accurately with the support of blockchain. This will improve inventory management, improve the accuracy of delivery times and lengthen product shelf lives for the benefit of consumers and retailers alike.

Problems faced in this study included difficulty in collecting substantial data through interviews from relevant individuals. It was difficult to secure interviews that would provide the necessary data due to the nascent research topic and few companies that could provide relevant data. In order to overcome this difficulty, further research was required to identify companies that linked to the research topic. It was also necessary to think carefully about access available to other parts of the beef supply chain, and this was where the butcher was identified and brought into the study. In order the extract the necessary data from the interviewees, questions had to be posed concisely and often with further explanation. This was difficult as the individuals interviewed had very different levels of knowledge on the topic already, but analysis after the interviews and linking to the literature ensured that enough conclusions could be made to support the study.

The research question, to understand the feasibility of blockchain application in relation to beef supply chain waste reduction was answered by the combination of interviews and current literature review. Types of waste in a beef supply chain were identified and the cause of each type of waste was identified. For each type of waste, the extent to which blockchain could mitigate the waste was discussed, and that answered the question about whether blockchain could be applied to a beef supply chain and also if blockchain could be used to reduce waste in a beef supply chain. Furthermore, acceptance of blockchain technology in a beef supply chain was touched upon, as well as current processes and the future of blockchain application. Therefore, the gap in the literature around waste reduction in a beef supply chain and also waste reduction by using blockchain technology, has been narrowed.

The link between blockchain and the reduction of waste in a beef supply chain was researched in this study, as this research had not been carried out previously. Further studies should include more research on the acceptance of blockchain technology on a widespread level for beef supply chains, as this arose as an important point in this study. Further studies should also include the feasibility of blockchain to reduce waste in other meat and food substance supply chains, as blockchain is a technology that is only increasing in popularity and the benefits it brings are only increasing in importance. Finally, research must be done on how to increase the awareness, education and adoption of streamlining technology amongst the various stakeholders in the supply chains.

Stakeholders in beef supply chains must be aware of the importance of including blockchain technologies in future strategies and they must also be aware that waste reduction is a topic that is increasing in importance as supply chains look to become more sustainable as well as more cost-effective.

REFERENCES

Abeyratne, S., & Monfared, R. (2016). Blockchain Ready Manufacturing Supply Chain Using Distributed Ledger. *International Journal of Research in Engineering and Technology*, *05*(09), 1–10. doi:10.15623/ijret.2016.0509001

Arc-net. (2018). *Arc-net and PwC join forces to use blockchain in combating food fraud.* Available at: https://arc-net.io/blog/arc-net-pwc-blockchain-food-fraud/

Authenticate. (2018). *Blockchain Technology for the Global Food Supply Network: Transformation or Hype?* Whitepaper.

Bryman, A., & Bell, E. (2011). *Business Research Methods*. New York: Oxford University Press.

Creasey, S. (2017). *The potential for blockchain in the food industry - just-food global news*. Aroq Limited.

Davies, G. (1992). The two ways in which retailers can be brands. *International Journal of Retail & Distribution Management*, *20*(2). doi:10.1108/09590559210009312

DecisionNext. (2018). *The Missing Link in the Food Chain: Blockchain.* Whitepaper.

FAO. (2011). Global food losses and food waste – Extent, causes and prevention. Rome: FAO.

Fearne, A., Hornibrook, S., & Dedman, S. (2001). The management of perceived risk in the food supply chain: A comparative study of retailer-led beef quality assurance schemes in Germany and Italy. *The International Food and Agribusiness Management Review*, *4*(1), 19–36. doi:10.1016/S1096-7508(01)00068-4

Food Standards Agency. (2018). *FSA trials first use of blockchain.* Available at: https://www.food.gov.uk/news-alerts/news/fsa-trials-first-use-of-blockchain

Fridgen, G., Radszuwill, S., Urbach, N., & Utz, L. (2017). Cross-Organizational Workflow Management Using Blockchain Technology – Towards Applicability, Auditability, and Automation. *51st Annual Hawaii International Conference on System Sciences.*

Garrone, P., Melacini, M., & Perego, A. (2014). Opening the black box of food waste reduction. *Food Policy*, *46*, 129–139. doi:10.1016/j.foodpol.2014.03.014

Good Logistics. (2017). *Blockchain Next on Food Supply Chain Menu.* Available at: https://www.foodlogistics.com/technology/article/12382466/blockchain-next-on-food-supply-chain-menu

gov.uk. (2018). *China lifts ban on exports of British beef.* Available at: https://www.gov.uk/government/news/china-lifts-ban-on-exports-of-british-beef

Haddud, A., DeSouza, A., Khare, A., & Lee, H. (2017). Examining potential benefits and challenges associated with the Internet of Things integration in supply chains. *Journal of Manufacturing Technology Management*, *28*(8), 1055–1085. doi:10.1108/JMTM-05-2017-0094

Hammerich, T. (2018). *Blockchain and the Internet of Food.* Future of Agriculture.

Hobbs, J.E. (2003). *Consumer demand for traceability.* Academic Press.

Improving efficiency, generating better returns & tackling environmental impacts in beef supply chains. (2017). WRAP. Available at: http://www.wrap.org.uk/sites/files/wrap/Tackling%20hotspots%20in%20the%20beef%20value%20chain_Published%20March%202017%20(2).pdf

Järvensivu, T., & Törnroos, J. (2010). Case study research with moderate constructionism: Conceptualization and practical illustration. *Industrial Marketing Management*, *39*(1), 100–108. doi:10.1016/j.indmarman.2008.05.005

Johnson, E. (2018). *Blockchain: shaking up logistics*. Epicor.

Koonce, L. (2017). *Blockchains & Food Security in the Supply Chain*. Academic Press.

Korpela, K., Hallikas, J., & Dahlberg, T. (2017, January). Digital Supply Chain Transformation toward Blockchain Integration. *Proceedings of the 50th Hawaii International Conference on System Sciences*. 10.24251/HICSS.2017.506

Management, S. (2017a). *Case study: Walmart*. Available at: https://www.cips.org/en/supply-management/analysis/2017/june/case-study-walmart/

Management, S. (2017b). *Six things wrong with blockchain*. Available at: https://www.cips.org/en/supply-management/analysis/2017/december/six-things-wrong-with-blockchain-/

Management, S. (2018). *Blockchain for supply chains: From hype to reality*. Available at: https://www.cips.org/supply-management/analysis/2018/june-/blockchain-hype-reality/

Management, S. (2018). *Unilever and Sainsbury's testing blockchain for supply chain data*. Available at: https://www.cips.org/supply-management/news/2017/december/unilever-and-sainsburys-testing-blockchain-for-small-farmers/

Material Handling and Logistics (MHL News). (2017). *Blockchain Can Improve Visibility in Food Supply Chain*. Available at: http://www.mhlnews.com/technology-automation/blockchain-can-improve-visibility-food-supply-chain

Medium. (2018). *The Amazon of China has Adopted Blockchain to Track high-end Beef Imports*. Available at: https://medium.com/@Michael_Spencer/the-amazon-of-china-has-adopted-blockchain-to-track-high-end-beef-imports-bf927671191a

Mena, C., Adenso-Diaz, B., & Yurt, O. (2011). The causes of food waste in the supplier–retailer interface: Evidences from the UK and Spain. *Resources, Conservation and Recycling*, *55*(6), 648–658. doi:10.1016/j.resconrec.2010.09.006

Mishra, N., & Singh, A. (2016). Use of twitter data for waste minimisation in beef supply chain. *Annals of Operations Research*, 1–23.

National Food Crime Unit. (2017). *Food Crime Annual Strategic Assessment - A 2016 Baseline*. Food Standards Agency and Food Standards Scotland.

Papargyropoulou, E., Lozano, R., Steinberger, J.K., & Wright, N., & bin Ujang, Z. (2014). The food waste hierarchy as a framework for the management of food surplus and food waste. *Journal of Cleaner Production*, *76*, 106–115. doi:10.1016/j.jclepro.2014.04.020

Parfitt, J., Barthel, M., & Macnaughton, S. (2010). Food waste within food supply chains: Quantification and potential for change to 2050. *Philosophical Transactions of the Royal Society of London. Series B, Biological Sciences*, *365*(1554), 3065–3081. doi:10.1098/rstb.2010.0126 PMID:20713403

Ripe.io. (2018). *Ripe – Let's Talk About the Quality of our Food*. Available at: http://www.ripe.io

Roberts, J. (2012). Organizational ignorance: Towards a managerial perspective on the unknown. *Management Learning*, *44*(3), 215–236. doi:10.1177/1350507612443208

Saberi, S., Kouhizadeh, M., Sarkis, J., & Shen, L. (2018). Blockchain technology and its relationships to sustainable supply chain management. *International Journal of Production Research*, 1–19.

Saunders, M., Lewis, P., & Thornhill, A. (2012). *Research methods for business students* (6th ed.). Harlow, UK: Pearson.

Steiner, J., Baker, J., Wood, G., & Meiklejohn, S. (2015). *Blockchain: the solution for transparency in product supply chains*. Project Provenance Ltd.

Taylor, D. H. (2006). Strategic considerations in the development of lean agri-food supply chains: A case study of the UK pork sector. *Supply Chain Management*, *11*(3), 271–280. doi:10.1108/13598540610662185

Tian, F. (2016, June). An agri-food supply chain traceability system for China based on RFID & blockchain technology. In *Service Systems and Service Management (ICSSSM), 2016 13th International Conference on* (pp. 1-6). IEEE.

Traill, B. (1997). Structural changes in the European food industry: Consequences for innovation. In *Products and Process Innovation in the Food Industry* (pp. 38–60). Boston, MA: Springer. doi:10.1007/978-1-4613-1133-1_2

Treiblmaier, H. (2018). The impact of the blockchain on the supply chain: A theory-based research framework and a call for action. *Supply Chain Management*, *23*(6), 545–559. doi:10.1108/SCM-01-2018-0029

Vitasek, K. L., Manrodt, K. B., & Abbott, J. (2005). What makes a lean supply chain? *Supply Chain Management Review*, *9*(7), 39–45.

Wolf, M. (2017). *How Could Blockchain Be Used With Food*. Smart Kitchen Show.

World Health Organization. (2017). *WHO's first ever global estimates of foodborne diseases find children under 5 account for almost one third of deaths*. Available at: http://www.who.int/mediacentre/news/releases/2015/foodborne-disease-estimates/en/

WRAP. (2007). *Understanding the Food Waste*. Research Summary.

Yin, R. K. (2013). *Case study research: Design and methods*. Sage publications.

Chapter 4
Exploring the Blockchain Technology Application in the Chinese New Retail Business Model

Yuhong Li
Jiuzhou Industrial Holdings Group Co., Ltd., China

Nachiappan Subramanian
University of Sussex, UK

ABSTRACT

While the new retail is revolutionizing retailing in China, the supply chain of the new retail has a few problems or potential risks which will decrease the customer satisfaction level. However, the implementation of one of the cutting-edge technologies—the blockchain—can revolutionize the supply chain of the new retail in China. This qualitative piece uses multiple interviews to find out the specific outcomes blockchain will make for the supply chain of the new retail in China. The major contribution is to fill the gap of the academic literature as well as the business application and such as the new retail in China to increase supply chain security and efficiency with blockchain.

DOI: 10.4018/978-1-5225-9078-1.ch004

INTRODUCTION

The new retail format is introduced by the Alibaba founder Jack Ma that combines ecommerce business model, physical supermarket and restaurant to offer 360 degree experience to customers with the support of extensive data and technology (Declan, 2017). The new retail uses consumer centric model which is an amalgamation of online and offline format. Online aspect customizes order and improve effectiveness whereas customers can order fresh food or vegetables online and the delivery support systems will move the stuff in next 30 minutes if the houses are located within 5 km radius; the offline part works on efficiency reduce complexity in handling diverse products using technology. The other attractiveness of the store is the layout, which combine traditional supermarket and restaurants where customers can, choose fresh vegetables and seafood to eat in addition to grocery shopping. For example, the Alibaba-backed HEMA supermarket is one of the new retail stores, which have already set 65 stores across China with fine-dining grocery and 3000 products from over 100 countries (Declan, 2017). Furthermore, integration of O2O (Online to Offline) strategy is done through data and technology where detailed data of shopping behavior of customers is captured in the final payment using Alipay. This process will enable HEMA to personalize the retail experience (Declan, 2017).

The new retail is fascinating to experience. however, it has two potential issues such as food security and achieving efficiency. In terms of security issue, China as a country feeds 20% population of the world (China Ministry of Agriculture, 2017) on 7% of total farmlands in the planet (Carter, Zhong, & Zhu, 2012) has issues related to excessive use of chemical fertilizer which increased three-fold in the past decade (Cui, Shoemaker, 2018). In addition, genetically modified food could be another threat for the food industry in China. Hence, there is an urgent need for the food industry as well as new retail in China to add more assurance to protect the security of the food and make the supply chain more transparent. In terms of efficiency there are several non-value-added activities that needs to curb such as documentation and bureaucratic processes (Heutger & Kückelhaus, 2018). One way to handle these issues is through blockchain technological application.

Blockchain has deeper and wider usage than what people have thought and it is expected to revolutionize the business applications. Blockchain was developed by Satoshi Nakamoto in 2008 essentially as an alternate to conventional money transfer and used the cryptocurrency bitcoin in a wider network with the record of transactions (Tapscott, 2016). Recently, lots of managers or investors trust blockchain because it is an open global infrastructure where there is no middle man or third parties so that companies and individual can make transactions with low transaction cost and time dealing with third parties (Underwood, 2016). Furthermore, the blockchain uses the technology of distributed ledger and consensus process to ensure all the transactions between companies and individuals can be shared in a connected computer network, which is available for verification by the users on the network. On the other side, blockchain can also be private and permissioned which allow certain group of people in a particular company, industry or supply chain to work together and maintain the information as well as transactions private (Underwood, 2016).

One of the most significant applications is the implementation of blockchain in supply chain management, which will bring benefits to all the members in the supply chain including suppliers, producer, retailer and customers, and at the same time, save huge amount of cost and waste. In addition, supply chain management possess additional risks such as delays and disruptions. For instance, the delays in material flow will have impact on the efficiency of supply chain and results in the customer satisfaction level which is important for companies or logistics organizations (Chopra & Sodhi, 2004). Hence, it is necessary to overcome these risks, and this study will focus on how implementing blockchain can help supply chain to be more reliable and efficient.

The purpose of this study is to address the application of blockchain in supply chain that could enhance security, transparency and efficiency. Specifically the study answers the following research question 'How blockchain can improve the security, transparency and efficiency of new retail supply chains in China?'

LITERATURE REVIEW

This section provides a critical review of academic studies relevant to the topics new retail format, supply chain issues, and blockchain.

New Retail in China

The retail industry in China have substantially changed in terms of scale and capability after the economic reform that took place in 1978. This reform transformed the country into one of the largest consumer markets. However, Edward (2017) mentioned few challenges posed by ecommerce giants such as Taobao and JD with their 24-hour online sales of Alibaba in 'singles day' which is on November 11 of every year. As per the data from Jon (2017), Alibaba sold over $25 billion of products on single's day in 2017, which is a global achievement in terms of national sales. The data shows the shopping behavior of consumers in China and their preference to buy online to avail discount and to receive the home delivery. This trend instigated several offline retail giants to integrate online and offline sales with some innovation in creating unique experience to customers.

Most of the retailers extensively rely on technologies such as mobile payment, wireless internet, and artificial intelligence to grow their market share. Interestingly, some retailers developed ubiquitous touchpoint to connect customers and their social groups (Edward, 2007). Jason (2018) also observed integration of online and offline channels to develop consumer-centric closed loop business model to unlock the growth potential. A few retailers also attempted to offer unique experience to customer by making them to stay longer in store (Edward, 2007).

On the other hand, the new retail format's philosophy is to maximize the customer experience and make consumers as co-producers rather than being only customers (Jason, 2018). Hence, retailers were forced to stimulate the need for consumers and let them to co-create the brand. These changes mandates supply chain to be more flexible and efficient to adjust to the sales as well as involve emerging technologies such as blockchain. More importantly, Jason (2018) mentioned the engagement of customers at the early stage of product development in designing and planning so as to improve the higher customer loyalty. However, this also increases the complexity and uncertainty for the supply chain with the increase in personalization (Jason, 2018).

Issues in Traditional Retail Chain

Supply chain delay is one of the biggest problems retailers face, and Chopra (2004) indicated the reason for supply chain delay is due to involvement of high levels of handling or inspections during border crossings between countries and change in transportation modes during shipping. Kehoe and Gindner (2017) argued the inclusion of paper-based processes are still common which causes delay and obstructs visibility. This is quite common in retail where many handoffs are required from wholesalers and distributors before it goes to local and global stores (Jason, 2018).

There are some other ethical problems because of lack of transparency. Dickson (2016) stated that both buyers and customers are not sure the true value of the products or service due to non- transparent supply chain that are unable to track environmental issues. According to Blowfield (2003), the UK based Premier Brands was one of the first mainstream companies who are concerned about the ethical sourcing in supply chain for quality management where company takes responsibility for others. However, Maloni (2006) mentioned that besides these proactive measures there are some other ethical issues relating to procurement in the supply chain including environment, and labor practices. The issues are related to child labor, forced labor, health and safety, and poor working conditions, which is very hard for retailers to check in the existing supply chain. Besides, Maloni (2006) also stated the ethical problems in procurement such as bias towards suppliers, allowing personalities decisions to influence buying decisions, and failure to provide prompt responses to customers.

The lack of transparency can cause other problems such as information sharing and unsuccessful assessment of customers. Jason (2018) stated that the lack of transparency of supply chain make suppliers and retailers difficult to collect and gather information of customer behavior. Cecere (2014) also mentioned that there are information losses and barriers in every step of the supply chain. Lamming (2004) also suggested that customers should see suppliers as a critical resource to create competitive advantage in terms of supply chain transparency.

Blockchain

As per Tapscott (2016), blockchain has three specific characteristics which makes them special, decentralized, public, and encrypted every member in the network can view the transactions instead of a single institution charged with auditing transactions and keeping record. In addition, there are private and public keys in encryption to ensure virtual security. From the perspective of Swan (2015), blockchain should be viewed as another avatar which transformed world such as internet and information technology applications. Underwood (2016) mentioned the similarity between blockchain and internet which are open and have a global infrastructure.

In terms of Chinese context, the government unveiled the Internet plus in 2015, which aim to integrate mobile internet, cloud computing, big data and O2O with traditional industries in order to promote the increase of economic transformation (Wang, 2016). Like the era of internet plus which make many traditional industries got second birth, blockchain is described as the innovation accelerator based on capability of third platform of technology by Underwood and Michael (2016). From this point of view, Swan (2015) envisioned blockchain technology has the capacity to reconfigure all aspects in the society and its operation.

In purchasing, smart contract is activated when a pre-set or set condition agreed by parties (Francisco & Swanson, 2018). It is good to note that Swan (2015) provided the data to avoid contract disputes that account for litigation in the United States (44%) and United Kingdom (57%) with automated mechanism like smart contracts. The smart contracts have several elements, which make it distinct, autonomy, self-sufficiency, and decentralization to ensure trust between parties (Swan, 2015). However, blockchain is still in an early stage of development where there are many limitations both internal and external including technical, perception of practitioners, and government regulation (Swan, 2015). Underwood (2016) also stated that there are three key challenges for blockchain: data transfer, integration of existing systems and security. Few compatible issues are standardized database and centralized business mode (Swan, 2015).

Block Chain and Supply Chain

Sexena (2018) stated that supply chain across industries will benefit after adopting the blockchain because both supply chain and blockchain are globally spread, multi-layered parties with transfer of assets, material and information in the contract. In terms of logistics, Heutger (2018) indicated the usefulness of blockchain in developing trust in logistics globally. Even more, Lubowe and McDermott (2016) mentioned that the demand data of blockchain can eliminate the bullwhip effect, one of the biggest challenges of supply chain due to the transparency of supply chain. We will discuss a few benefits blockchain can offer to deal with supply chain issues.

Cost

As per Brody (2017) inventory cost is substantial roughly it is 20%-40% of total supply chain cost. Certainly blockchain will enable to track and manage resource will increase the accuracy of the forecast and reduce inventory cost. Potentially smart contract will inform logistics carrier and trigger the automatic digital invoicing through banking system, which will reduce the transportation capital requirement and simplify the financial operations. Similarly, Ward (2018) also shared a cash perspective where digital currency will trigger payment from retailer to producer and enhance total working capital. Cost reduction possibility is been confirmed by Heutger and K點kelhaus (2018) in their recent study. Further, Ward (2018) mentioned Open Port (OPN), an asset built on smart contract through blockchain technology will slim the margin of transporters thereby reducing the transportation cost. According to Kakavand (2016), the custodial function of blockchain like recordkeeping and auditing will reduce the transactional cost for both parties, which is mentioned in the study by Satyavolu and Sangamnerkar (2016).

Food Chain Security

There are long debates in support of building supply chain security, an early study by Peleg-Gillai (2006) justified the investment on developing security will not be a financial burden, but will lead to operational improvement, cost reduction and higher revenue. Saxena (2018) stated the non-tampering advantage of blockchain through cryptographic protection that cannot be changed or modified without evidence. Moreover, Ward (2018) complemented the capability of proprietary tokens that will offer additional security and less risks from external threats like improper data access. Few others backed the great capability of blockchain such as not possible for strangers to hack the data (Wu & Li, 2018; Sadouskaya, 2017). However, Sadouskaya

(2017) argued that evaluation of security need of a threat model which is different in each blockchain model.

Transparency

Transparency in supply chain is related to the availability of information in supply chain network (Francisco & Swanson, 2018). Microsoft (2018) shared how blockchain is uniquely positioned to offer trust and transparency in supply chain. Heutger and K點kelhaus (2018) described the mechanism in which stored data and information from variable source can enable stakeholders to access and maintain accuracy. However, Wu and Li (2017) argued that both public and private ledgers have certain limitation that public ledger is not attractive for all industries, while private ledgers have limitations on technology.

Furthermore, Francisco and Swanson (2018) are confident that blockchain can take supply chain transparency into the next level by offering traceability and tracking of data and to overcome some of the limitations it have in terms of ethical issues such as child labor and unethical use of rainforest resources. Sadouskaya (2017) mentioned that adopting blockchain into supply chain can track products from origin to the customers with secured access to the stakeholders participating in the network.

Efficiency

Efficiency is essential for supply chain, Ward (2018) offered suggestion how block chain can enable efficiency to the in terms of payments and clearance processes between suppliers. Heutger and K點kelhaus (2018) also mentioned about blockchain improvement to track the lifecycle and ownership of products. Besides, blockchain can enable to overcome some frictions in supply chain such as difficulty to track provenance of goods in order to gain more efficiency (Heutger & K點kelhaus, 2018).

METHODOLOGY

The study uses "interpretivism" philosophy to understand the difference between human role and social actors, and interpretivism researchers often assume that only social constructions such as language, shared meanings, or consciousness are connected to the reality (Myer, 2008). Hence, the aim of this study is to understand the application of blockchain techniques in supply chain. The inductive approach will be used in this study. The inductive approach is from simple cases or examples to generate ideas and conclusions, through which the researchers can explore the theory through data collection from organizations (Saunders & Tosey, 2013). Besides, the inductive research approach can also permit alternative explanation of what is going on (Saunders & Tosey, 2013) such as the facts and situation about blockchain as well as the supply chain market in order to build and generate the theory. The inductive research approach needs qualitative data collection. It is beneficial to use inductive research approach to develop ideas and frameworks about the supply chain combined with blockchain because there are only a few primary data is available in this topic. Hence, the inductive research approach will facilitate the understanding of the benefits as well as challenges of blockchain applications in supply chain.

The studies focused on leading global technology services company to analyze the strategy and how to implement blockchain in supply chain. The case study method allows the exploration and understanding of the complicated theory within organization and the context to analyze (Zanial, 2007). Moreover, Hill and Ackiss (1945) mentioned the case study method can help researchers to bridge the gap between community study and stereotype survey, so that the researchers can explore additional details about blockchain out of the information in the literature. It would be better if the research use multiple case study to study the blockchain applications in supply chain in the context of China. The information could be more accurate and reliable if the data is accessible for several business giants of the new retail such as HEMA, Yonghui, and 7Fresh so that the researcher can analyze the demand and the attitude of them for implementing the blockchain. However, the source of information is unfortunately limited, in addition only a few information could be collected for this study.

Research Choices and Time Horizon

The study used qualitative mono-method. As per Saunders and Tosey (2013) the research choices mono-method are suitable for single data collection in terms of qualitative or quantitative way. Hence, the findings of this research will bases on single qualitative method of data collection. The case company chosen is a leading technological provider and study gathered data using open interviews. This study is a cross-sectional in the area of blockchain and supply chain. Cross-sectional study is suitable to answer questions in a particular period of time, and to make use of strategies in survey or case study (Saunders & Tosey, 2013). The study is to understand the impact of blockchain on new retail supply chains.

Research Data Collection

The qualitative data is retrieved from the interview with the procurement manager who as several years of experience in both procurement and technology application. is engaged in the combination with the blockchain and supply chain and the manager in the industry of new retail in China. Hence, this research will collect the primary data through the face to face interview with semi-structured. This method is better for researchers to focus on the application of the blockchain in the supply chain management in particular companies and form the conclusion, some electronic method such as the phone call interview and email interview are not available because the information gathered through these methods will be not enough.

At the beginning of the research, there seems to be some difficulty in gathering primary data. However, we got hold of a procurement manager who is involved in applying blockchain in his company and agreed to participate in the interview. We followed the ethical review process and maintained confidentiality. We also struggled a bit to collect data from the Chinese companies with respect to new retail business model. The authors tried to collect from the leading ecommerce and technology companies. Interestingly, we got hold of one practitioner in the leading ecommerce company which is involved in establishing new retail format.

The semi-structured interview contains list of semi open questions on specific topic. The interview questions included questions related to security aspects in procurement of supply chain using blockchain, how does global technology provider implement blockchain to its procurement and processes to provide security and efficiency, the possible barriers or difficulties after implementing blockchain to the supply chain. We took notes of the response since we are not allowed to record or take photos. Although we got minimal data that is not abundant to capture the entire scenario, however it was useful to understand the viewpoint of technological providers experience.

FINDINGS

This section will summarize the results of the primary data gathered by the authors in the data collection process. We document the opinion of the interviewees on the specific topic of blockchain with supply chain, their general strategy for the implementation on blockchain into supply chain, the benefits and challenges of the strategy, and the opinion of blockchain with supply chain from the perspective of the new retail in China. Key findings are reported and discussed in comparison to the relevant theories

The data collection took place between July and September 2018, the interviewee is an experienced manager in procurement department of leading global technological service firm who is leading the projects of the blockchain with supply chain. The respondent mentioned a few benefits after adopting blockchain and business application of blockchain with supply chain based on their experience. The interviewee also mentioned the challenges and limitations about the blockchain in the field of supply chain, such as the regulation issues and trust. However, the information collected is not complete due to some confidential issue that some of the material that cannot be triangulated. It would be better to have additional data and application method to support the theories as well as the conclusion.

Global Technological Service Provider Perspective

The technology company X is offering a solution to enhance traceability and transparency of the supply chain in order to make it more secure and more efficient (IBM blockchain, 2018). Dispersive data collection and record keeping of the supply chain leads to incomplete and unreliable certificates and manifests. Also, there is increasing demand for the information transparency for the supply chain regarding where and how the products are made. Besides the information, the supply chain with high level of complexity need more trust between members to function properly. However, the facts of distrust historically have decreased sharing information and data between supply chain members. The IBM blockchain with supply chain was created to make these situations better.

Company X's blockchain uses a shared ledger and the smart contract to deal with the problems in the procurement function in supply chain where blockchain can validate each transaction and process so that everybody in the company or the supply chain can see and access the progress of the flow. Secondly, the technology can reflect the geographical flow of the products and how were they treated so that companies can check the inventory level, investigate the certificates of the industry, and on some occasions even the dangerous components of products (IBM blockchain, 2018).

However, there are a few limitations in blockchain implementation. First, the improvement of transparency of the supply chain can influence the accuracy of information at the early stage of the project. Second, blockchain implementation in supply chain needs the cooperation of customs and governmental agencies. Taxes and duty payment plays essential role due to non-availability of unified tax policy. Furthermore, the cost of the blockchain is another problem because small to medium sized companies still cannot afford to manage the investment on this, or lack of the skills to implement. Finally, due to early stages there are some issues related to proper standard and poor regulations.

Food Chain Security

The technology company have established the point of contact contingent labor and Know Your Suppliers to help the procurement department to achieve the higher level of supply chain security. The POC-Contingent is the two-time sheet which is meant to work out the problems with suppliers. There is a trust contract in the POC-Contingent where suppliers can register and share the information or the transaction, so there will be no physical sheet with blockchain. It can also record the time of transaction. Know your suppliers can also increase the security level greatly. The project Know Your Suppliers is the trusted source of digital identity of suppliers which benefits both suppliers and buyers by simplifying the supplier onboarding and validation process. It is simple to register and can be linked to the ERP system available in the companies to save time and increase trust and security.

For buyers, the Know Your Suppliers can allow immediate access to the real-time data, that can minimize supplier risks and simplify the process. First the access avoids delay and flaws between buyers and suppliers and speeds up the decision making of procurement manager easier. Second, the supplier risk can be eliminated greatly after using Know Your Suppliers. The procurement manager, as a buyer, can check the situation of suppliers with the smart contract in this system. The financial statement includes the management situation, the origin of the product they sold, involvement of number of the employee at various levels are transparent for buyers. The other benefits include real-time updates by 95%, with 70%-90% reduction of cycle time, and 50% reduction of cost for onboard new supplier.

For suppliers, the Know Your Suppliers can eliminate the need for suppliers to provide same information due to the smart contract, so that the information can be seen at the same time and the transparency is guaranteed. It can also create the ecosystem of suppliers and buyers in which they can benefit each other in a more secure and efficient way.

Supply Chain Efficiency

Global supply chains are slowing down due to the complexity and sheer volume of point to point communication across coupled land transportation providers, freight forwarders, customs, government ports, and ocean carriers. However, pilot project between IBM and Maersk have announced a joint venture to streamline shipping and global logistics with an efficient, secure global trade digitization platform based on blockchain (IBM blockchain, 2018). IBM address this with a distributed permission platform accessible by the supply chain ecosystem to exchange real-time data and handle necessary documents.

IBM also employ blockchain technology to create a global tamperproof system for digitizing trade workflow and tracking shipment end to end, which will eliminate frictions including point to point communications, track millions of containers journey per year and integrate with customs authority in selected trade lanes. The tamperproof system can digitalize some distinct entity including the signature and documents needed for producers, authority ports and customs and importers. Each member can submit the documents, the real-time registration progress through computers or mobile devices, after which the submission is visible to all participants in the supply chain. The action can initiate the smart contract in the blockchain instantly to the approval of the next participant, so there will be no delays or missed steps.

Besides that, the tamperproof system can also trigger the remediation actions such as supplier substitution or price adjustment immediately to reduce the potential crisis due to the delay of weather or labor issues. So, the tamperproof system can offer secure data exchange, and a tamperproof repository for those shipping documents and events, which will significantly reduce delays and fraud as well as save billions of costs annually (IBM blockchain, 2018).

New Retailer Perspective

As the second largest ecommerce platform in China, company Y has already put the attention on blockchain to improve both its logistics and financial sectors with the cooperation with government. The company Y has launched the blockchain-as-a-service (BaaS) platform which use the blockchain as a tool to develop several applications such as government taxation, charity donation, financial settlement, logistics, and big data security.

For the supply chain of the new retail, company Y has revealed the new system with blockchain to track the beef imported from Australia. They partnered with one of Australian companies InterAgri to bring beef products such as Angus beef to Chinese consumers, which are realized by the help of blockchain to make the whole supply chain especially production process traceable. This will not only incredibly increase the food safety for consumers, but also the customer experience and satisfaction because they will know some essential information such as the places where livestock were raised and fed, where and how the meat were produced and processed, and in which circumstance they were transported. Besides, JD has also adopting blockchain of joining a transportation blockchain alliance to create the shared ledger in blockchain to improve its logistics.

Contributions

The current literature shows the characteristic of blockchain which is decentralized and public which is different to other centralized third parties who record and keep data and information. New retail in China have some issues in supply chain such as the high cost, less transparency of food products and less efficiency which need the support of blockchain technology.

With the system using blockchain like the tamperproof system from IBM, companies in the new retail can maximize the supply chain efficiency to transport products immediately avoiding paperwork to guarantee freshness of the perishable products. The blockchain can significantly improve the food chain security for companies in the new retail. The study highlighted the linkage between blockchain application and supply chain with the new retail format.

These efforts can pave way for future researchers to identify more practical methods or tactics in supply chain with blockchain. Furthermore, the author also mentions several feasible strategies for companies in new retail in China to implement blockchain. The four issues discussed in the chapter are shown in Figure 1.

One simple proposition of this study is about the type of supply chain, for example private blockchain can enhance the food security, and public blockchain can increase supply chain efficiency for new retail business model. This proposition needs further validation.

Figure 1. Block chain and supply chain issues

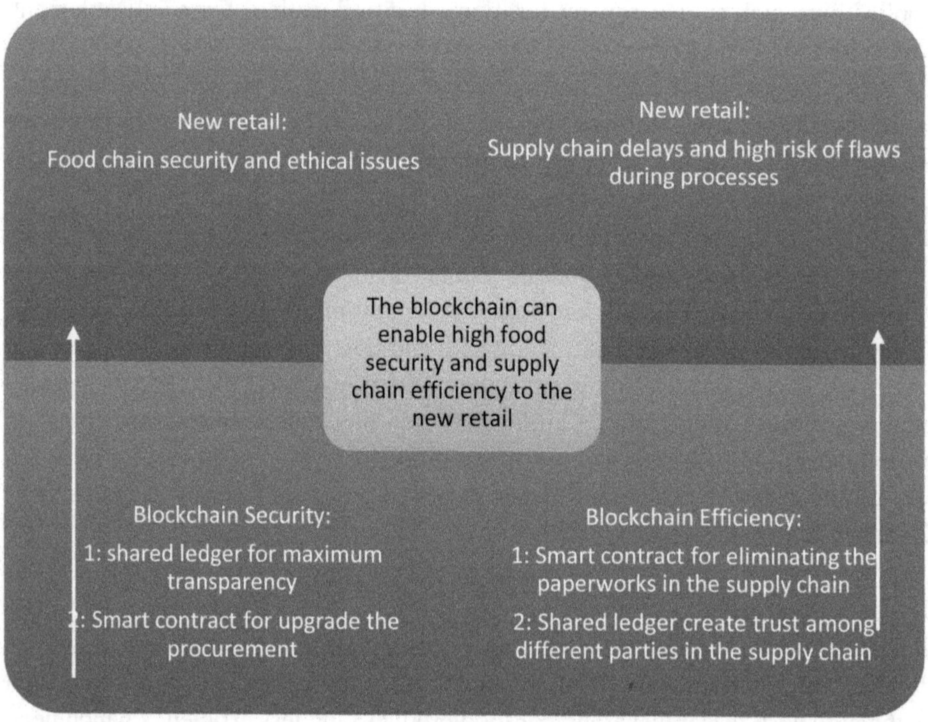

CONCLUSION

The objective of this study is to investigate the benefits of adopting blockchain in the supply chain management. Specifically this study investigated the application of new retail format a. Hence, this study aims to explain the opportunities and challenges for implementing blockchain in new retail format supply chains. Even though we discussed the major advantages of blockchain there are several limitations such as essentiality of cooperation of all stakeholders in members in the supply chain. The study addressed the four issues with the limited data but future studies can verify the findings using large-scale empirical data.

REFERENCES

Blowfield, M. (2003). Ethical Supply Chains in the Cocoa, Coffee and Tea Industries. *Greener Management International*, (43).

Bowersox, D.J., & Closs, D.J. (1996). *Logistical Management—The Integrated Supply Chain Process*. McGraw-Hill Companies Inc.

Brody, P. (2017). *How blockchain is revolutionizing supply chain management*. Academic Press.

Carter, C. A., Zhong, F., & Zhu, F. (2012). Advances in Chinese agriculture and its global implications. *Applied Economic Perspectives and Policy*, *34*(1), 1–36. doi:10.1093/aepp/ppr047

Cecere, L. (2014). Supply chain visibility in business networks. *Supply Chain Insights Report*, 1-19.

China Ministry of Agriculture. (2017). *China Agricultural Yearbooks*. Beijing, China: China Agriculture Press.

Chopra, S., & Sodhi, M. S. (2004). Supply-chain breakdown. *MIT Sloan Management Review*, *46*(1), 53–61.

Clemons, E. K., Jin, F., Wilson, J., Ren, F., Matt, C., Hess, T., & Koh, N. (2013, January). The role of trust in successful ecommerce websites in China: Field observations and experimental studies. In *System Sciences (HICSS), 2013 46th Hawaii International Conference on* (pp. 4002-4011). IEEE.

Cui, K., & Shoemaker, S. (2018). *A look at food security in China*. Academic Press.

Declan, E. (2017). *What Jack Ma's New Retail looks like in action.* Available at: https://www.enterpriseinnovation.net/article/what-jack-mas-new-retail-looks-action-967015705

Dickson, B. (2016). *Blockchain has the potential to revolutionize the supply chain.* Aol Tech.

Ding, Lannes, Zhu, Gao, Peng, Song, & Jiang. (2018, March). *Embracing China's New Retail.* Bain report.

Francisco, K., & Swanson, D. (2018). The supply chain has no clothes: Technology adoption of blockchain for supply chain transparency. *Logistics*, *2*(1), 2. doi:10.3390/logistics2010002

Hackius, N., & Petersen, M. (2017). Blockchain in logistics and supply chain: trick or treat? In *Proceedings of the Hamburg International Conference of Logistics (HICL)* (pp. 3-18). epubli.

Heutger, M., & Kückelhaus, M. (2018). *Blockchain in Logistics.* Academic Press.

Hill, M. C., & Ackiss, T. D. (1945). The "Insight Interview" Approach to Race Relations. *The Journal of Social Psychology*, *21*(2), 197–208. doi:10.1080/00224545.1945.9714166

IBM. (2016). *An example of using blockchain to increase safety and reveal product provenance in food supply chains.* Retrieved from https://www-03.ibm.com/press/us/en/photo/50817.wss

IBM Blockchain. (2018). *One truth across the supply chain network.* Available at: https://www.ibm.com/blockchain/industries/supply-chain

Kakavand, H., Kost De Serves, N., & Chilton, B. (2016). *The Blockchain Revolution: An Analysis Of Regulation And Technology Related To Distributed Ledger Technologies.* Available at: http://www.fintechconnectlive.com/wp-content/uploads/2016/11/Luther-Systems-DLA-Piper-Article-on- Blockchain-Regulation-and-Technology-SK.pdf

Kehoe, L., & Gindner, K. (2017). *When two chains combine, supply chain meets blockchain.* Academic Press.

Lamming, R.C., Caldwell, N., & Phillips, W. (2004). Supply chain transparency. *New et al*, 191-208.

Lubowe, D., & McDermott, B. (2016). *Trust in trade, towards stronger supply chains.* Academic Press.

Maloni, M. J., & Brown, M. E. (2006). Corporate social responsibility in the supply chain: An application in the food industry. *Journal of Business Ethics*, *68*(1), 35–52. doi:10.100710551-006-9038-0

Microsoft. (2018). *How block chain will transform the modern supply chain.* Academic Press.

Myers, M. D. (2008). *Qualitative Research in Business & Management.* SAGE Publications.

Peleg-Gillai, B., Bhat, G., & Sept, L. (2006). *Innovators in supply chain security: better security drives business value.* Washington, DC: Manufacturing Institute.

Russell, J. (2017). *Alibaba smashes its Single's Day record once again as sales cross $25 billion.* Available at: https://techcrunch.com/2017/11/11/alibaba-smashes-its-singles-day-record/?guccounter=1

Sadouskaya, K. (2017). *Adoption of Blockchain Technology in Supply Chain and Logistics.* Academic Press.

Satyavolu, P., & Sangamnerkar, A. (2016). Blockchains Smart Contracts: Driving the Next Wave of Innovation Across Manufacturing Value Chains. *Cognizant 2020 Insignts.*

Saunders, M., Lewis, P., & Thornhill, A. (2009). *Research methods for business students.* Pearson Education.

Saunders, M., & Tosey, P. (2013). *The layers of Research Design.* Academic Press.

Saxena, S. J. S. (2018). *Harness Abundance the Blockchain Way In Supply Chain Management.* Academic Press.

Schary, P.B., & Skjt-Larsen, T. (2001). *Managing the Global Supply Chain.* Handelsh 鳶skolens forlag.

Su, T. Y., & Lei, X. H. (2008). Research on supply chain cost reduction based on process and time analysis. *IEEE International Conference on Industrial Engineering and Engineering Management,* 1625–1629. 10.1109/IEEM.2008.4738147

Swan, M. (2015). *Blockchain: Blueprint for a new economy.* O'Reilly Media, Inc.

Tapscott, D., & Tapscott, A. (2016). *Blockchain revolution: how the technology behind bitcoin is changing money, business, and the world.* Penguin.

The Economist. (2015). *Blockchains: The great chain of being sure about things.* Author.

Tse, E., & Wang, J. (2017). *How China is leading the 'new retail' revolution.* Available at: https://www.scmp.com/comment/insight-opinion/article/2125831/how-china-leading-new-retail-revolution

Underwood, S. (2016). Blockchain beyond bitcoin. *Communications of the ACM, 59*(11), 15–17. doi:10.1145/2994581

Wang, Z., Chen, C., Guo, B., Yu, Z., & Zhou, X. (2016). Internet plus in China. *IT Professional, 18*(3), 5–8. doi:10.1109/MITP.2016.47

Ward, M. (2018). *Building the Blockchain Logistics Protocol.* Academic Press.

Wu, H., Li, Z., King, B., Ben Miled, Z., Wassick, J., & Tazelaar, J. (2017). A distributed ledger for supply chain physical distribution visibility. *Information, 8*(4), 137. doi:10.3390/info8040137

Yin, R. K. (2015). *Qualitative research from start to finish.* Guilford Publications.

Zainal, Z. (2007). Case study as a research method. *Jurnal Kemanusiaan*, (9), 1-6.

Chapter 5
The Challenges and Solutions of Cybersecurity Among Malaysian Companies

Puteri Fadzline Tamyez
University Malaysia Pahang, Malaysia

ABSTRACT

The objective of this chapter is to analyze the challenges faced by Malaysian companies in cybersecurity and to determine solution for Malaysian companies to overcome challenges in cybersecurity. The data were collected from the expert people in cybersecurity fields using interview sessions. The finding confirmed that the awareness and budget are very important in other to implement the element of cybersecurity in the company. Cybersecurity is good and desired as a protection for an organization in developing strategic planning to gain more profitability and increase the productivity of goods and services. This research will be beneficial for the organization because it will provide the solution for the company to overcome the cybersecurity issues. From this research, an organization can have potential to enhance competitiveness and understand the problem occur, then do the improvement by implementing cybersecurity.

DOI: 10.4018/978-1-5225-9078-1.ch005

INTRODUCTION

Industry 4.0 invites tremendous advantages for companies towards business sustainability. It has nine pillars altogether, namely, internet of things, big data, supply chain, cloud computing, horizontal and vertical integration, autonomous robot, addictive manufacturing, cyber security, simulation and augmented reality. However, the major challenge in facing this digitalization era is on cyber security (Jay Lee, 2016). Privacy and security of the data will always be top security measures that any organization should take. We live in a world where information are secured in digital or a cyber-form. Data from multiple sources has different formats gives difficulty for analysts to integrate the data (Ibrar, 2017). Lack of monitoring and protection against unauthorized changes or alteration will create unwanted changes in data information. Most companies faces inadequate of development phase (Ibrar, 2017), thus makes it important in order to limit the risk of application related assaults or attacks.

The inadequacy of development phase as brought small and medium size enterprise (SMEs) face different risks as compared to large companies, as these organizations have limited and minimum human and monetary resources to apply information technology (IT) and cyber security systems (Heikkila, 2016). Most SMEs own traditional security mechanisms, which could not accommodate the technology of the Internets of Things (IoT) due to limited resources (Ibrar, 2017). Apart from that, they need to overcome the inadequacy of security budget and low security alertness among the workers. Lack of employee training and recovery planning has contributed to its low security alertness among employees (Heikkila, 2016). This finding further elaborates that only a few companies report on the provision of safety-related training for all employees. However, the lack of deployment process shall cause problems in managing the cyber security (Ibrar, 2017).

The inadequacy of security software's upgradability and patch ability is also one of the issues in cyber security. A number of companies usually do not put much effort in upgrading security software due to lack of resources. The low and inadequacy of physical security will allow and provide an unrecognized user to enter the data or devices using Universal Serial Bus (USB) port. This may cause companies to face many problems. Another issue is trust. Network interactions with systems that have lower standard security will invite more trust issues. Data transference is mostly carried out by wireless network, which increases the probability of miss-data problem to occur (Ibrar, 2017). This may affect in aspects of incomplete or false information. Thus, cyber security is crucial in all industries to make sure all of their data were being safely secured. This research attempts to answer the accompanying inquiry in aspects of the challenges faced by Malaysian company in cybersecurity and the solution for Malaysian companies to overcome challenges in cybersecurity.

BACKGROUND OF STUDY

Most industries are affected by technological change and innovation or rather called industrialization revolution (Jay Lee, 2016). This revolution is due to mechanization in the first industrial revolution, the use of electricity is 2nd industrial revolution and electronics and automation is industrial revolution 3. The revolution not only affected the production itself, but also the labour market and education system as well (Lasi, 2014). Due to development of digitalization and robotics, the industry faces the next industrial revolution, known as Industry 4.0. These new emerging technologies have a huge impact on people's education and scope of work (Katharina M., 2015).

Only qualified and highly educated workers will be able to control this technology. Digital supply chain is smart, worth-driven network that is the current path to automation and analysis in order to generate unique forms of interest and business value (Keliang Zhou, 2015). Cyber security as one of the nine pillars is exposed to external and insider cyber threats with complicated and sophisticated cyber security landscape (Wells, 2016). It is known as an activity or process, ability or capability, or state whereby information and communications systems and the information contained therein are protected from and or defended against damage, unauthorized use or modification, or exploitation" (DHS, 2014). Other than that, cyber security involves reducing the risk of malicious attack to software, computers and networks (Dan Craigen, 2014). This includes tools used to detect break-ins, stop viruses, block malicious access, enforce authentication, enable encrypted communications, and on and on. Thus, the goal of this study is twofold; to analyse the challenges faced by Malaysian companies in cyber security and to determine solution for Malaysian companies to overcome challenges in cyber security.

Cybersecurity has become a general concern for all citizens, professionals, politicians, and, more generally all decision makers. It has also become a serious concern for societies that must protect against cybersecurity attacks with both preventive and reactive measures, which imply a lot of monitoring, and must simultaneously preserve the freedom and avoid general surveillance. Cybersecurity can be defined as computer security, also known as cyber security or IT security, is the protection of computer systems from the damage to their hardware, software or information, as well as from disruption or misdirection of the services they provide (Roca et al., 2019).

Security in general includes both cybersecurity and physical security. Cybersecurity consists in ensuring three basic and essential properties of information, services, and IT infrastructures well known as the CIA triad as confidentiality, integrity, and availability (Fischer, 2016). Therefore, getting an information system means blocking unauthorized entities such as users, processes, services, machines from accessing, altering, or providing inaccessible computer data, computing services, or computing infrastructure. Other property, such as legality which means original proof of information, privacy, or protection against illegal copying may also be listed.

LITERATURE REVIEW

Challenges of Cyber Security

One of the challenges is there are many types of attacks against information systems. The threats are therefore numerous. The attacks can target the hardware, the network, the system, or the applications for example through the malicious actions of a malware, or even the users themselves like social engineering, phishing and others. The attacker can be an insider or an outsider. According to Roca et al., (2019), these attacks can be classified as observation attacks, perturbation attacks, and a new field known as hardware-targeted software attacks. Attacks against information systems do not usually involve the hardware layer but exploit software vulnerability. This new type of attack is especially dangerous as it makes hardware attacks possible at a distance, as opposed to classical side channel attacks. This task requires expertise at the hardware, firmware, and operating system levels.

Other types of challenge in security threat for IoT are cloning device and sensitive data disclosure (Naik, 2017). Cloning device is a foreign hardware which acts as the right device unlike in reality it is not. Bad data can quickly cause problem to the server and this require higher budget to fix. Sensitive Data Disclosure occurs when an application does not adequately protect sensitive information. Information need to be encrypted during transit as data originates from various information.

The challenge in managing cyber security is data integrity. To secure data integrity in IoT environment, it is found to be quite difficult due to the large of information and data. Data from multiple sources has variety of concepts and formats, which makes it difficult for analysts to integrate this data. Besides that, lack of monitoring and protection against unauthorized changes or alteration will make unwanted changes in data information (Ibrar, 2017). Malicious data searchers are always looking for new ways to steal the data particularly during peak hours where organizations might lack the internal capability and mechanisms to manage and secure the data (Jaime, 2016). Ransom ware attacks will affect the entire landscape of security services. It will control the entire of framework and will permit constrained access for client cooperation. From 2005 until March 2016, approximately 7600 ransom ware assaults were accounted for by Internet Crime Center.

As in many other fields, there is a well-known adage about security that says that the main threat lies between the chair and the keyboard. This adage may be exaggerated, and at the very least it deserves further study, but it must be recognized that the users are indeed sometimes a source of security problems (Fischer, 2016). Firstly, the user can be the target of the attack. In addition, the user can try to avoid using the available protection mechanisms due to the excessive complexity of use and finally the user level of education and training is too insufficient. The employee therefore not aware of the real risks or on the contrary overestimate the user. In either case, the user does not know what mechanisms are to be used when. This also can happen when the user interfaces of software systems are not well designed for the purpose of security, even if design principles that are suitable for other applicants are applied (Woodward, 2015).

On the other hand, lack of security investment among the companies and alertness among the employee also contribute to the issues in cyber security (Heikkila, 2016; Roca et al., 2019). When there is security failure caused by security attacks, organization usually do not consider it as a very serious attack that might happen again. Besides that, traditional security mechanisms are incapable to recover IoT devices as most of the devices have battery constraints and restricted assets (Yaaqob, 2017). Additionally the lack of employee training and recovery planning also contribute to the risk of cyber security (Heikkila, 2016). Empirical work reveals that only a few companies report on the provision of safety-related training for all employees.

Next is the lack of security software upgradability for protection (Yaaqob, 2017). This is due to limited resource of equipment, tools and system. Software upgradability is also costly for an organization. On one hand, most organization has lack capability to organise unstructured data (Jaime, 2016). The vast array of data makes it difficult to detect problems because the data from various sources with different formats, thus makes it difficult for the analyst to interpret and integrate the unstructured data. On the other hand, integration among the stakeholder of any industry 4.0 organization may cause obstacles in language (Lane Thames, 2017). There are many types of industry 4.0 environment such as from diverse technologies and different types of subject matter experts where only few common language are used in standard or processes that align with the company objectives and goals.

Solutions for Cyber Security Issues

Tools to Secure Against Cyber Threats

Rather than using several type of security protection such as antivirus, firewall, strong passwords, protecting Wi-Fi connection to secure the data, there are other few ways protect the data information and improve the level of security.

Digital Signature

This is a practice that is conceivable to protect electronic data such that the original of the data, and additionally the honesty of the data, can be checked. This method of ensuring the inception and the respectability of the data is likewise called authentication. A computerized mark is just a system that can be utilized for various verification purposes. For an E-record, it comes practically near the conventional written by hand marks (Iqbal, 2016). The client can create key match by utilizing the particular crypto programming. Presently, Microsoft Internet Explorer and Netscape enable the client to make their own particular key combination. Any individual may make an application to the Certifying Authority for issue of Digital Signature Certificate.

Encryption

This effective and imperative strategy for security in the computer system is to encode the sensitive records and messages in travel and capacity. Generally, there are four gatherings of individuals that are utilized and added, namely; the specialty of cryptography, military, discretionary corps and diarists. The military has the most sensitive part and has shaped the field (Iqbal, 2016).

Security Audits

This practice is an effective evaluation of the security of an institution data framework by estimating on how the arrangement is going well in create up the criteria. It is to discover the vulnerabilities that an association is looking for its IT framework. A thorough audit regularly overviews the security of the structure's physical setup and condition, programming, information dealing with systems, and customer home (Iqbal, 2016).

Cyber Forensics

Cyber forensics is the basic solution in the investigation of computerized violations. Digital legal sciences are the revelation, examination, and remaking of confirmation extricated from any component of computer frameworks, systems, media, and peripherals that enable specialists to unravel a wrongdoing (Iqbal, 2016). The primary concern with computer forensics involves imagery storage media, restoring files were deleted, finding slacks and free space, and maintain the data that being collected for prosecution purposes. Another issue is forensic networking, where it is consistent forensic in terms of the network (Sadeghil & Waidner, 2016).

Cryptography

Cryptography is based on strong mathematical reasoning and aims to guarantee more the nature of confidentiality only. Cryptography provides tools for protection integrity and legitimacy of the message for example avoid deep amounts financial transactions are changing, to ensure not being rejected in aspects dispatchers cannot denies being the author of the message and unnamed (Roca et al., 2019). This chapter is organized into three sections:

- Cryptographic Primer is the most basic building block; such primitive permitting to encrypt or digitally sign messages;
- Cryptography Scheme usually builds primitive to provide more powerful safety goals, ensuring integrity and legitimacy of size messages;
- Cryptographic protocols rely on schemes to achieve more complex security aims, for example, to establish a safe communication channel that can be used confidential and explicit message exchange.

Crytapanalysis

The purpose of academic cryptanalysis is to understand the threats to security of the existing primitives in order to be ahead of malicious (Roca et al., 2019). It provides an empirical measure of security thanks to a thorough and never-ending scrutiny, searching for possible weaknesses. The knowledge of state-of-the-art cryptanalysis is thus the backbone for the design of secure primitives.

Manufacturing Risk

Manufacturing that associates with business is significantly prone to cyber security attacks due to its very nature. Collaboration with suppliers associate an increasing reliance on IoT technology to trace finished merchandise and material shipments throughout the producing method and this leaves several individual points at risk of potential attacks. Additionally, its production is heavily dependent on information, together with internal, customer, payroll, checking account, and provider data. Finally, recent international power struggles have concerned the producing business in many ways. Producing companies created up half the businesses targeted by the high-profile malware and ransom ware attacks (Shelzer, 2018). According to Jay Lee (2016), the solution is to build up an automation CPI with a time machine observing action so a virtual testing algorithm can be applying for any control activity to the framework. Cyber security requires a smart-driven approach that is used for Defence in Depth. It requires regular improvement of network and proactively response and actions to face the probable attacks.

To overcome the cyber security issues, it needs to fully apply the safety security by design (Yaaqob, 2017). The IoT systems need to moderate ransom ware during the whole lifecycle of utilization execution. This is to ensure the security verification and registration of devices in IoT networks. This is followed by educating and increasing the alertness and awareness of security issues among employees. This can be consistent through lifelong learning such as introducing the basic security concepts, practice and tools. Recuperation plans to guarantee a minimum economic loss in the case of a security failure and do the security planning for all workers. Ultimately, security policies, security risks investigation, security principles and activity models must continuously be up to date and need to monitor and maintenance regularly (Heikkila, 2016).

Security objectives, security policies, availability, integrity and confidentiality need to be clearly understood by all employees. Security principle or policy describes the responsibility of the management to security task, responsibilities and methods in implement in the security domain (Heikkila, 2016; Jaime, 2016). Apart from that, there is a need to have standard formats such as Machinery Information Management Open System Alliance (MIMOSA) databases as suggested by Jaime (2016) in order to ensure the trustworthiness of data. The security threat mechanism is known as Advanced Persistent Threat Mechanism (ATP) (Penang Skill Development Centre, 2017). This mechanism is using malware to detect sensitive data.

There are numerous risks that incorporate risky tasks, pernicious code with information uprightness to keep disclosure of unknown weaknesses vulnerabilities. Thus, it will create more troublesome for the programmers to figure out the code. Additionally, integrity mechanism is to ensure the consistency and accuracy of data while anonymity is the service of keeping and hiding data sources. It assists in guaranteeing information classification and security. Wireless communication security is to ensure the secure of configuration when communicate across wireless network and Integrity monitoring tools also can be implemented to keep the alert while completing their job duties (Yaaqob, 2017).

METHODOLOGY

The method to collect the data in this study is a qualitative method. Observations and interviews were used as an approach to have a better knowledge regarding the topic. In addition, researchers also have use literature studies such as content analysis, case studies and phenomenological studies. The populations for this research are the company in Malaysia that applied the cybersecurity in the manufacturing companies or services while the research sample consists of top management in the Malaysian companies and professional people that are using cybersecuirty. The number of respondents depends on saturation of data from the respondents but the targeted respondents are in industries which are manager companies to answer the first research objective.

For this research, researcher was used purposive sampling method. This purposive sampling method was selected require to the characteristics of a population. For this sampling used the expert purposive samplings to provide various vary of cases relevant to a specific development or event. Other than that, researchers also have use snowball sampling to know the existing about the research studies from the acquaintances.

Interview sessions were conducted with different participants. The purpose for the interview is to obtaining specific estimation on the research topic and intended answer the question to the research topic. The interview was through telephone, email and face to face. In the interview protocol contain a headline, the description and instruction to the interviewer, crucial questions to ask, investigate for major query, transition info for the interviewer, space to record on what the interviewers are mention and the space where investigators noted a reflective note(National Center for Postsecondary, 2003).

The semi-organized interview approach was, consequently, most proper to address these issues. Despite the fact that the questioner has a prepared list of inquiries that calls interview guide, the interview procedure is adaptable and the interviewee can react openly in his/her words (Bryman, 2004). Open-inquiries in the in-depth interviews enable respondents to clarify their perspective and comprehension of supportability and economical business rehearses, without being restricted by biased classifications provided. The questioner can allow the respondents to clarify their answers (Zikmund, 2015).

Reliability and validity are from members checking and different sources. The researcher sent the transcript to the respondents in other to get the approval validation of data while different sources are from journal or past articles. The semi-structured interview data were digitally recorded and later transcribe using the denaturalized convention (Maclean, Mechthild, & Alma, 2004). Through this adaptability, thematic analysis enables detailed data exploration (Clarke, 2016). This analysis helps the researcher to discover the patterns and construct specified researcher query. For this research Atlas.ti software was applied.

RESULTS AND DISCUSSIONS

Three participants were selected from different level managerial where two respondents are from top management and another respondent is an officer. Table 1 indicates the background of the respondents.

Table 1. Background of Participants

Respondent ID	Designation
A	Head Strategic Research And Advisory Department Strategy Research Division
B	Branch Manager
C	ICT Security Officer

Minor Findings

Awareness of the Importance Cyber Security in Workplace

Cyber security is a global phenomenon representing a complex socio-technical challenge for governments and private companies with the involvement of individuals. Although cyber security is one of the most important challenges faced by governments today, the visibility and public awareness remains limited. All respondents agreed on the importance of the implementation of cyber security in any organizations due to the trend of Industry 4.0.

As claimed by respondent A:

Cyber security is important because it involves the protection of information that is accessed and transmitted via the internet or more generally through any computer network. From technology perspective, organization need to use the appropriate mix of such resources as encryption techniques, firewall, access controls and intrusion detection systems.

This is aligned with Goutam (2015) who stated that cyber security is now considered as important part of individuals and families, as well as organizations, governments, educational institutions and our businesses. The findings of the current study is also consistent with those of Walter (2018) who found that an appropriate learning on online behaviour and system protection results decreases the vulnerabilities with safer online environment.

The average unprotected computer connected to the Internet can be compromised in moments. Thousands of infected web pages are being discovered every day. Hundreds of millions of records have been involved in data breaches. New attack methods are launched continuously. The importance of information security should be highlighted as a necessary approach to protect data and systems.

According to respondent B, "*the protection and security of employees' work and personal lives are no longer separate.*"

They have been intertwined with evolving trends of social networks, the internet of things, and unlimited connectivity.

"This is because cyber security is no longer just the responsibility of the company IT department". "It is now the responsibility of every employee, not just to protect their work assets but their personal data as well", as stated by respondent C.

Failure to do that will make a risk or problem in the company. Other than that, the trend for cyber threat and security kept changing in an instant.

Role of Cyber Security in Organizations

100% of respondents agreed that the role of cyber security is to protect confidentiality, and ensure integrity with availability. This is aligned with the works of Gordon (2006), who stated that cyber security is to protect the privacy of information as claimed by all of participants. Integrity is described as protecting the accuracy, reliability and validity of information. It is also to ensure that authorized users can access information on a timely basis. All the e-business models, business to business, business to consumer and business to government rely on secure transmission and storage sensitive information.

"The role of cyber security is to establish trust so that the buyer and sellers are willing to participate in electronics transactions" as stated by respondent A.

According to respondent B:

The function of cyber security is to protect the assets in our organization such as intangible assets of organizations such as computer software, patents, intellectual property (IP), company information, licensing agreements, employment contracts and trade secrets are mainly stored, managed and shared on network systems.

This is aligned with literature that states a breach of security controls by hackers or sometimes unauthorized employees which results in data privacy issues (Yaaqob, 2017). Unauthorized persons or agencies can gain access to sensitive personal information leading to identity theft and misrepresentation. Cyber security could also prevent any system failure that will cause physical damages and harm to the humans, machine, process, data and others (Ahmad et al., 2017; Goutam, 2015). This is aligned with respondent C statement that the role of cyber security is *"to secure customer data and company records"*. Organizations manage records of finances, employees, physical assets, passwords and access codes to sensitive information or to carry out specific tasks on network systems. Sensitive information must be secured from unauthorized modification or access by both employees and hackers. Knowing the crucial role of cyber security, thus, all department areas that consists of automated system, computer and networks that has information data must be protected.

MAJOR FINDINGS

The Challenges of Cyber Security Faced by Malaysian Companies

Cyber security which everyone is talking about has, without a doubt, become one of the most significant threats to global business in year 2018 and beyond. The characteristics of cyber-crime are a threat to all businesses regardless of size of companies. When you have an organisation that employs others, holds confidential client data, accounts, personnel files, business plans, confidential project information. If any of it is breached, it will signify that the company is vulnerable.

The challenges can be divided to five themes; namely technology, human or people, process, budget and awareness. This is based on respondent A, who claimed that, "*technology is used to support the communication that is always changing*". Thus, with the increasing use of digital technologies such as cloud, big data, mobile, IoT and artificial intelligence with the growing connectivity of everything, come greater challenges on the level of security, compliance and data protection and regulations which should be effectively tackled.

Respondent B claimed that:

with the growth of technology cyber threats and cyber-attacks, it has took new shapes in the form of next-generation ransom ware, web attacks and others. Scammer or hackers are always step ahead from us.

This is accordance to the opinion of respondent C, "*the security parameter has changed.*"

The second theme covers human or people, which refers to employees, employers, suppliers and customers. Organisations that do not produce user security policies or train the users in recognised good security practices will be vulnerable to many of these risks. "*Human or people need to know how to use and implement technology track or risk while using the systems*" as stated by respondent A.

This is followed by respondent B and C; "*the level of knowledge and expertise of employee are also important as it needs to ensure that the system will run smoothly.*"

The present findings seem to be consistent with other research such as by Heikkila (2016), who reported that the lack of employee training qualified professionals is one of the challenges to implement learning environment that can give real access o industrial control system. Process is reported to be one of the challenges in implementing cyber security elements. This is due to the difficulty in handling the procedure of implement information security. It requires numerous stages and requires involvement of all departments in the company.

As mentioned by respondent A:

Process means the systematic process or SOP to runs the program or systems, thus, people need to know how to use and implement the technology element track or risk when using the system.

If users or people are not trained in securing the usage of their organisation's ICT systems or the functions of a security control, it may accidentally misuse the system, potentially compromise the security control and further affect the confidentiality, integrity and availability of the information held on the system. Hence, a proper and systematic SOP should be designed and established in certain parts of the information security in an organization. This supports literature which describes the lack of improper or unsafe operation and malicious code modifications (Heikkila, 2016).

However, in the perspectives of budget, it is also considered as a challenge in establishing cyber security system in organizations. All of the respondents agreed that implementing cyber security requires a huge amount of investment.

As stated by respondent B: "*financial or budgets also needs to be strong enough when we have information security systems in company.*"

It corroborates with the findings of Heikkila (2016), who found that the lack of security investment among companies is one of the problem that is faced by companies to implement the systems.

According to respondent B and C: "*in terms of financial, the company should put priority in the information security budget under the company's strategic plan.*"

This is to avoid the company from having lack of budget in aspects of security. Hackers or malware always have new ways to breach the security systems and if this happen the company will always compactible to overcome the issues if the company have stable financial. This finding is in agreement with Jaime (2016)'s findings which showed that the malicious data seekers will always find new methods to steal the data especially during the peak hours.

There are similarities between budget by Jaime (2016) and those described by Naik (2017). Bad data can quickly cause problem to the server and requires higher budget to be fixed (Naik, 2017). Sensitive data disclosure occurs when an application does not adequately protect sensitive information. Hence, information needs to be encrypted during transit. Awareness and training program is crucial to disseminate information to all users, namely; employees, consumers, and including managers. In the case of an Information Technology (IT) security program, it should also be appropriately to communicate on security requirements and appropriate behaviour (Bada, 2015). An awareness and training program can be effective, if the material is interesting, current and simple enough to be followed. Thus, it is critical for all employees to be aware of their personal security responsibilities and the requirement to comply with corporate security policies. This can be achieved through systematic delivery of a security training and awareness programme that actively seeks to increase the levels of security expertise and knowledge across the organisation as well.

As stated by respondent B:

Awareness is the first step that an organization needs to do in other to give knowledge to the employee" while respondent C claimed that, "awareness is the first protection that user can avoid from hackers' activity."

If users are not aware of any special handling or the reporting requirements for particular classes of sensitive information, the organisation may be subjected to legal and regulatory sanctions.

The Implementation of Cyber Security

There are several ways to implement the cyber security in the organization. In this aspect we found five themes, such as "*build a team of IT and non-IT staff which conduct top to bottom security audits, update software and systems, provide new and continuing security education, conduct risk analysis and test systems*" as agreed by all respondents.

Firstly, "*it is crucial to build a team of IT and non-IT staff*" as stated by all respondents. This team will determine the security policies.

According to all the respondents, "*conduct top to bottom security audits to review the security practices and policies of IT systems.*"

Thus, it is beneficial organizations to involve all employees in the process of implementing cyber security (Bada, 2015). This theme supports the idea of Hans de Brujin (2017) who reported that if non-technical people in the organization could not understand the importance of certain policy or investment for cyber security, the process will not be well implemented.

The majority of respondents agreed that the procedure of implementation begins with "*starting with the basic elements such as accounts, passwords, workstations, and mobile devices and then work up to network hardware, then servers and applications.*"

Update software and systems was not different from what other IT vendors do when the vulnerability security was discovered. As mentioned by respondent B:

Adopts a 'push' methodology, forcing new security updates onto a user's device when they connect to the network, instead of a 'pull' methodology because we don't know when a user need their device and these updates can solve a problem.

These finding accords with literature, where creating policies and procedures for automatic updates software and systems can help ensure the proper security measures are in place and running optimally (Shelzer, 2018).

Respondent B and C suggested that "*the company will have a good security protection if they have an advanced systems and software.*" Thus, companies must provide new and continuing security education for all employees to ensure that security policies and practices stay fresh in employees' minds and can understand any policy additions or changes.

All three respondents conclude that; "*the first thing to do is the session awareness to all staff and then setup one technical team to have specific training on cyber security.*"

The findings corroborate with Jaime (2016) works that there is a need of awareness by employees regarding the issues of information security. As respondent B suggest; "*one way can review whether the employees are demonstrate vulnerability of confidential data is to do social engineering*". This social engineering can be as simple as someone give a password to another person or it could be a user who is pulling up a website at work and surrenders passwords or other vital information that ultimately gets into the wrong hands.

Other than that, is by conducting a risk analysis to identify potential areas of weakness, failure, or compromise in workplace. All respondents found that it is "crucial *to do risk analysis which can determine on what department are critical to have security protection.*"

The three respondents also suggest that there are three processes of risk analysis conditions for its acceptance. The process of risk analysis is as follows:

1. Firstly, "*group the results based on high, medium, and low risks in similar meaning, examine business services and categorize which ones are critical, important, or optional in implementation the system.*"
2. Secondly, it needs to test system whether it have good impacts to the organization or not. In other to test the system *"it is important to perform regular data backups.*" This also can physically secure the information assets.

According to 50% of respondent stated that; "*all software, hardware and network need to be secured*".

This is aligned with literature Shelzer (2018) who stated that cyber security requires knowing what devices authorized and unauthorized are connected and accessing information. This knowledge will usually require a full security assessment and ongoing accountability procedure setups.

RO2: Solutions to Overcome Challenges in Cyber Security

Type of improvement in cyber security aspects can be described as three themes which are awareness, training and budget or financial investment. It is very crucial to have awareness among all the employee and employer, a good training in other to enhance employee skills and stable financial for implement cyber security in an organization

Theme 1: The awareness of cyber security issues.

All organizations from large and small businesses to healthcare providers, academic institutions, and government agencies can experience data breaches or be targets of cybercrime, which can result in stolen intellectual property, theft of personal identity information, a disruption to the way do the business. A lot of campaign regarding awareness cyber security can be implemented.

For instance, "*awareness that focuses on ways to help create good cyber security awareness in the workplace which this campaign included on how to use secure methods of communication, beware of scams, minimize storage sensitive information and others*" as claimed by respondent A.

As concluded by 100% of respondents "*it is crucial to increase the level of awareness about cyber security among employees and third party.*"

This is aligned with Heikkila (2016) who suggests implementing the education and increase alertness and awareness of security issues among all employees. The company needs to maintain the user awareness of the cyber risk faced by the organisation. This is because without exception, all users should receive regular refresher training or reminder on the cyber risks to the organization which are to the employee and individuals. The company also need to promote an incident reporting culture. According to Maria et al, (2015) the organization should enable establish a security culture that empowers staff to voice their concerns about the poor security practices and security incidents without fear of recrimination.

Theme 2: Training program.

Employees can do harm to the business by visiting infected websites, responding to phishing emails, using business email through public Wi-Fi and more. Therefore it is important for companies to offer cyber security training. Training for employee is one of the solutions for the company to overcome the challenges and problem faced in implementing the cyber security. As stated by the three respondents: *"It is crucial to increase the number of staff training."*

Top management should monitor the effectiveness of security training. This is in accordance to the findings of Shelzer (2018) who suggest establishing mechanisms in other to test the effectiveness and value of the security training provided to all staff.

"This should be done through formal feedback by including questions in the staff survey on security training and the organization's security culture" as stated respondent C.

Those areas that regularly feature in security reports or achieve the lowest feedback ratings should be targeted for remedial action. Respondent B added: *"Support the formal assessment of Information Assurance (IA) skills."*

Staff in security roles should be encouraged to develop and formally validate their information assurance (IA) skills through enrolment on a recognized certification scheme for IA Professionals. Some security related roles such as system administrators, incident management team members and forensic investigators will require specialist training. Training to educate staff sometimes be enough to defend against some of the most common types of attacks levelled against businesses.

Theme 3: Budget or financial investment

Financial is one of the resources that quite challenging in cyber security. The cost needed to implement cyber security is usually high and requires a lot of investment from the company. Other than that, budget for cyber security is a challenging process, in part due to the limited tasks. It is a series of interrelated and persistent processes (Lawrence A. Gordon, 2006). As one of the respondents stressed that,

In terms of budget focus more on the risk department and need to allocate more investment and budget in other to face the new cyber threat.

The companies are suggested to purchase cyber security insurance. This can help the financial costs of breaches of personal data, payment card information, and intellectual property, as well as damage to brand reputation. This tactic is to ensure they have the financial resources necessary to respond to and remediate security incidents:

In aspect of budget the company must allocate the element of cyber security budget not in the IT department but in the earlier start project or the whole departments and in the strategic plan" as claimed by respondent A while respondent B stated

that "the organizations can employ managed security services to make sure that cyber security programs are managed in a cost-effective manner.

Doing so can also help ensure that companies have access to highly trained cyber security talent within their budget constraints.

Theme 4: Steps to secure confidential data that are being connected with the technology.

It is a well-known fact that passwords and usernames used by the majority of data users are weak (Wells, 2016). This makes it easy for hackers to get access to the information systems and compromise sensitive data of a business entity or government agency.

"Hardware authentication is very important in other to secure the technology that have been used" as stated by all respondents.

According to respondent B:

Technology gurus or cyber-guru have developed a solution in the user authentication process with a new Core vPro processor that belongs to the sixth generation of processors.

The core vPro can combine different hardware components with enhanced factors simultaneously for user identity validation purposes.

CONCLUSION AND IMPLICATIONS

The discussion of this research was discussed based on two objectives; (1) to analyse the challenges faced by Malaysian companies in cyber security and (2) to determine solution for Malaysian companies to overcome challenges in cyber security.

It is the best practice for an organization to use the same level to judge the challenges against information assets as it would be a legal, regulatory, financial or operational risk. This is achievable by incorporating information risk management information throughout the organization, which is actively supported by the board of directors, senior managers and authorized Information Assurance (IA) structure. Determining and communicating organizational attitudes and approaches to risk management is important. Thus, this study has shown that there are five challenges in the implementation of cyber security in Malaysian companies, namely; awareness, people, budget and technology.

Unfortunately, the usage by employees of organisation's Information and Communications Technologies (ICT) brings with various risks and challenges. It is critical for all employees to be aware of their personal security responsibilities and the requirement to comply with corporate security policies. This can be achieved through systematic delivery of an awareness programme that actively seeks to increase the levels of security expertise and knowledge across the organisation as well as a security-conscious culture.

Any employee may release personal or sensitive information to others. Not only that, unsatisfied users may try to abuse the system-level privileges or force other users to gain access to systems that are not permitted. Similarly, they can try to steal or physically disable the computer resources. These findings suggest that an awareness training program can be efficient if the programme is interesting, current and simpler. It was also shown that an awareness of cyber security in employee can give benefits to the organization in other to avoid cyber threats.

Ultimately, employee education and skills play major roles in running the cyber security system. Employee needs to be trained on information security to enhance the employee knowledge and the skills to secure the data. If employees are not trained to secure the usage of organisation's ICT systems or the functions of a security control, they may accidentally misuse the system, potentially compromise security control in terms of the confidentiality, integrity and availability of the information held on the system. Thus, this study has found that generally huge investment is required to implement cyber security in the company. The company should consider the budget in the early stage which stems from the strategic plan of companies.

Other than that, the increasing digital connectivity and automation practically of all processes in the business world throughout the entire value chain has led to the creation of agility. This has led to the development of a very high level of threat and significantly raises the risk of cyber security. Hardware authentication can be especially important when it comes to the Internet of Things, where the network of connected devices ensures that any device that seeks to be connected has the rights for connectivity to that particular network.

In today's world, due to the Industrial Revolution 4.0, it is essential to have information security in all aspects of operation and departments. To compete in the market situation today, it is important for companies to implement cyber security in the organization. It is time to build a better and more secured system that protects the confidential data for initiatives and better work. From the respondents' point of view, there is an urgency to strengthen cyber security in any organization. This must be considered in any company's strategic planning. However, each technology strategy has to suit to capabilities and resources available within the organization covering aspects of the capital, resources and capabilities of people to doing work.

Additionally, the ability of organizational employees to carry out assigned tasks aggressively will increase mutual profit. Therefore, an advanced level of protection security is also crucial to reduce the probabilities of cyber threats and ultimately ensure the successful operation of the organization.

REFERENCES

Ahmad-Reza Sadeghi, C. W., & Waidner, M. (2016). *Security and Privacy Challenges in.* Fraunhofer Institute for Secure Information Technology, Darmstadt, Germany.

Barbour, R. S. (2008). *Introducing Qualitative Research: A student guide to the craft of doing qualittive research.* London: Sage.

Baxter, P. (2008). Qualitative Case Study Methodology: Study Design and Implementation for Novice Researcher. *Qualitative Case Study Methodology: Study Design and Implementation*, *13*(4), 544–549.

Bogdan, R. C. (2007). *Qualitative research for education: An introduction to theories and methods* (5th ed.). Boston: Pearson.

Braun, V., & Clarke, V. (2006). Using thematic analysis in psychology. *Qualitative Research in Psychology*, *3*(2), 77–101. doi:10.1191/1478088706qp063oa

Brikci, N. (2007, February). *A Guide to Using Qualitative Research Methodology.* Retrieved from Google: https://cloudfront.ualberta.ca/-/media/science/research-and-teaching/teaching/qualitative-research-methodology.pdf

Bryman. (2004). *Social Research Method* (2nd ed.). Oxford, UK: Oxford University Press.

Dan Craigen, N. D.-T. (2014, October). Defining Cybersecurity. *Science Direct*, 1-9.

Deloitte. (2015). *Challenges and Solutions for the Digital Transformation and use exponential technologies.* Zurich: The Creative Studio at Deloitte.

Fischer, E. A. (2016, August 12). Cybersecurity Issues and Challenges. *Congressional Research Service*, 12.

Gizem Öğütçü, Ö. M. (2016, February). Analysis of personal information security behavior and awareness. *Science Direct, 56*, 83-93. Retrieved from https://www.sciencedirect.com/science/article/pii/S0167404815001406

Goutam, R. K. (2015). Importance of Cyber Security. *International Journal of Computers and Applications, 17*. Retrieved from https://pdfs.semanticscholar.org/5cfb/7a5bd2e6c181e8a69ebd49b1dadb795f493b.pdf

Hans de Bruijn, M. J. (2017, January). Building cybersecurity awareness: The need for evidence-based. *Science Direct, 34*, 1-7. Retrieved from https://www.sciencedirect.com/science/article/pii/S0740624X17300540

Hasnah Haron, S. N. (2017). A Hanbook for Business Research Methods. Kuala Lumpur: Pearson Malaysia Sdn Bhd.

Insight Active Learning. (2014, November 26). *Cyber Security and Related Issues: Comprehensive Coverage*. Retrieved from InsightIAS: http://www.insightsonindia.com/2014/11/25/cyber-security-related-issues-comprehensive-coverage/

Iqbal, S. (2016, May 31). On Cloud Security Attacks: A Taxonomy and Instrusion Detection and Preventation as a Service. *Science Direct*, 98-120.

Jaime, C. (2016). The challenges of cybersecurity frameworks to protect data required for the development of advanced maintenance. *Science Direct*, 223.

Jansen, C. (2016). Developing an Operating Industrial Security Services to Mitigate Risks of Digitalization. *Science Direct*, 133-137.

Jansen, D.-I. D.-P. (2016). Developing and operating industrial security services to mitigate risks of digitalization. *Science Direct*, 133-137.

Jay Lee, B. B. (2016). Introduction to cyber manufacturing. *Science Direct*, 13.

Jugder, N. (2016). *The thematic analysis of interview data: an approach used.* University of Leeds. Retrieved from http://hpp.education.leeds.ac.uk/wp-content/uploads/sites/131/2016/02/HPP2016-3-Jugder.pdf

Katharina, M. D. G. (2015). *Qualification For The Actory Of The Future*. Acta Universitatis Cibniensis – Technical Series.

Lane Thames, D. S. (2017). *Cyberrsecurity for Industry 4.0*. Springer Nature. doi:10.1007/978-3-319-50660-9

Lasi, H. F. (2014). Business & Information Systems Engineering. *Industry*, *4*(0), 239–242.

Lawrence, A., & Gordon, M. P. (2006). *Managing Cyber Security Resources A cost Benefit Analysis*. McGraw-Hills Companies.

MacLean, L. M., Meyer, M., & Estable, A. (2004). Improving Accuracy of Transcript in Qualitative Research. *Qualitative Health Research*, *14*(1), 113–123. doi:10.1177/1049732303259804 PMID:14725179

Manuel Dominguez, M. A. (2017). Cybersecurity training in control systems using real equipment. *Science Direct*, 12180-12184.

Maria Bada, J. R. (2015). Cyber Security Awareness Campaigns. *Research Gate*, *3*, 15.

Marjo Heikkila, A. R. (2016). Security Challenges in Small- and Medium-Sized Manufacturing Enterprise. *Science Direct*, 23.

Michael Mylrea, S. N. (2017). An Introduction to Buildings Cybersecurity. *Google Scholar*, 978-982.

Naik, S. (2017). Cyber Security - IoT. *IEEE International Conference On Recent Trends in Electronics Information & Communication Technology*, 764.

National center for postsecondary improvement. (2003). *National Center for Postsecondary Improvement*. Retrieved from https://web.stanford.edu/group/ncpi/unspecified/student_assess_toolkit/sampleInterviewProtocol.html

Oliver, D. G., Serovich, J. M., & Mason, T. L. (2005). Constraints and Opportunties with Interview Transcription: Towards Reflction in Qualitative Research. *Social Forces*, *84*(2), 1273–1289. doi:10.1353of.2006.0023 PMID:16534533

Paul van Schaik, D. J. (2017, October). Risk perceptions of cyber-security and precautionary behaviour. *Science Direct, 75*, 547-559. Retrieved from https://www.sciencedirect.com/science/article/pii/S074756321730359X

Penang skill development centre. (2017). *Introduction to the Industry 4.0*. Penang: Author.

Ponelis, S. R. (2015). Using Interpretive Qualitative Case Studies for Exploratory Research in Doctoral Studies: A Case of Information System Research in Small and Medium Enterprise. *International Journal of Doctoral Studies*, *10*, 535–550. doi:10.28945/2339

René Waslo, T. L. (2017, March 21). *Industry 4.0 and cybersecurity*. Retrieved from Deloitte Insight: https://dupress.deloitte.com/dup-us-en/focus/industry-4-0/cybersecurity-managing-risk-in-age-of-connected-production.html

Roca, S. K.-L.-D.-V. (2019). *Cybersecurity Current Challenges and Inria's research directions*. Le Chesnay Cedex, France: Inria.

Ruggunan, D. S. (2015). Introduction to Qualitative Analysis: Thematic Analysis. *Discipline of Human Resources*, 1-38. Retrieved from http://smitg.ukzn.ac.za/Libraries/General_Docs/Introduction-to-Qualitative-Analysis.sflb.ashx

Shelzer, R. (2018, April 17). *Cybersecurity and Manufacturing: What's at Risk and How to Protect Your Facility.* Retrieved from Global Electronic Services, Inc: https://blog.gesrepair.com/cybersecurity-and-manufacturing-whats-at-risk-and-how-to-protect-your-facility/

Veal. (2006). *Research Methods for Leisure and Tourism* (3rd ed.). Harlow, UK: Pearson Education.

Wells, J. C. (2016). Cyber-physical security challenges in manufacturing systems. *Manufacturing Letters*, 74-77.

Woodward, P. A. (2015, July 20). Cybersecurity vulnerabilities in medical devices: a complex environment and multifaceted problem. *US National Library of Medicines National Institute of Health,* 305-316. Retrieved from https://www.ncbi.nlm.nih.gov/pmc/articles/PMC4516335/

Yaaqob, I. (2017). The rise of ransomware and emrging security challenges in the Internet of Things. *Science Direct*, 453-454.

Yulia Cherdantseva, P. B. (2016, February). A review of cyber security risk assessment methods for SCADA systems. *Science Direct, 56*, 1-27. Retrieved from https://www.sciencedirect.com/science/article/pii/S0167404815001388

Zhang, Z. (2016). Social media security and trustworthiness. *Science Direct*, 1-11.

Keliang Zhou, T. L. (2015). Industry 4.0: Towards Future Industrial. *Science Direct*, 2147 - 2152.

Zikmund, W. (2015). *Business Research Methods* (16th ed.). Fort Worth, TX: Dryden.

Section 2

Smart Manufacturing and Supply Chain

Chapter 6
Multi-Objective Optimization of Economic and Environmental Aspects of a Three-Echelon Supply Chain

Rajaram R.
Tata Consultancy Services, India

Jawahar N.
Ramco Institute of Technology, India

S. G. Ponnambalam
https://orcid.org/0000-0003-4973-733X
University Malaysia Pahang, Malaysia

Mukund Nilakantan Janardhanan
University of Leicester, UK

ABSTRACT

It is very relevant in today's competitive world for suppliers to ensure that customer-demanded products are made available. Customers expect to obtain a product that has benefits and are available within an acceptable price and time. It is necessary for companies to optimally use their ability to satisfy customers' specified needs. Researchers and industries are working on developing green supply chain concept in the last few years due to environmental concerns. The objective of this chapter is to propose a three-echelon supply chain model that optimizes economic and environmental objectives simultaneously. The objectives considered are minimizing the total supply chain cost and minimizing CO_2 *emission of the supply chain network. The proposed model falls into NP-hard category. Multi-objective genetic algorithm is proposed to solve the proposed model and illustration is provided to explain the use of the proposed model. A procedure that could be followed to find the best possible solution based on user's choice among the Pareto front solutions is also explained.*

DOI: 10.4018/978-1-5225-9078-1.ch006

INTRODUCTION

Due to increasing awareness among industries on environmental concerns, lot of researchers and industries are working on green supply chain concept in the last few years (Nurjanni et al., 2014). As defined in supply chain management, for every product reaching the end user there would be a cumulative effort of various elements in the supply chain such as suppliers, manufacturing plants and customers (retailers/distributors). Now-a-days, the main focus in supply chain management (SCM) is to achieve overall competitive advantage (Marinagi et al., 2014). Major focus has been on objectives that are related to economy and many researchers attempted it by means of maintaining optimum inventory level (Ghodsypour & O'brien, 2001, Reddy et al., 2011), maximizing channel profit (Nachiappan et al., 2007), improving customer service performance (Nozick & Turnquist, 2001) and determining optimal distribution plan (Jawahar & Balaji, 2009, Hamedi et al., 2009). In the last few year's times, the industries and researchers have understood that competitive advantage cannot be only derived by focusing on economic aspects alone. Sustainability is another aspect which needs to be focused in achieving the global competitive advantage, resulting in both economical advantage and clean environment. This is very much applicable for SCM as well and few research work has been reported focusing on this element (Nurjanni et al., 2014). To be different from each other in market, industries realize the importance of 'green image' very seriously and recent research works focus on addressing the environmental aspects such as minimizing carbon emission (Zhang et al., 2015, Aksoy et al., 2014), minimizing energy consumption and maximizing the utilization of eco- friendly resources (Aksoy et al., 2014, Subulan et al., 2015, Abdallah et al., 2013) can be classified under Green Supply Chain Management (GSCM).

The environmental pollution is one of the major issues while transporting goods between the elements of the supply chain (SC). The emission from the vehicle during transportation leads to greater environmental problems. Among the different modes of transportation such as air, sea, rail and roadways, road freight transportation is widely used for movement of goods because of high flexibility, cheaper investment funds and availability. According to Jofred and Öster (2011), 30% to 40% of CO_2 emission are directly resulted from freight transportation in roadways. Hence, every industry working with SC mode aims to be sustainable in the business and attempts to implement plans to reduce the environmental impact as a result of the transportation and processing activities, besides concentrating on the goal of making profit. Many recent work are being reported with models considering both economy and environment aspects (Wang et al., 2011, Govindan et al., 2014, Pop et al., 2015, Validi et al., 2014, Chibeles-Martins et al., 2016, Urata et al., 2015). Within a country, road transportation is considered to the most common and cost effective. In countries like China and India, many industries such as automobile manufacturing industry, tire manufacturing industry, and industrial belt manufacturing sector, majorly rely for distribution of goods using road transportation.

Based on the above concerns, this chapter's major contributions are:

- A bi-criteria model for the sustainable supply chain with the objectives of minimization of SC cost to be competitive in market (economic) and minimization of CO2 emission for pollution (environmental) for a three-echelon supply chain.
- This problem considered can be classified as belonging to Sustainable Supply Chain model (SSCM). The proposed problem considers two objectives, hence it belongs to Multi-criteria optimization problem. The problem considered is to a Multi-criterial optimization problem and following approaches problems can be used to solve them:
    - **Case (a):** Converting the problem as a single objective problem:
        - Combine multiple objectives into a single objective by assigning weight to each objective and solve the problem.
        - Consider one of the objectives as the objective criterion and others to be set as constraints and solve the problem.
    - **Case (b):** Pareto optimal front, which retains multiple solutions for the different combinations of objectives solutions. The Pareto solution set once generated can be an easy reference solution set comprising of various non-dominated (i.e. quality) solutions. From the set of Pareto front solutions, a best suited solution for the user could be chosen.
- In this book chapter, the mathematical formulation of the model is proposed and it consists of non-linear functions and binary decision variables, and falls under the category of NP-Hard problem. On this consideration, this chapter proposes a Multi Objective Genetic Algorithm (MOGA) to generate the Pareto optimal set for the developed SSCM. To the authors knowledge this is for the first time multi objective GA is proposed to solve the considered problem.

The rest of the chapter is organized as follows: section 2 reviews the relevant SC literature based on economy perspective, environmental perspective and combined economic and environmental perspective. In section 3; working environment, problem definition and mathematical formulation for SSC model is described. Section 4 discusses elaborately the proposed MOGA with an illustrative problem. Section 5 discusses the details on the determination of the supplier selection and vehicle allocation for supplier and customer end based on user requirements on cost and CO_2 emission. Section 6 concludes the present work.

LITERATURE REVIEW

Economic and environmental objectives have been dealt in different manners in the supply chain (SC) literature in the last few years. This section explores the literature on supply chain problems reported in literature with features of the models, their objective functions and approaches proposed by researchers in solving them.

Economy Oriented Objectives

The research works in SCM focusing on economic objectives are presented in this section.

Ghodsypour and O'brien (2001) developed mixed integer non-linear programming model to solve multiple sourcing problem. In their model budget, quality and service are included. The objective of the model is to minimize the total annual purchasing cost which includes ordering cost, holding cost and purchasing price. Demand, capacity and quality are included as constraints. The proposed non-linear programming model is solved using Excel solver and the model is illustrated with numerical problems. Nozick and Turnquist (2001) developed a model which helps to minimize the cost and to maximize the customer service. They used linear functions to determine the optimal inventory level, location of distribution centers, and level of services. Rau et al., (2003) formulated a multi-echelon SC inventory model to achieve the minimization of joint total cost. The decision such as number of deliveries and order lot size is taken. Nachiappan et al., (2007) proposed vendor managed inventory (VMI) mode of operation based two-echelon single vendor–multiple buyers supply chain (TSVMBSC) model. The sales quantity and sales price determine the channel profit. A genetic algorithm based methodology is used to solve the considered problem. Jawahar and Balaji (2009) proposed a two-stage fixed charge transportation problem of a supply chain which involves manufacturing plant, distribution centers and retailers. The objective is to minimize the total cost of distribution. A genetic algorithm based methodology is used to solve the problem and its solution quality is compared with approximate and lower bound solutions. Hamedi et al., (2009) formulated a mixed integer nonlinear programming model to determine the optimal distribution plan which leads to minimum distribution cost for natural gas supply chain. The proposed model is solved in hierarchical manner, that is, each step section of the problem is solved and it is carry forward to next step as an input to attain the results. Reddy et al., (2011) considered a single echelon-two stage distribution inventory optimization model with warehouse and retailers with limited capacity. The objective of the model is to minimize the total cost of supply chain by providing an optimal inventory level for warehouse and retailer. The model is formulated as a linear programming problem and validated with the help of confectionary industry data. Gumasta et al., (2012) proposed an inventory model mapped with transportation model for perishable goods. The decision such as maximizing the profit by determining the optimal order quantity followed by minimizing the transportation cost by determining number of vehicles involved and capacity to be carried out by that vehicle is to be taken. Panda et al., (2015) have analyzed the coordination among three stages of supply which includes manufacturer, distributor and retailer. In this work they proposed contract-bargaining process to distribute channel profit among the supply chain members.

Mohammed and Wang (2017) presented a study on developing a cost-efficient supply chain model with an objective of minimizing total cost of transportation, number of transportation vehicles and delivery time. The developed model is based on a meat supply chain. They developed a multi-objective probabilistic programming model and tested the effectiveness and applicability of the developed model.

Environment Oriented Objectives

Few relevant research reported on SCM concentrating on environmental objectives are presented in this section.

Aksoy et al., (2014) constructed a fuel consumption and CO_2 emission calculation model considering vehicle technical specifications, transportation distance and vehicle load in green supply chain environment. The result shows that fuel consumption and CO_2 emission amount has direct relationship with vehicle type and travelled distance. The proposed algorithm can be applied to any kind of vehicle type. Subulan et al., (2015) proposed a model considering a multi echelon logistic network design problem considering the environmental issues. Different recovery options such as recycling, remanufacturing and energy recovery are concentrated in this model. The model is formulated as mixed integer linear programming and solved using fuzzy goal programming approach. Saffar and Razmi (2015) considered CO2 emissions in production and remanufacturing processes. They proposed a model to minimize total cost for designing a green supply chain network. However in their model, effects of transportation on environmental issues were not considered. Vidović et al., (2016) proposed a two echelon vehicle routing problem as mixed integer linear programming for recycling network design. The decision such as location of collection points and routes for the vehicle to be taken, which results the supply chain in a profit manner. The model is solved by a heuristic algorithm.

The Multi Criteria Decision Making models (MCDM) has been used extensively in solving supply chain management problems with focus on environmental impact objectives. The MCDM models are used for to evaluate the supply chain management practices such as green purchasing, green manufacturing, green warehousing, green transportation etc. and their performances. Jayant and Azhar (2014) used Interpretive

Structural Modeling (ISM) methodology to identify the most influential barriers for implementing green supply chain management (GSCM). Rostamzadeh et al., (2015) proposes a quantitative evaluation model for measuring the uncertainty of green supply chain management indicators and applies an approach named Vlsekriterijumska Optimizacija I Kompromisno Resenje (VIKOR) to solve the green multi-criteria decision making problem. Govindan et al., (2015) used the decision-making trial and evaluation laboratory (DEMATEL) method with intuitionistic fuzzy sets to evaluate the important relationship among GSCM practices and performances. Sazvar et al., (2018) presented a detailed review of literatures reported on GSCM. Following section presents the most relevant literature reported where both economic and environmental objectives are combined.

Economy and Environment Combined Objective

In SCM, the economic and environmental aspects are considered independently. Recently, the studies are focusing on models to address both economic and environmental aspects simultaneously. The most important research works concentrating on both objectives simultaneously are described in this section.

Wang et al., (2011) formulated a multi-objective mixed integer programming model to solve supply chain network design problem. The decision such as location of facilities, supplier selection and transportation routes are taken. The model is solved using ILOG CPLEX solver. Pishvaee and Razmi (2012) developed a bi-objective optimization method to minimize total cost and environmental impact of a supply chain network. They utilized a life cycle assessment (LCA) method to assess and quantify the environmental impact. A multi-objective fuzzy optimization was also proposed. Govindan et al., (2014) formulated a model considering a two echelon supply chain network to solve multiple-vehicle location–routing problem. The decisions such as number and location of facilities, amount of products to be delivered and the optimal route for distribution is considered. The model is solved by metaheuristic algorithm called MHPV, a hybridization of adapted multi-objective particle swarm optimization (MOPSO) and the adapted multi-objective variable neighborhood search algorithm (AMOVNS). Pop et al., (2015) proposed an efficient reverse distribution system to solve sustainable supply chain network design problem. The decision such as determination of routes and size of shipments on the routes are decided. This problem is solved by several classical heuristic algorithms such as Classical Reverse Distribution Algorithm (CrdA), Classical High Reverse Distribution Algorithm (CHrdA) and Dynamic Reverse Distribution Algorithm (DYrdA). Govindan et al., (2015) proposed a bi objective mixed integer programming model to solve five echelon supply chain network design and order allocation problem. The products are dispatched from distribution centers to retailers by two ways either by direct shipment or by cross docking. The model is solved by MOHEV algorithm, which is the hybridization of adapted multi-objective electromagnetism mechanism algorithm (AMOEMA) and adapted multi-objective variable neighborhood search (AMOVNS). Chibeles-Martins et al., (2016) proposed a mixed integer linear multi objective optimization model for supply chain design and planning. The decision such as capacity level and location of the supply chain players (factories, warehouses and distribution centers), inventory level and material flow are considered. The model is solved by simulated annealing algorithm. Urata et al., (2015) developed a mixed integer programming model to solve global supply chain network design problem. In this model the demand for finished products and the reduction ratio of the environmental impact is considered. The determination of supplier location and factory location are taken as a decision. Huang et al., (2016) proposed a game theoretic model which includes greenhouse gas emission management among multi-level supply chain coordination. The genetic algorithm-based methodology is applied to solve the model. The decision such as optimizing product line design, transportation modes, supplier selection and pricing decision are taken as a result. (Tsao et al., 2018)

Based on above mentioned literatures, the CO_2 emission model during transportation is modeled differently. In this chapter, the CO_2 emission during transportation is calculated using the model by (Aksoy et al., 2014) which *is* based upon transportation distance, vehicle load and vehicle technical specifications for different truck types. Based on the literature review, research on supply chains with respect to economical and environmental aspects can be summarized:

- The performance evaluation of SC networks is mostly based on cost/profit oriented and risk faced by the supply chain members.
- Green supply chain model considering the environmental objectives are on the environmental degradations, which are carbon Emission, energy Consumption and waste generation.
- Constructing fuel consumption and CO_2 emission calculation models have been given importance.
- The Multi Criteria Decision Making Models (MCDM) have been extensively used in GSCM.
- Modeling approaches based on dynamic programming, goal programming and genetic algorithm have been rarely used in the existing research on SSCM, especially while formulating a model for SSCM (Ansari & Kant, 2017).
- Sustainability is the growing concern in supply chain management. The tradeoff between economic factors and environment factors are considered widely in SSCM.

On this consideration, this chapter mainly focuses on modeling a three-echelon sustainable supply chain and using a multi-objective genetic algorithm to minimize the total cost of SC and to minimize the CO_2 emission of the supply chain Network which has received less attention among researchers and very relevant.

PROBLEM DESCRIPTION

The SSCM considered in this paper is a three-echelon SC network comprising of suppliers, manufacturing plant and customers (i.e. retailers or distributors). This section discusses on the environment and the mathematical formulation of the SSCM.

Model Environment

Figure 1 presents the supply chain network of the considered problem. SC network is centered with a manufacturing unit that procures the required raw materials from S number of suppliers (indexed as 's') located at different origin /distance, produces P different models (indexed as 'p') of a product and delivers them to the C number of customers (indexed as 'c') located at different destination/distance.

Each customer 'c' has a specific demand for each model 'p' (order quantity of model 'p' by the customer 'c') and is denoted as D_{cp}. Equation 1 gives the total quantity (Q_p) of each model to be produced:

$$Q_p = \sum_{c=1}^{C} D_{cp} \tag{1}$$

Each model 'p' requires different combination of raw materials in different quantities and the quantity required per one unit of product is Q_{mp} and is known $\forall p$. The weight of each model (W_p) depends on the quantity (Q_{mp}) of material 'm' involved in producing the model 'p'. The different type of raw materials M (indexed as 'm') to be procured becomes the union of material combinations of all the models. Equation 2 gives the total quantity (Q_m) of each material (i.e. $\forall m$) to be procured for producing the quantity of Q_p.

$$Q_m = \sum_{p=1}^{P} \left(Q_p \times Q_{mp} \right) \tag{2}$$

Figure 1. Supply chain network

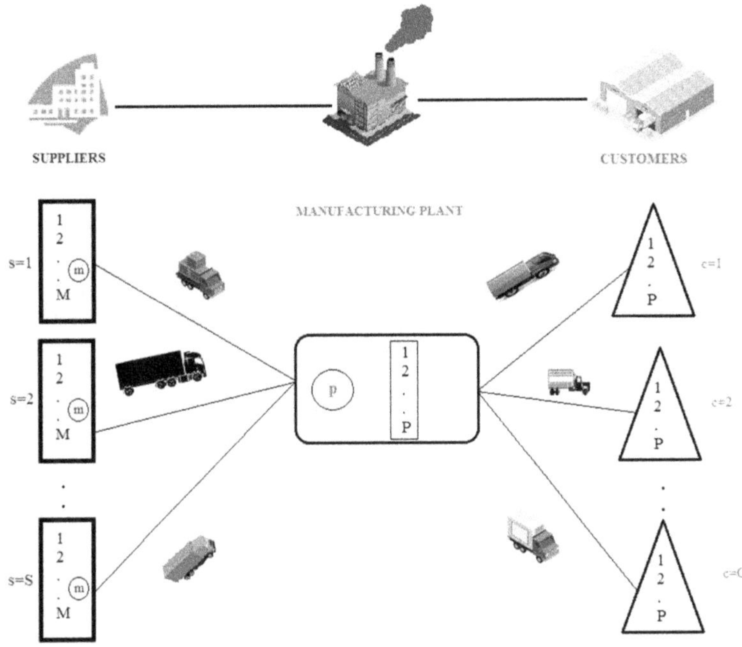

The cost of material 'm' varies with supplier 's' and is denoted by C_{ms}. However, it is not necessary to assure that all suppliers are capable of supplying all M materials. When a supplier 's' does not have material 'm', then the cost is given as a larger number 'L' to isolate from purchase decision. Besides, the policy of the manufacturing plant is to buy the required total quantity of each material (Q_m) as a single lot. Hence, the raw material cost (RMC) for all M materials can be estimated using Equation 3.

$$RMC = \sum_{m=1}^{M}\sum_{s=1}^{S}\left(C_{ms} \times Q_m \times \delta_{ms}\right) \tag{3}$$

where,

δ_{ms} = Binary variable indicating the supplier 's' for material 'm'

$$\delta_{ms} = \begin{cases} 1 & if\ supplier\ 's'\ supplies\ material\ 'm' \\ 0 & otherwise \end{cases}$$

The manufacturer procures raw material 'm' from supplier 's' in a particular type of vehicle 'v,' among the 'V_s' types available at his place. The vehicles are available at different capacity (C_{vs}) range. The transportation cost involved during the procurement (TC_{pro}) is based on vehicle selected for supplier and the vehicle cost elements of varying cost U_{vs} per unit and fixed cost F_{vs} per trip. Equation 4 shows transportation cost during procurement.

$$TC_{pro} = \sum_{s=1}^{S}\left[\sum_{v=1}^{V_s}\left(\delta_{vs}\right)\times\left\{\sum_{m=1}^{M}\left(Q_m\times\delta_{ms}\times U_{vs}\right)+\left(F_{vs}\times\frac{\left(Q_m\times\delta_{ms}\right)}{C_{vs}}\right)\right\}\right] \tag{4}$$

where,

δ_{vs} = Binary variable indicating the vehicle type 'v' from supplier 's'

$$\delta_{vs} = \begin{cases} 1 & if\ vehicle\ 'v'\ selected\ for\ supplier\ 's' \\ 0 & otherwise \end{cases}$$

The term $\frac{\left(Q_m\times\delta_{ms}\right)}{C_{vs}}$ rounds next higher integer to evaluate the number of trips required for the selected vehicle. The total procurement cost (PC) is calculated by the sum of raw material cost and the transportation cost during procurement as in Equation 5.

$$PC = \sum_{m=1}^{M}\sum_{s=1}^{S}\left(C_{ms}\times Q_m\times\delta_{ms}\right)+\sum_{s=1}^{S}\left[\sum_{v=1}^{V_s}\left(\delta_{vs}\right)\times\left(\sum_{m=1}^{M}\left(Q_m\times\delta_{ms}\times U_{vs}\right)+\left(F_{vs}\times\frac{\left(Q_m\times\delta_{ms}\right)}{C_{vs}}\right)\right)\right] \tag{5}$$

The manufacturing cost varies with model 'p'. It is addressed as MC_p and known $\forall p$. Hence the cost of production (CP) for all the models to satisfy the requirement of the entire customer (i.e. Q_p) can be determined using Equation 6.

$$CP = \sum_{p=1}^{P}\left(MC_p\times Q_p\right) \tag{6}$$

The manufacturer can hire a particular type of vehicle 'v' in any numbers among the 'V_c' types available at his place for shipping the models. Each vehicle is in different capacity range (C_{vc}). The distribution cost involved during the delivery (DC) of models from plant to customer is evaluated based on vehicle selected for customer and the vehicle cost elements are varying cost U_{vc} per unit and fixed cost F_{vc} per trip and is shown in Equation 7.

$$DC = \sum_{c=1}^{C} \left[\sum_{v=1}^{V_c} \left(\delta_{vc} \right) \times \left(\sum_{p=1}^{P} \left(D_{cp} \times W_p \times U_{Vc} \right) + \left(F_{vc} \times \frac{\left(D_{cp} \times W_p \right)}{C_{vc}} \right) \right) \right] \tag{7}$$

where,

δ_{vc} = Binary variable indicating the vehicle type 'v' from customer 'c'

$$\delta_{vc} = \begin{cases} 1 & if\ vehicle\ 'v'\ selected\ for\ customer\ 'c' \\ 0 & otherwise \end{cases}$$

The total cost (TC) of supply chain is the sum of procurement cost, cost of production and distribution cost and it is calculated using Equation 8.

$$\begin{aligned} TC = & \sum_{m=1}^{M} \sum_{s=1}^{S} \left(C_{ms} \times Q_m \times \delta_{ms} \right) + \sum_{s=1}^{S} \left[\sum_{v=1}^{V_s} \left(\delta_{vs} \right) \times \left(\sum_{m=1}^{M} \left(Q_m \times \delta_{ms} \times U_{vs} \right) + \left(F_{vs} \times \frac{\left(Q_m \times \delta_m^s \right)}{C_{vs}} \right) \right) \right] \\ & + \sum_{p=1}^{P} \left(MC_p \times Q_p \right) + \sum_{c=1}^{C} \left[\sum_{v=1}^{V_c} \left(\delta_{vc} \right) \times \left(\sum_{p=1}^{P} \left(D_{cp} \times W_p \times U_{Vc} \right) + \left(F_{vc} \times \frac{\left(D_{cp} \times W_p \right)}{C_{vc}} \right) \right) \right] \end{aligned} \tag{8}$$

The emission during procurement (E_{pro}) involves that the CO_2 emission $\left(CO_2^{vs} \right)$ from the vehicle 'v' which is selected for the supplier 's' during procurement and the load transmitted by that vehicle. Equation 9 determines the emission during procurement

$$E_{pro} = \sum_{s=1}^{S} \left[\sum_{v=1}^{V_s} \left(CO_2^{vs} \times \delta_{vs} \right) \times \left(\sum_{m=1}^{M} \frac{\left(Q_m \times \delta_{ms} \right)}{C_{vs}} \right) \right] \tag{9}$$

The emission from the plant (E_{plant}) is evaluated based on the CO_2 emission (CO_2^p) which is released during the production of each model 'p' and is known $\forall p$. Equation 10 calculates E_{plant}.

$$E_{plant} = \sum_{p=1}^{P}\left(Q_p \times CO_2^p\right) \tag{10}$$

The emission occurs during delivery (E_{deli}) is calculated based on CO_2 emission (CO_2^{vc}) from the vehicle which is selected for distributing the model to customer and the load to be carried by that vehicle and it is determined as in Equation 11.

$$E_{deli} = \sum_{c=1}^{C}\left[\sum_{v=1}^{V_c}\left(CO_2^{vc} \times \delta_{vc}\right) \times \left(\sum_{p=1}^{P}\frac{\left(D_{cp} \times W_p\right)}{C_{vc}}\right)\right] \tag{11}$$

CO_2^{vs} and CO_2^{vc} are calculated based upon the capacity of vehicle (C_{vs}, C_{vc}) transportation distance from supplier to plant d_s and from plant to customer d_c and vehicle technical specification(Aksoy et al., 2014). The total emission (TE) of supply chain in grams is the sum of emission from the vehicle which released during procurement (E_{pro}), delivery (E_{deli}) and emission from the plant (E_{plant}) and it is calculated using Equation 12.

$$TE = \sum_{s=1}^{S}\left[\sum_{v=1}^{V_s}\left(CO_2^{vs} \times \delta_{vs}\right) \times \left(\sum_{m=1}^{M}\frac{\left(Q_m \times \delta_{ms}\right)}{C_{vs}}\right)\right] + \sum_{p=1}^{P}\left(D_p \times CO_2^p\right)$$
$$+ \sum_{c=1}^{C}\left[\sum_{v=1}^{V_c}\left(CO_2^{vc} \times \delta_{vc}\right) \times \left(\sum_{p=1}^{P}\frac{\left(D_{cp} \times W_p\right)}{C_{vc}}\right)\right] \times 1000 \tag{12}$$

Problem Definition

The problem aims at the optimum determination of the type of vehicle (truck) for both supplier (δ_{vs}) and customer (δ_{vc}) end transportation and the supplier selection (δ_{ms}), for the minimum total cost of supply chain and the CO_2 emission without affecting the required capacity of shipping a raw material 'm' from supplier 's' to the manufacturing plant for producing the model 'p' and then shipping the model 'p' from manufacturing plant to customer 'c' with the following information available:

- Number of suppliers (S), number of vehicles (V_s, V_c), number of materials (M) and number of models (P).
- Demand of customer on the model (D_{cp}), weight of each model (W_p), Quantity of material required for producing the model (Q_{mp}) and cost of material from supplier (C_{ms}).
- Unit cost of shipping U_{vs} and fixed cost associated with its distribution F_{vs} for procuring the raw material 'm' from supplier 's' to manufacturing plant ($\forall_m$ and $\forall_s$).
- Unit cost of shipping U_{vc} and fixed cost associated with its distribution F_{vc} for shipping the different model 'p' from manufacturing plant to customer 'c' ($\forall_p$ and $\forall_c$).
- CO_2 emission $\left(CO_2^{vs}\right)$ associated with its distribution for procuring the raw material 'm' from supplier 's' to plant ($\forall_m$, $\forall_s$ and $\forall_v$).
- CO_2 emission $\left(CO_2^{vc}\, QUOTE\right)$ associated with its distribution for shipping the model 'p' from plant to customer 'c' ($\forall_p$, $\forall_c$ and $\forall_v$).

Mathematical Formulation

Based on the relations discussed in Section 3.1, the formulation of a mathematical model for the SSCM considered in this paper is presented.

The notations used are as follows:

Identifiers

c = Customer identifier (c= 1 to C)
m = Material identifier (m= 1 to M)
p = Model identifier (p= 1 to P)
s = Supplier identifier (s= 1 to S)
v = Vehicle identifier (v= 1 to V)

Parameters

C = Number of customers *c*
C_{ms} = Cost of material *m* from supplier *s*
C_{vs} = Capacity of the vehicle *v* for supplier *s*
C_{vc} = Capacity of the vehicle *v* for customer *c*
CO_2^{vs} = CO_2 emission from vehicle *v* from supplier *s* to plant

CO_2^{vc} = CO_2 emission from vehicle *v* from plant to customer *c*
CO_2^{p} = CO_2 emission for producing model *p*
d_s = Distance between supplier *s* to manufacturing plant
d_c = Distance between manufacturing plant to customer *c*
D_{cp} = Demand of customer *c* on the model *p*
F_{vs} = Fixed cost of transportation of material *m* from supplier *s*
F_{vc} = Fixed cost of transportation of model *p* from manufacturing plant to customer *c*
M = Number of materials *m*
MC_p = Manufacturing cost for producing the model *p*
P = Number of models *p*
Q_m = Quantity of each material *m*
Q_{mp} = Quantity of material m for producing model *p*
Q_{ms} = No. of units of material *m* from supplier *s* to manufacturing plant
Q_p = Quantity of each model *m*
S = Number of suppliers *s*
TC = Total Cost of Supply Chain
TE = Total CO_2 emission from the supply chain
U_{vs} = Unit cost of transportation of material *m* from supplier *s* to manufacturing plant
U_{vc} = Unit cost of transportation of model *p* from manufacturing plant to customer *c*
V_s = Number of Vehicle type *v* for supplier *s*
V_c = Number of Vehicle type *v* for customer *c*
W_p = Weight of each model *p*

Decision Variables

δ_{ms} = Binary variable that specifies the material *m* from supplier *s*
δ_{vs} = Binary variable that specifies the vehicle *v* from supplier *s* to plant
δ_{vc} = Binary variable that specifies the vehicle *v* from plant to customer *c*

The mathematical formulation for SSCM is as follows:

$$Minimize\ TC$$
$$=\sum_{m=1}^{M}\sum_{s=1}^{S}\left(C_{ms}\times Q_m\times\delta_{ms}\right)+\sum_{s=1}^{S}\left[\sum_{v=1}^{V_s}\left(\delta_{vs}\right)\times\left(\sum_{m=1}^{M}\left(Q_m\times\delta_{ms}\times U_{vs}\right)+\left(F_{vs}\times\frac{\left(Q_m\times\delta_{ms}\right)}{C_{vs}}\right)\right)\right]$$
$$+\sum_{p=1}^{P}\left(MC_p\times Q_p\right)+\sum_{c=1}^{C}\left[\sum_{v=1}^{V_c}\left(\delta_{vc}\right)\times\left(\sum_{p=1}^{P}\left(D_{cp}\times W_p\times U_{Vc}\right)+\left(F_{vc}\times\frac{\left(D_{cp}\times W_p\right)}{C_{vc}}\right)\right)\right] \quad (13)$$

$$Minimize\,TE = \sum_{s=1}^{S}\left[\sum_{v=1}^{V_s}\left(CO_2^{vs} \times \delta_{vs}\right) \times \left(\sum_{m=1}^{M}\frac{\left(Q_m \times \delta_{ms}\right)}{C_{vs}}\right)\right] + \sum_{p=1}^{P}\left(Q_p \times CO_2^{p}\right)$$

$$+\sum_{c=1}^{C}\left[\sum_{v=1}^{V_c}\left(CO_2^{vc} \times \delta_{vc}\right) \times \left(\sum_{p=1}^{P}\frac{\left(D_{cp} \times W_p\right)}{C_{vc}}\right)\right] \times 1000 \tag{14}$$

Subject to:

$$\sum_{s=1}^{S}\delta_{ms} = 1; \forall m \tag{15}$$

$$\sum_{p=1}^{P}\left(\sum_{c=1}^{C}D_{cp} \times Q_{mp}\right) = \left[\left(Q_m\right) \times \left(\sum_{s=1}^{S}\delta_{ms}\right)\right];\ \forall m \tag{16}$$

$$\sum_{v=1}^{V_s}\delta_{vs} = \max\left(\delta_{1s}, \ldots \delta_{Ms}\right); \forall s \tag{17}$$

$$\sum_{v=1}^{V_c}\delta_{vc} = 1; \forall c \tag{18}$$

$$\delta_{ms} \in \left\{0,1\right\}; \forall m, \forall s \tag{19}$$

$$\delta_{vs} \in \left\{0,1\right\}; \forall v, \forall s \tag{20}$$

$$\delta_{vc} \in \{0,1\}; \forall v, \forall c \tag{21}$$

Equation 13 shows the objective function of total cost (TC). Equation 14 shows the objective function of total emission (TE). Constraint 15 shows that each material 'm' is supplied by only one supplier 's'. Constraint 16 shows that based on the demand of the material the supply of material from supplier to plant is carried out. Constraint 17 represents that one vehicle type is allocated to supplier 's' when the material is procured from that supplier. Constraint 18 represents that it will be assured one vehicle type to be allotted for customer 'c' during shipment of goods. Constraint 19, 20 and 21 represents the binary requirements of decision variables.

METHODOLOGY

SSCM is formulated as a zero-one integer programming problem. In multi-objective optimization problem, there is an existence of non-dominated solutions called pareto optimal solutions or trade off solutions. Integer programming packages, such as CPLEX, can be used efficiently for small and medium size problems but for large size problem it takes more computational time (Harris et al., 2014). Metaheuristic approaches are useful to obtain a set of trade-off solutions for large size problems. The SSCM model proposed in this paper can be classified as NP hard (Wang et al., 2011). Hence, a metaheuristic algorithm is required to obtain near optimal solution (Chibeles-Martins et al., 2016). In this, we propose to use genetic algorithm (GA) to solve the considered problem. GA is an evolutionary search algorithm which generates optimal solutions to the problem using techniques inspired by natural evolutionary progression (Kalyanmoy, 2001). GA follows the principle of Darwinian evolution theory i.e. fittest individuals has high chances for survival (Yeh & Chuang, 2011). GA is good in handling huge search space to find the optimal solutions. Flexibility, simplicity and the robust response to changing circumstance of the genetic algorithm are its practical advantages attracting researchers to use this algorithm to solve complex problem such as the considered on in this book chapter (Sadegheih, 2006). Due to complexity of the problem, this problem considered is solved used GA.

Description

Figure 2 shows the flowchart of the proposed MOGA. The description of various modules are provided in this section.

Data Input and Parameter Initialization

The following data input and parameter initialization are to be given in this module. The basic inputs for SSCM are:

- Number of suppliers (S), number of customers (C), number of vehicle type for suppliers (V_s), number of vehicle type for customers (V_c), number of materials (M), number of models (P), capacity of the vehicle for suppliers (C_{vs}), capacity of the vehicle for customers (C_{vc}).
- Demand of customer on the model (D_{cp}), weight of each model (W_p), Quantity of material required for producing the model (Q_{mp}) and cost of material from supplier (C_{ms}).
- Unit cost of transportation of material from supplier to manufacturing plant(U_{vs}) and Fixed cost of transportation of material from plant to customer (F_{vs}).
- Unit cost of transportation of model from manufacturing plant to customer (U_{vc}) and Fixed cost of transportation of model from plant to customer (F_{vc}).

The parameter initializations of MOGA are population size (pop_size), number of generation (max_gen), probability of crossover (P_c) and probability of mutation (P_m).

Generation of Initial Population

The genetic algorithm starts with randomly generated set of chromosomes called initial population. Each chromosome represents the decision on binary variable to specify the supplier for each material (δ_{ms}), integer variable that specifies the vehicle from supplier to plant (δ_{vs}) and integer variable that specifies the vehicle from plant to customer (δ_{vc}). The first S×M number of genes of the chromosome represents the decisions of δ_{ms} and the alleles take the binary values that are generated randomly. The next S genes are for vehicle type determination at supplier end (i.e. δ_{vs} decisions) with its value randomly generated within range of 1 to V_s. The last C genes have the decisions of δ_{vs} which also has a random value in the range of 1 to V_c. Equation 22 gives the length (l) of the chromosome. Figure 3 shows the genetic representation of a chromosome with δ_{ms}, δ_{vs} and δ_{vc} decisions.

$$l = \left(S \times M\right) + S + C \tag{22}$$

For instance, consider a problem with S=4, M=7, C=3. The length of chromosome is 35 (i.e. (4×7) + 4 +3).

Figure 2. Procedure of the proposed MOGA

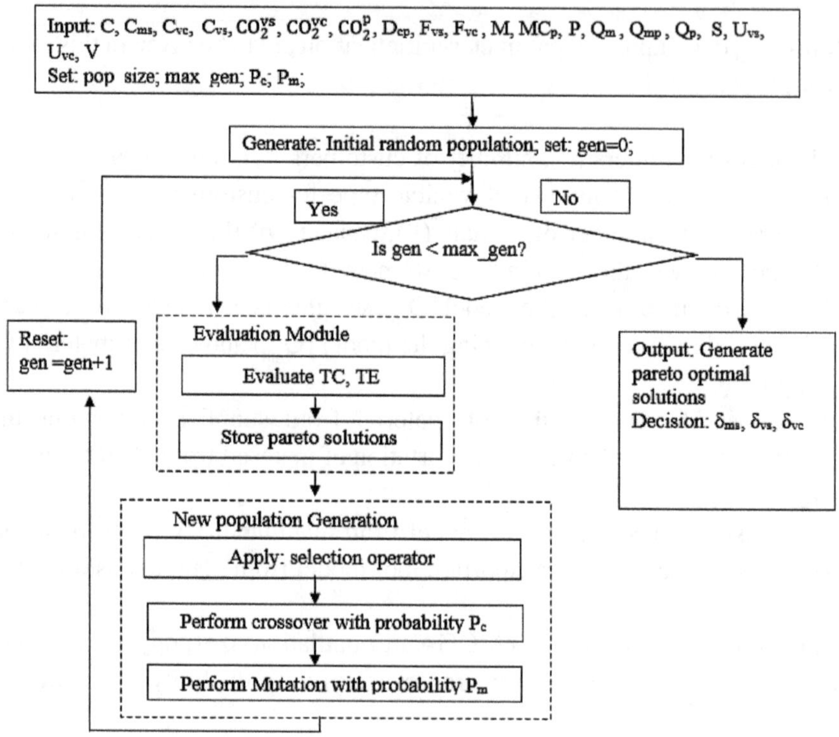

Figure 3. Genetic representation of chromosome

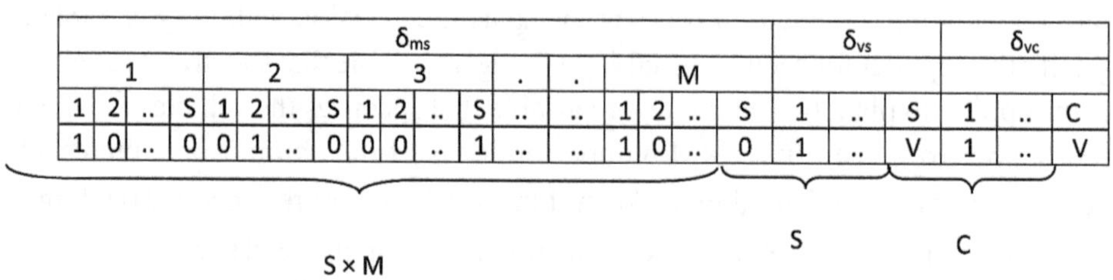

Termination Condition

The number of times the process of evaluation and new population generation is based upon the size of the problem. The number of generation (max_gen) is considered as a termination condition. If the termination condition is not satisfied, then the population of solution is modified to create the better population.

Evaluation Module

The chromosomes in the population are evaluated using equation (13) and (14), to get total cost (TC) and total emission (TE) respectively. After TC and TE evaluation, the non-dominated solutions within the population of each generation are stored.

New Population Generation

At each step, the current population is used to create the children that make up the next new generation. The new population generation will tend to cause the problem to converge on good solutions. The new population generation is carried out by selection, crossover and mutation operations. The reproduction or selection operation is carried out to make multiple copies of good solutions and eliminate bad solutions in the current population, while population size remains to be same. In this paper, roulette wheel or proportionate selection method is used. Crossover is done to explore the search space. Crossover is applied to pair of individuals selected randomly from the population which are created after the reproduction/ selection operation. The probability of crossover (P_c) is an important parameter for doing crossover operation and it is specific to the problem. The random number is generated between 0's and 1's. The chromosomes which get random number less than P_c will undergo crossover operation to form new child. If the condition is not satisfied, then the pair of selected chromosomes will be stored as it is. The chromosomes selected for crossover is subjected to multi-point crossover, which has three random cut points at each decision. At the cut points, the segments of the strings are interchanged so that the new set of strings are formed. The mutation is helpful to keep the diversity in the population. Mutation alters the genetic character of child chromosome by changing a random a bit in the string with a probability of mutation (P_m). A random number is generated between 0's and 1's for each chromosome. The chromosome which gets random number less than P_m will be mutated. If the random number generated for the chromosome is not less than P_m then the corresponding chromosome is stored without doing mutation. The random material is selected and for that existing supplier is swapped with any one of the remaining suppliers for the decision on δ_{ms}. For decision on δ_{vs} and δ_{vc}, the random supplier and customer are selected and replaced with any one of the vehicle types.

Output Module

The non-dominated set or pareto set from each generation which is stored by following the section 3.4, from that on comparing TC and TE with other solutions the best pareto optimal solutions are generated. The chromosome that corresponds to pareto optimal solutions are obtained as the output and the decision such as δ_{ms}, δ_{vs}, δ_{vc} can be made easier.

Numerical Illustration

As there is no benchmark problem available for the model proposed in this paper, an example problem is generated to suit the model and the details are presented in this section.

Step 1: Data Input and Parameter Initialization

Table 1 shows the data input and parameter initialization

Step 2: Generation of Initial Population

Table 13 gives the random initial population in which each chromosome represents the decision on δ_{ms}, δ_{vs}, and δ_{vc}. The first 28 (SxM=4x7) number of genes representing the decision of δ_{ms}. The alleles take the binary values that are generated randomly such that each four genes will have the presence of 1's at one place only, indicating that each material is procured by only one supplier. The next 4(S) genes determine the decision of δ_{vs}, that is the vehicle type selected for supplier which are allowed to generate randomly within the range of 1 to 4 (V_s). The last 3(C) genes indicate the decision of δ_{vc} that is vehicle type selected for customer which are allowed to generate randomly within the range of 1 to 4 (V_c). The total length of the chromosome is 35 (i.e. 28+4+3). Similarly 20 (population size) strings are generated randomly.

Step 3: Termination Check Module

The number of times the process of evaluation and new population generation is kept to be 1000 for this problem, which is considered as a termination condition.

Step 4: Evaluation Module

In the evaluation module, the populations are evaluated based on objective function TC and TE. After that each chromosome fitness values (TC and TE) are compared with every other chromosome in the current population to check that if it is dominated by any other chromosome in that population.

Step 5: Genetic Operations

Table 1. Data input and parameter initialization

S. No.	Particulars	Data Source	Data Value
1	Number of suppliers S	Random generation	4
2	Number of customers C		3
3	Number of models P		4
4	Number of materials M		7
5	Number of vehicle type for suppliers V_S		4
6	Number of vehicle type for customers V_C		4
7	Demand of customer on model D_{cp}	Random generation	Shown in table 2
8	Quantity of material for producing model Q_{mp}	Random generation	Shown in table 3
9	Cost of material C_{ms}	Random generation	Shown in table 4
10	Capacity of vehicle for supplier C_{vs}	Random generation	Shown in table 5
11	Unit cost of transportation of vehicle for supplier U_{vs}	Random generation	Shown in table 6
12	Fixed cost of transportation of vehicle for supplier F_{vs}	Random generation	Shown in table 7
13	CO_2 emission from the vehicle for supplier CO_2^{vs}	Adopted and can be calculated from Aslı Aksoya et al., (2013)	Shown in table 8
14	Capacity of vehicle for customer C_{vc}	Random generation	Shown in table 9
15	Unit cost of transportation of vehicle for customer U_{vc}	Random generation	Shown in table 10
16	Fixed cost of transportation of vehicle for customer F_{vc}	Random generation	Shown in table 11
17	CO_2 emission from the vehicle for customer CO_2^{vc}	Adopted and can be calculated from Aslı Aksoya et al., (2013)	Shown in table 12

Table 2. Demand of customer on model (D_{cp})

Customer (C)	Models (P)			
	1	**2**	**3**	**4**
1	4	3	2	1
2	0	2	1	3
3	3	1	0	0

Table 3. Quantity of material for producing model (Q_{mp})

Model (P)	Material (M) (Kg)						
	1	2	3	4	5	6	7
1	2.25	1.5	2	0.8	0.4	0.5	0.55
2	1.5	1.25	1.5	0.5	0.35	0.4	0.5
3	2	1.8	1.5	0.6	0.3	0.4	0.4
4	2.25	1.5	2	0.8	0.4	0.5	0.55

Table 4. Cost of material (C_{ms})

Supplier (S)	Material (M) ($)						
	1	2	3	4	5	6	7
1	650	200	50	L	40	50	L
2	625	L	35	30	35	L	20
3	L	225	L	35	20	25	L
4	600	250	55	25	45	30	15

In this L is given as a large number i.e 1000.

Table 5. Capacity of vehicle for supplier (C_{vs})

Vehicle (V_s)	1	2	3	4
Capacity (C_{vs}) (Kg)	20	35	40	55

Table 6. Unit cost of transportation of vehicle for supplier (U_{vs})

S	V_s			
	1 ($)	2 ($)	3 ($)	4 ($)
1	30	35	30	25
2	60	70	60	50
3	25	15	15	10
4	65	75	65	55

Table 7. Fixed cost of transportation of vehicle for supplier (F_{vs})

S	V_s			
	1 ($)	2 ($)	3 ($)	4 ($)
1	50	45	45	55
2	50	45	45	55
3	50	45	45	55
4	50	45	45	55

Table 8. CO_2 emission from the vehicle for supplier $\left(CO_2^{vs}\right)$

S	V_s			
	1 (Kg)	2 (Kg)	3 (Kg)	4 (Kg)
1	1.5	3	2.75	3.5
2	3	6	5	7
3	0.75	1.5	1.75	2
4	3.5	6.5	5.5	7.5

Table 9. Capacity of vehicle for customer (C_{vc})

Vehicle (V_s)	1	2	3	4
Capacity (C_{vc}) (Kg)	35	45	50	75

Table 10. Unit cost of transportation of vehicle for customer (U_{vc})

C	V_c			
	1 ($)	2 ($)	3 ($)	4 ($)
1	45	50	50	55
2	90	100	100	110
3	95	105	105	115

Table 11. Fixed cost of transportation of vehicle for customer (F_{vs})

C	V_c			
	1 ($)	2 ($)	3 ($)	4 ($)
1	40	45	40	50
2	40	45	40	50
3	40	45	40	50

Table 12. CO_2 emission from the vehicle for customer $\left(CO_2^{vs}\right)$

C	V_c			
	1 (Kg)	2 (Kg)	3 (Kg)	4 (Kg)
1	3	3.25	4	3.75
2	6	7.5	8	7.75
3	6.5	8	8.5	8

GA operations such as selection, crossover and mutation are carried out for each generation.

Step 6: Output Module

Steps 3 to 5 are repeated until the termination condition is satisfied. Pareto solutions obtained in all 1000 generations are stored. Table 13 shows the pareto optimal solutions set. The plot of the pareto front is shown in Figure 4 (see Appendix).

Parameter Tuning

The performance of GA is studied with the example problem presented in section 4.2. Parameter settings shown in Table 14 is used to identify the best genetic parameters, population size (pop_size), crossover probability (P_c) and mutation probability (P_m). The problem is solved with different combinations of parameter settings by three trials of each combination to find out the consistence pareto optimal solutions set. Best solutions are obtained for the following parameters: pop_size of 50, crossover probability-0.8 and mutation probability -0.15.

Table 13. Pareto optimal solutions set

S. No.	Material							Supplier				Customer			TC ($)	TE (grams)
	1	2	3	4	5	6	7	1	2	3	4	1	2	3		
1	4	1	1	3	3	3	1	4	0	4	4	1	1	1	5507525	7219981
2	4	1	1	3	3	3	1	4	0	2	4	1	1	1	5519535	7219693
3	4	1	1	3	3	3	4	4	0	2	4	2	1	1	5537721	7219643
4	4	1	1	3	3	3	2	4	3	2	4	2	1	1	5537823	7219637
5	4	1	1	3	3	3	1	1	0	2	4	1	1	1	5545755	7218761
6	4	1	1	3	3	3	1	4	0	4	4	1	1	2	5548686	7218569
7	4	3	1	3	3	3	1	4	0	2	4	1	1	1	5556218	7217943
8	4	3	1	3	3	3	1	1	0	2	4	1	1	1	5570358	7217441
9	4	1	1	3	3	3	1	1	0	2	4	1	1	2	5586916	7217349
10	4	3	1	3	3	3	2	4	4	4	4	2	1	2	5591436	7217053
11	4	3	1	3	3	3	2	4	2	4	4	2	1	2	5591641	7217051
12	4	3	1	3	3	3	1	4	0	2	4	1	1	2	5597379	7216531
13	1	1	1	3	3	3	1	4	0	2	0	1	3	1	5605980	7214283
14	1	1	1	3	3	3	1	4	0	4	0	1	1	2	5613986	7213202
15	1	3	1	3	3	3	4	4	0	4	4	1	3	1	5618996	7213130
16	1	3	1	3	3	3	1	4	0	2	0	1	1	1	5621518	7212577
17	1	3	1	3	3	3	4	4	0	2	4	2	1	1	5639703	7212526
18	1	3	1	3	3	3	2	4	2	2	0	2	1	1	5639908	7212518
19	1	3	1	3	3	3	1	1	0	2	0	1	1	1	5651715	7211504
20	1	3	1	3	3	3	1	4	0	2	0	1	1	2	5662679	7211164
21	1	3	1	3	3	3	1	4	0	2	0	2	1	2	5680403	7211097
22	1	3	1	3	3	3	1	1	0	2	0	1	1	2	5692876	7210091
23	1	3	1	3	3	3	1	1	0	2	0	1	3	2	5714021	7210048
24	1	3	1	3	3	3	1	2	0	2	0	1	1	2	5723670	7209902
25	1	3	1	3	3	3	1	1	0	2	0	4	1	2	5727815	7209880
26	1	3	1	3	3	3	2	1	1	2	0	1	2	2	5736160	7209767
27	1	3	1	3	3	3	1	1	0	2	0	2	2	2	5753371	7209688
28	1	3	1	3	3	3	1	2	0	2	0	1	2	2	5766441	7209566
29	1	3	1	3	3	3	4	2	0	2	1	2	2	2	5784625	7209517
30	1	3	1	3	3	3	2	2	4	2	0	4	2	2	5801739	7209367

Table 14. Parameter settings

population size, Pop_size	20 and 50
crossover probability, P_c	0.7,0.8 and 0.9
mutation probability, P_m	0.05,0.1 and 0.15

Decision Making Using Pareto Front Solutions

Pareto front solutions obtained are helpful to make decision according to user's priority on the objectives. This section explores how the user can take decision according to their requirement from the Pareto front solutions. The three cases could be: (1) minimum cost variation with emission constraint, (2) minimum emission with limitation on cost variation, and (3) restriction on both emission and cost variation. Figure 5, (see Appendix), reveals all the above cases using the Pareto front solutions which are obtained for the example problem detailed in Section 4.2.

Case 1: If the user set the limit for emission within 7216000 grams (shown as line AA'), point 13 is the right choice satisfying the emission constraints whose minimum emission is 7214283 grams and minimum cost to be $ 5605980.

Case 2: If the user wants to take the decision when he/she set the limit for cost less than $ 5600000 (shown as line BB'), point 12 is right choice satisfying cost constraints whose minimum cost is $ 5597379 and minimum emission is to be 7216531 grams.

Case 3: When the user set the limit for cost within $ 5750000 (shown as line DD') and emission within 7210000 grams (shown as line CC'), points 24, 25 and 26 becomes the set of choices.

Besides, there is a chance for non-existence of solutions in the Pareto optimal solutions set satisfying the user requirements. When the user set the limit for cost to be within range of $ 5550000 (shown as line EE') and emission within 7212000 grams (shown as line FF'), there is no solution exists for this condition. Similarly, user can choose the limits on the supply chain cost and CO_2 emission and then identify the solution among the pareto solutions within that region in the solution space. A user can choose a best solution from the pareto front solutions based on his/her priority A solution reveals on supplier, vehicle allocation for supplier and customer with related cost and CO_2 emission.

CONCLUSION

In this paper, a mathematical model for the sustainable supply chain is formulated with two objectives (minimum cost and minimum emission). A MOGA based heuristic methodology is proposed to solve the formulated model and it is illustrated with an example. As a result, the Pareto optimal solution sets are obtained for the objectives of minimum total cost and minimum emission. This Pareto solutions set are helpful to make decision from which supplier the materials are going to be procured, which type of vehicle is to be selected to distribute the raw material from supplier to plant and which type of vehicle is to be selected to deliver the models from plant to customer. Sensitivity analysis on the proposed GA is done to identify the best parameters of population size, crossover probability and mutation probability. The problem considered is a realistic problem which will have to be solved by supply chain managers in their day to day activities. The study gives the manager a useful tool to optimize the supply chain with an objective of minimizing the total supply chain cost and minimizing CO_2 emission of the supply chain network. By using GA algorithm, the managers are able to obtain the solution in short computation time which helps in real time decision making. One of the major disadvantages of the study is difficulty in the collection of the real data. In future, the complexity of the problem by incorporating uncertainty in the demand and testing it using the hybrid versions of metaheuristic algorithms. The model could also be extended using environmental and greenness factors such as noise pollution, accident risk and time assessment factors etc.

REFERENCES

Abdallah, T., Diabat, A., & Rigter, J. (2013). Investigating the option of installing small scale PVs on facility rooftops in a green supply chain. *International Journal of Production Economics*, *146*(2), 465–477. doi:10.1016/j.ijpe.2013.03.016

Aksoy, A., Küçükoğlu, İ., Ene, S., & Öztürk, N. (2014). Integrated emission and fuel consumption calculation model for green supply chain management. *Procedia: Social and Behavioral Sciences*, *109*, 1106–1109. doi:10.1016/j.sbspro.2013.12.595

Ansari, Z. N., & Kant, R. (2017). A state-of-art literature review reflecting 15 years of focus on sustainable supply chain management. *Journal of Cleaner Production*, *142*, 2524–2543. doi:10.1016/j.jclepro.2016.11.023

Chibeles-Martins, N., Pinto-Varela, T., Barbosa-Póvoa, A. P., & Novais, A. Q. (2016). A multi-objective meta-heuristic approach for the design and planning of green supply chains-MBSA. *Expert Systems with Applications*, *47*, 71–84. doi:10.1016/j.eswa.2015.10.036

Ghodsypour, S. H., & O'Brien, C. (2001). The total cost of logistics in supplier selection, under conditions of multiple sourcing, multiple criteria and capacity constraint. *International Journal of Production Economics*, *73*(1), 15–27. doi:10.1016/S0925-5273(01)00093-7

Govindan, K., Jafarian, A., Khodaverdi, R., & Devika, K. (2014). Two-echelon multiple-vehicle location–routing problem with time windows for optimization of sustainable supply chain network of perishable food. *International Journal of Production Economics*, *152*, 9–28. doi:10.1016/j.ijpe.2013.12.028

Govindan, K., Jafarian, A., & Nourbakhsh, V. (2015). Bi-objective integrating sustainable order allocation and sustainable supply chain network strategic design with stochastic demand using a novel robust hybrid multi-objective metaheuristic. *Computers & Operations Research*, *62*, 112–130. doi:10.1016/j.cor.2014.12.014

Gumasta, K., Chan, F. T., & Tiwari, M. (2012). An incorporated inventory transport system with two types of customers for multiple perishable goods. *International Journal of Production Economics*, *139*(2), 678–686. doi:10.1016/j.ijpe.2012.06.020

Hamedi, M., Farahani, R. Z., Husseini, M. M., & Esmaeilian, G. R. (2009). A distribution planning model for natural gas supply chain: A case study. *Energy Policy*, *37*(3), 799–812. doi:10.1016/j.enpol.2008.10.030

Harris, I., Mumford, C. L., & Naim, M. M. (2014). A hybrid multi-objective approach to capacitated facility location with flexible store allocation for green logistics modeling. *Transportation Research Part E, Logistics and Transportation Review*, *66*, 1–22. doi:10.1016/j.tre.2014.01.010

Huang, Y., Wang, K., Zhang, T., & Pang, C. (2016). Green supply chain coordination with greenhouse gases emissions management: A game-theoretic approach. *Journal of Cleaner Production*, *112*, 2004–2014. doi:10.1016/j.jclepro.2015.05.137

Jawahar, N., & Balaji, A. (2009). A genetic algorithm for the two-stage supply chain distribution problem associated with a fixed charge. *European Journal of Operational Research*, *194*(2), 496–537. doi:10.1016/j.ejor.2007.12.005

Jayant, A., & Azhar, M. (2014). Analysis of the barriers for implementing green supply chain management (GSCM) practices: An interpretive structural modeling (ISM) approach. *Procedia Engineering*, *97*, 2157–2166. doi:10.1016/j.proeng.2014.12.459

Jofred, P. & Öster, P. (2011). *CO2 Emissions from Freight Transport and the Impact of Supply Chain Management: A case study at Atlas Copco Industrial Technique*. Academic Press.

Kalyanmoy, D. (2001). *Multi objective optimization using evolutionary algorithms*. John Wiley and Sons.

Marinagi, C., Trivellas, P., & Sakas, D. P. (2014). The impact of information technology on the development of supply chain competitive advantage. *Procedia: Social and Behavioral Sciences, 147*, 586–591. doi:10.1016/j.sbspro.2014.07.161

Mohammed, A., & Wang, Q. (2017). Developing a meat supply chain network design using a multi-objective possibilistic programming approach. *British Food Journal, 119*(3), 690–706. doi:10.1108/BFJ-10-2016-0475

Nachiappan, S., Gunasekaran, A., & Jawahar, N. (2007). Knowledge management system for operating parameters in two-echelon VMI supply chains. *International Journal of Production Research, 45*(11), 2479–2505. doi:10.1080/00207540601020478

Nozick, L. K., & Turnquist, M. A. (2001). Inventory, transportation, service quality and the location of distribution centers. *European Journal of Operational Research, 129*(2), 362–371. doi:10.1016/S0377-2217(00)00234-4

Nurjanni, K. P., Carvalho, M. S., & Da Costa, L. (2014). Green supply chain design with multi-objective optimization. *Proceedings of the 2014 International Conference on Industrial Engineering and Operations Management Bali*, 7-9.

Panda, S., Modak, N., Basu, M., & Goyal, S. (2015). Channel coordination and profit distribution in a social responsible three-layer supply chain. *International Journal of Production Economics, 168*, 224–233. doi:10.1016/j.ijpe.2015.07.001

Pishvaee, M. S., & Razmi, J. (2012). Environmental supply chain network design using multi-objective fuzzy mathematical programming. *Applied Mathematical Modelling, 36*(8), 3433–3446. doi:10.1016/j.apm.2011.10.007

Pop, P. C., Pintea, C.-M., Sitar, C. P., & Hajdu-Măcelaru, M. (2015). An efficient Reverse Distribution System for solving sustainable supply chain network design problem. *Journal of Applied Logic, 13*(2), 105–113. doi:10.1016/j.jal.2014.11.004

Rau, H., Wu, M.-Y., & Wee, H.-M. (2003). Integrated inventory model for deteriorating items under a multi-echelon supply chain environment. *International Journal of Production Economics, 86*(2), 155–168. doi:10.1016/S0925-5273(03)00048-3

Reddy, K. B., Narayanan, S. & Pandian, P. (2011). Single-Echelon Supply Chain Two Stage Distribution Inventory Optimization Model for the Confectionery Industry. *Journal of Applied Mathematical Sciences, 5*, 2491-2504.

Rostamzadeh, R., Govindan, K., Esmaeili, A., & Sabaghi, M. (2015). Application of fuzzy VIKOR for evaluation of green supply chain management practices. *Ecological Indicators, 49*, 188–203. doi:10.1016/j.ecolind.2014.09.045

Sadegheih, A. (2006). Scheduling problem using genetic algorithm, simulated annealing and the effects of parameter values on GA performance. *Applied Mathematical Modelling, 30*(2), 147–154. doi:10.1016/j.apm.2005.03.017

Saffar, M., & Razmi, J. (2015). A new multi objective optimization model for designing a green supply chain network under uncertainty. *International Journal of Industrial Engineering Computations, 6*(1), 15–32. doi:10.5267/j.ijiec.2014.10.001

Sazvar, Z., Rahmani, M., & Govindan, K. (2018). A sustainable supply chain for organic, conventional agro-food products: The role of demand substitution, climate change and public health. *Journal of Cleaner Production, 194*, 564–583. doi:10.1016/j.jclepro.2018.04.118

Subulan, K., Taşan, A. S., & Baykasoğlu, A. (2015). Designing an environmentally conscious tire closed-loop supply chain network with multiple recovery options using interactive fuzzy goal programming. *Applied Mathematical Modelling, 39*(9), 2661–2702. doi:10.1016/j.apm.2014.11.004

Tsao, Y.-C., Thanh, V.-V., Lu, J.-C., & Yu, V. (2018). Designing sustainable supply chain networks under uncertain environments: Fuzzy multi-objective programming. *Journal of Cleaner Production, 174*, 1550–1565. doi:10.1016/j.jclepro.2017.10.272

Urata, T., Yamada, T., Itsubo, N., & Inoue, M. (2015). Modeling and balancing for costs and CO2 emissions in global supply chain network among Asian countries. *Procedia CIRP, 26*, 664–669. doi:10.1016/j.procir.2014.07.107

Validi, S., Bhattacharya, A., & Byrne, P. (2014). A case analysis of a sustainable food supply chain distribution system—A multi-objective approach. *International Journal of Production Economics, 152*, 71–87. doi:10.1016/j.ijpe.2014.02.003

Vidović, M., Ratković, B., Bjelić, N., & Popović, D. (2016). A two-echelon location-routing model for designing recycling logistics networks with profit: MILP and heuristic approach. *Expert Systems with Applications, 51*, 34–48. doi:10.1016/j.eswa.2015.12.029

Wang, F., Lai, X., & Shi, N. (2011). A multi-objective optimization for green supply chain network design. *Decision Support Systems*, *51*(2), 262–269. doi:10.1016/j.dss.2010.11.020

Yeh, W.-C., & Chuang, M.-C. (2011). Using multi-objective genetic algorithm for partner selection in green supply chain problems. *Expert Systems with Applications*, *38*(4), 4244–4253. doi:10.1016/j.eswa.2010.09.091

Zhang, C., Gu, P., & Jiang, P. (2015). Low-carbon scheduling and estimating for a flexible job shop based on carbon footprint and carbon efficiency of multi-job processing. *Proceedings of the Institution of Mechanical Engineers. Part B, Journal of Engineering Manufacture*, *229*(2), 328–342. doi:10.1177/0954405414527959

APPENDIX

Figure 4. Pareto optimal solutions and pareto front

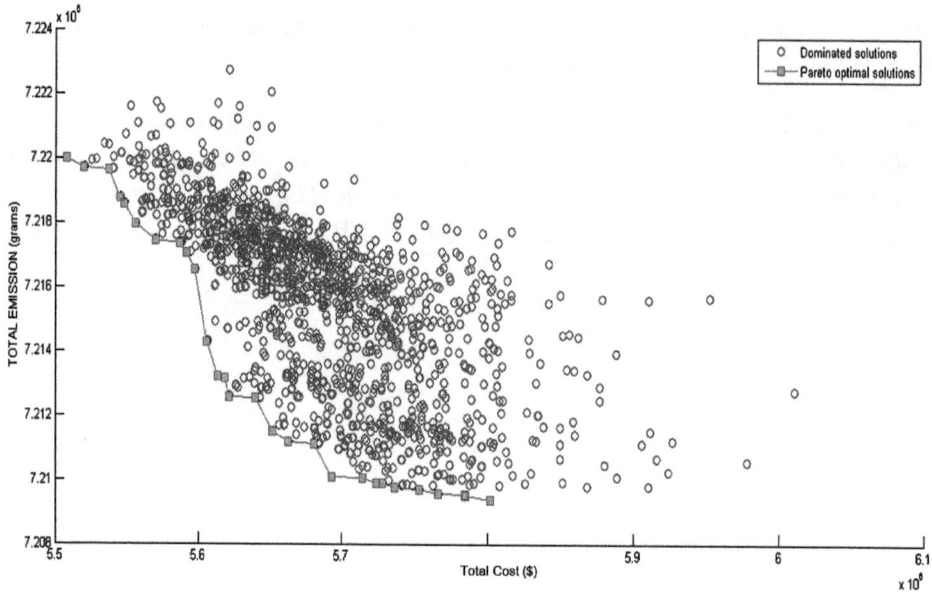

Figure 5. Model utility demonstration

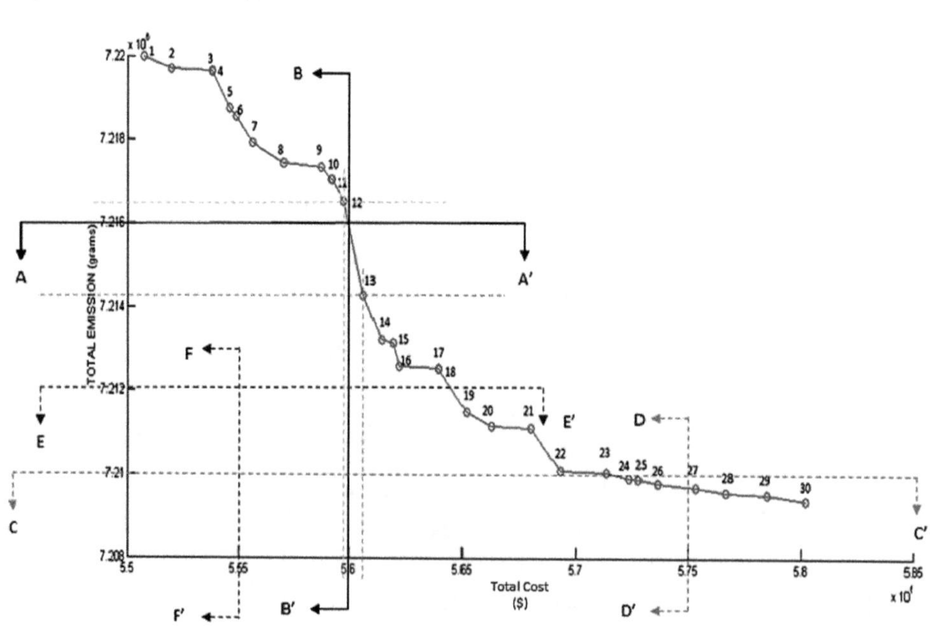

Chapter 7
Economic and Environmental Assessment of Spare Parts Production Using Additive Manufacturing

Atanu Chaudhuri
Aalborg University, Denmark

Dennis Massarola
Aalborg University, Denmark

ABSTRACT

This chapter aims to investigate the potential economic and environmental sustainability outcomes of additive manufacturing (AM) for spare parts logistics. System dynamic simulation was conducted to analyze the sustainability of producing a spare part used in a railways subsystem using a particular additive manufacturing (AM) technology (i.e., selective laser sintering [SLS]) compared to producing it using injection molding. The results of the simulation showed that using SLS for the chosen part is superior to the conventional one in terms of total variable costs as well as for carbon footprint. Compared to the conventional supply chain, for the AM supply chain, the costs of the supplier reduces by 46%, that of the railways company reduces by 71%, while the overall supply chain costs reduce by 61.9%. The carbon emissions in the AM supply chain marginally reduces by 2.89% compared to the conventional supply chain.

DOI: 10.4018/978-1-5225-9078-1.ch007

INTRODUCTION

Additive manufacturing (AM) technologies produce objects from computer-aided design (CAD) model data, usually adding layer upon layer, in contrast to conventional subtractive manufacturing methods that involve the removal of material from a starting work piece (Gibson et al., 2015).

Usage of AM in the spare parts supply chain has been studied by several authors (Ghadge et al., 2018; Li et al., 2017; Holmström et al., 2016; Khajavi et al., 2014). Use of AM for spare parts manufacturing is useful in industries with severe consequences for late deliveries and non-availability (Holmström et al., 2017). Spare parts produced using AM can support the maintenance process of advanced capital goods throughout their lifecycles, which often spans several decades (Knofius et al., 2016) without the need of having high inventory levels. Thus, producing spare parts using AM is also expected to have environmental sustainability implications

However, despite these prospective advantages, the sustainability impact of AM and specifically for producing spare parts using AM has not been explored sufficiently in the academic literature (Kellens et al., 2017). While AM could be an enabler and a driving force for improved industrial sustainability, the consequences of its implementation on the industrial system could lead to an alternative scenario in which less eco-efficient localised production combined with higher rate of product obsolescence can also result in an overall increase in resource consumption (Ford and Despeisse, 2016). Hence, further research on the implications of AM on supply chain performance, especially those in which quantitative evaluations are involved, is needed but is scarce in the literature (e.g. Liu et al., 2014; Li et al., 2017; Ghadge et al., 2018). Thus, this study aims to fill this research gap by contributing to the field of sustainable AM in its application to spare parts logistics. Therefore, the economic and environmental outcomes of AM in the structure of the spare parts supply chain, are assessed by performing a comparative study of the alternative manufacturing solutions from the perspective of a railways company. Specifically, the research tries to answer the following research questions:

RQ1: What are the sustainability implications of the implementation of AM in the structure of a spare parts supply chain?

RQ2: How does the adoption of AM technologies impact the economic and environmental sustainability performances in the spare parts supply chain of a railways company?

The flow of the chapter is shown in Figure 1 below:

Figure 1. Sections in the chapter

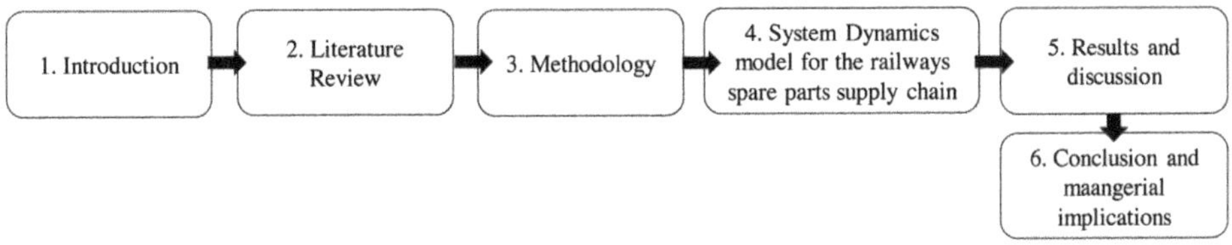

LITERATURE REVIEW

The relevant keywords and associated research string selected for the review included: (additive manufacturing OR 3D Printing OR 3DP OR rapid prototyping) AND (sustainability OR environmental sustainability OR sustainable manufacturing OR industrial sustainability). The selection process was not restricted to these keywords, given the wide range of terminology actively used. The search for relevant literature was conducted using Primo, the library database of Aalborg University, which includes journal articles from all major databases such as ABI-INFORM, EBSCO and SCOPUS. After the string research, the abstracts and keywords of all full papers were examined for relevance to the topics. Given the topic has obtained the attention of academics and practitioners only in recent years (Peng et al., 2018), no specific timeframe for the literature collection was set. The screening excluded editorials, commentaries and book reviews. Despite their relevance in additive manufacturing, studies highly focused on the issue of material yields and energy consumption in specific AM process categories, and research featuring on consumer applications of AM were deemed out of scope and not considered in this review.

Overview of AM Technology: Selective Laser Sintering

AM is a process of joining materials to make objects from three-dimensional (3D) model data, usually as ‘layer upon layer’, as opposed to subtractive manufacturing methodologies which take away layers of materials. Powder Bed Fusion is an AM process which use thermal energy to selectively fuse regions of powder bed. Among the Powder Bed Fusion technologies, Selective Laser Sintering (SLS) uses a laser to fuse particles of build materials layer by layer on top of each other (Gao et al., 2015). SLS can produce parts from any material that can be pulverized, including polymers, metals, ceramics, and glass. Post-curing is not required, the build time is fast, and complex parts can be manufactured.

Sustainability Implications of AM

As AM mimics biological processes by fabricating products layer-by-layer, it is considered less wasteful than conventional subtractive methods of production, displaying the potential to decouple economic and social value creation from the environmental implications of business activities (Ford & Despeisse, 2016). Among the potential sustainability benefits of AM, three are prominent: 1) improved resource efficiency in both production and use phases, as AM allows the redesign of manufacturing processes and products (Gibson et al., 2015); 2) extended product life through repair, remanufacture and refurbishment, and 3) more sustainable socio-economic patterns such as stronger person- product affinities and closer relationships between producers and consumers (Kohtala, 2015). Reconfigured value chains characterized shorter and simpler supply chains, more localized production, innovative distribution models, and new collaboration opportunities (Holmstrom and Gutowski, 2017).

Gebler et al., (2014) presented a comprehensive evaluation of AM industrial implications on the three sustainability dimensions, quantifying changes in life cycle energy and environmental impact on a global scale by 2025. The results obtained from the proposed model showed that AM may reduce the total primary energy demand by 2.54 to 9.30 EJ and CO2 emissions by 130.5 to 525.5 million tons in the time horizon between 2013 and 2025. Chen et al., (2015) assessed AM sustainability outcomes through a detailed case study and compared the environmental impact of SLS as a typical AM process and injection moulding processes as an example of conventional mass production process. According to the study, direct digital manufacturing has the possibility of combining the advantages of the other production paradigms and can have a positive impact on sustainable development. Ford and Despeisse (2016) qualitatively analysed the mechanisms through which AM can enable more sustainable models of production and consumption across the product and material life cycles. The research provided a broad review of previous studies on the sustainability of AM by exploring best practices observed in case studies in various industrial settings. Based on the case examples, Ford and Despeisse (2016) identified the sustainability advantages of AM, along with the challenges that must be overcome if the benefits of AM are to be realized. The study concluded by defining an agenda for further development, calling for increased research on specific single case studies and comparative case studies to provide richer insights into the effects of AM on sustainability in the particular contexts within which each of these occur. Kellens et al., 2017 provided an overview of currently available studies, posing specific focus on the environmental dimension of AM. The research encompassed all product life cycle stages from material production to the part manufacturing and use phase up to the waste treatment of the AM production waste. The study documents impact of improvement efforts for a wide range of AM processes, covering the core AM technologies. Subsequently, the environmental performance of AM processes is compared to the impact of conventional manufacturing processes, examining potential benefits and impacts of AM manufactured parts arising among other stages of the product life cycle. Cerdas et al., (2017) conducted lifecycle assessment of a conventionally manufactured eyeglass frame with those produced using AM in a distributed supply chain. The results indicated no potential environmental advantage between the two production systems but indicated that choice of energy source and materials used have significant environmental impact and have to be optimized. Li et al., (2017) compared conventional supply chain, centralised AM-based supply chain and distributed AM-based supply chain and showed that the spare part supply chain utilising AM was superior to the traditional one in sustainable performance. But, in the above study, a hypothetical situation was modeled. Peng et al., (2018) presented the background and status of the AM industry, stressing the need to conduct thorough research on its energy and environmental impacts.

Sustainability Outcomes of AM in the Spare Parts Supply Chain

The expected sustainability outcomes are discussed in the light of three major impacts: the reduction of the need for transportation through localized production and employment; the reduction of material consumption through decreased overproduction and simplified product operations; the extension of products' lifecycle by means of additive repairs and on-demand spare parts; (Holmstrom and Gutowski, 2017).

Reduction of the Need for Transportation Through Localized Production and Employment

Localizing the production of parts close to the point of assembly is seen as a mechanism to reduce the need for transportation. This assumption stems from the expectation that the transportation of raw materials in bulk is more efficient and environmentally sustainable than the transportation of finished parts which, in terms of volume, comprise of a high amount of air (Holmstrom & Gutowski, 2017). AM can reduce the number of production steps for production of items with complex geometries and intricate designs. Nevertheless, a trade- off is experienced between AM production rate and the need for post-processing, since faster processing leads to lower surface finish, thus requiring a larger need for post-processing operations (Gibson et al., 2015). In turn, the increased need for post-processing may require specialized manufacturing techniques and skills that might make the choice of localized production less desirable from an economic perspective (Holmstrom & Gutowski, 2017). The high acquisition price of AM machinery currently hinders the distributed deployment of AM systems (Khajavi et al., 2014). A prospective development which addresses the mentioned limitations, may be the establishment of localized manufacturing bureaus in which AM capacity is shared, allowing increased capacity utilization and reducing the cost for capacity buffer (Ford & Despeisse, 2016). In addition, a local hub may enable more efficient raw material transportation through shipment consolidation, but may also increase the need for transportation of finished parts (Holmstrom &Gutowski, 2017).

The Reduction of Material Consumption Through Decreased Overproduction and Simplified Product Operations

An AM-enabled on-demand parts manufacturing can help minimizing inventory waste and reduce the risk of part obsolescence deriving from changing product specifications or lack of demand. Such a make-to-order approach for production may potentially reduce material consumption, particularly in the product introduction phase, where engineering changes are more frequent, and in the end-of-life phase, where a higher risk for spare parts obsolescence is generally experienced (Ford & Despeisse, 2016). However, when a high number of parts are required within a short lead time, on- demand production is inhibited by the low production rate of AM. Slow speed of production also limits the opportunities granted by localization when lead-time requirements are tight (Holmstrom & Gutowski, 2017).

Extension of Products' Lifecycle by On-Demand Spare Parts

OEMs are required to fulfil high service requirements during each life cycle phase of the primary products, supporting previous generations of their products as well as their new products, magnifying the number of stock keeping units in after-sales inventories with clear (Khajavi et al., 2014). In practice, it is very expensive to maintain such variety of spare parts in inventory, thus OEMs generally provide support for the primary product in use only for a limited time.

The economics of AM make it ideal for on-demand production of one-off producing spare parts allowing to extend the time OEM can indefinitely support the primary products in use. However, the application of this practice is limited by the large number of parts that cannot be produced using AM without the necessary redesign effort to make them AM-based on- demand production. It is worth underlining that the sustainability outcome of products' lifecycle extension is not self-evidently positive, as it may keep less energy efficient equipment in use for a longer period slowing the renewal of the installed base (Gutowski et al., 2011).

Gaps From the Literature Review

The review showed that though there can be potential sustainability benefits of AM, those cannot be guaranteed. Hence, there is need to assess the economic and environmental sustainability implications for AM for specific parts and products. This research aims to address the above gap for specific spare parts used in railways industry.

METHODOLOGY

In the context of this research, system dynamics (SD) is identified as the suitable integrated sustainability assessment method to account for the dynamic and inherent complexity of sustainability assessment of complex manufacturing supply chains. Due to its unique features, SD is considered as a suitable approach to describe and analyse a supply chain system in which a wide range of materials and information flows and complex dynamic interactions occur (Angerhofer & Angelides 2000). Introduced by Forrester in 1961, SD is defined as a technique for the computational modelling of complex, nonlinear systems which aim is to understand the mechanisms through which systems function, along with the consequences that may follow as a result of the interdependence of system states. SD is an interdisciplinary approach based on the theory of system structures, which allows the representation of complex systems and the analysis of their dynamic behaviour over time (Sterman, 2000).

SD based framework has been used for comprehensive assessment of the sustainability (Musango et al., 2011). The SD approach has been widely adopted in the field of supply chain management (Spengler & Schröter, 2003) as well as in a number of different applications comprising, corporate planning and policy design, economic behaviour, biological and medical modelling and software engineering (Shamsuddoha, 2018). As the problem of sustainability assessment in manufacturing supply chains involves complex relationships between multiple variables and our objective is also to compare the sustainability performance of both conventional and AM supply chains, SD modeling is considered to be a suitable approach. It also has practical relevance as the results from SD simulations can help companies to decide whether to adopt AM in manufacturing a particular part considering both its economic and environmental implications.

The initial step in SD modelling is to determine the system structure consisting of positive and negative relationships between variables, feedback loops, system archetypes, and delays (Sterman, 2000). In SD, the conceptualization of the causal model is an important phase prior to the design of as exhaustive stock and flow model. Before the characterization of a SD model, the researcher is needed to sketch a diagram which captures his/her mental model of an organization and its processes. The mental model conceptualized is then converted into a causal loop diagram and, subsequently, in the final stock and flow simulation model. A causal loop diagram consists of variables linked by arrows which denote the causal influences among the variables. Each causal link is assigned a positive (+) or negative (-) polarity to denote how the dependent variable changes when the independent variable changes. The important loops are highlighted by a loop identifier which shows whether the loop is characterized by a positive (reinforcing) or negative (balancing) feedback. Most SD models contain dynamic feedback processes along with stock and flow structures, delays in time and non-linearity. These positive and negative loops communicate with one another to develop the real dynamic behaviour of a system (Sterman, 2000). In the context of this research, the initial causal loop diagram of the railways spare parts supply chain was developed followed by stock and flow model to capture stocks (accumulations) and flow information (rates) within a system structure (Sterman, 2000).

System Dynamics Model for the Railways Spare Parts Supply Chain

The spare parts supply chain of the railways company consists of numerous stakeholders operating globally attempting to meet supply with demand. Spare parts are developed, produced and tested by a large number of OEMs, who are also responsible for the delivery of the final product to the railways company. The assembly and maintenance operations are performed internally by the railways company. Each heavy maintenance location is specialized in definite maintenance operations. In these locations, vehicles undergo MRO services periodically, every 5 or 6 years. It is at these locations that the railways company holds inventory of finished parts for its rolling stock and, although a high degree of specialization is observed, some parts are found simultaneously in different maintenance locations. Along with the mentioned heavy maintenance locations, the company owns about 140 “short-term” service locations, where routinely maintenance operations, such as cleaning, are performed. When a failure occurs, defective parts are removed and replaced by operational ones, if available in stock. The part removed can either be immediately sent to the closest repair workshop or scrapped. Both the OEMs the and maintenance locations hold inventory to cope with the non-stationary demand for spare parts.

Simulation Scenario Development

For the production of its spare parts, the railways company currently relies on a wide network of external suppliers. In the same way, since it does not own AM machines, the production of AM printed parts is outsourced to external service bureaus. The railway company's portfolio of AM spare parts portfolio comprises of several polymer-based components characterised by a relatively high demand. Examples of these include, coat hooks, head rests, and a variety of plugs, plates and caps for electric and mechanical systems and frames. Metal components have also been printed. For these types of parts, demand is relatively lower and their increased utilization in operations is still under evaluation. Examples of metal elements are locking pawls, cross dampers and moulds employed in rapid tooling activities. As mentioned, repairable parts, are of particular importance for the railway industry, as they are typically heavily utilised and relatively expensive equipment. Thus, the supply chain involved in the provision of the plugs with safety clip, might be an interesting unit of analysis for the economic and environmental comparison of conventional manufacturing and AM-based systems. This type of plugs is part of the passenger information subsystem of trains, which demand usually arises due to components malfunction or after periodic inspection. Being standardized components required in a wide variety of vehicles. The demand for this component is 100 units per year. Hence, these parts are stored in inventories to allow a timely repair of the passenger information system upon failure. This specific component is also chosen due the availability of data for the conventional and AM based production of the part, allowing to provide a realistic case study with credible results.

The supply chain for the considered part is configured as a demand-pull supply chain, in which the only actors actively involved are the railways company and its supplier, which is the OEM of the plugs in both conventional and AM scenarios. In this model, when the failure of a subsystem occurs, defective parts are removed and replaced by operational ones picked from stock and directly utilized in the railways' maintenance workshops, leading to a decrease in the inventory level. The stock level is continuously reviewed so that purchasing orders are issued considering the difference between supply and demand for the part. The same mechanism takes place on the manufacturer side, where products are shipped outbound in response of the customer's orders and parts are produced considering the required delivery rate and the difference between supply and demand.

The causal loop diagram is further employed to guide the development of simulation model.

The underlying assumptions made to establish the simulation scenario are discussed here:

- The time horizon of the simulation is set as 260 weeks (5 years), assuming that, during this specific time span, the OEM has the capabilities to supply the required spare parts. Given the low daily demand observed for the plugs, the preferred unit for time set was "weeks" to provide the required level of aggregation.
- Spare parts demand rate due to subsystem failures is normally distributed and accumulates over time, increasing the aggregate demand for spare parts. On the other hand, when the failure occurred in a subsystem is solved, the usage rate of spare parts decreases, causing a reduction in the aggregate demand. The initial value for the demand is set at 0, assuming the equipment has no failure at the start of the simulation.
- The spare parts inventory increases due to inbound deliveries to the maintenance workshops and decreases due to the outbound flow of the parts required to fix the failures occurring to passenger information systems.
- From supplier's perspective, inventory is increased by the production of parts and decreased due to outbound deliveries to the railways company.
- The value of both spare parts inventories can change instantly, since the inbound and outbound flows as well as production and delivery flows can take place simultaneously. The amount of outbound spare parts from inventory is determined by the demand. The time spent in stock by the parts is given by the sum of the time interval between demand generation and inbound delivery of the spare parts to the warehouse, and the estimated number of inventory picks per year identified considering components with similar characteristics.
- Safety stock is maintained to avoid delays associated with long production times or any unexpected increase in demand. The considered cycle service level is 99%, and is estimated considering the service level established for similar articles in stock.
- The demand in transit status variable is defined to represent the number of spare parts that have been ordered and produced, but not yet stocked in the maintenance workshops. The variable is necessary to characterize the delay in the system caused by the production and delivery of the plugs.
- The ordered amount of spare parts expressed in both the railways company and the supplier order rates is determined considering, per each time step, the sum of the outbound amount of spare parts from stock and the difference between spare parts inventory and demand. When demand is higher than the available inventory, the ordered amount is certainly greater than the previous cycle and vice versa.
- Accessorial variables and constants are employed to facilitate the development of the model. Specifically, the accessorial variables are used to link different

status or rate variables, while constants represent the basic settings of the simulation model.

- The delays caused by information flows taking place between the actors of the supply chain system is not considered.

Having introduced the main assumptions underlying the system dynamic simulation model, this section presents the system dynamics model of the railways spare parts supply chain and the characteristics of the alternative scenarios.

Figure 2 presents the final version of the stock and flow model developed by building upon the previously discussed causal loop diagram and modelling assumptions. Although several assumptions have been made to facilitate the development and subsequent understanding of the model, the stock and flow diagram is considered to portray a realistic picture of the existing dynamic feedback processes of railways spare parts supply chain system. Particularly, the stock and flow structures, the delays in time and nonlinearities which take place in supply chains, are able to be characterized thanks to the adoption of system dynamics as the chosen tool for analysis.

Figure 2. Stock and flow model of the supply chain system

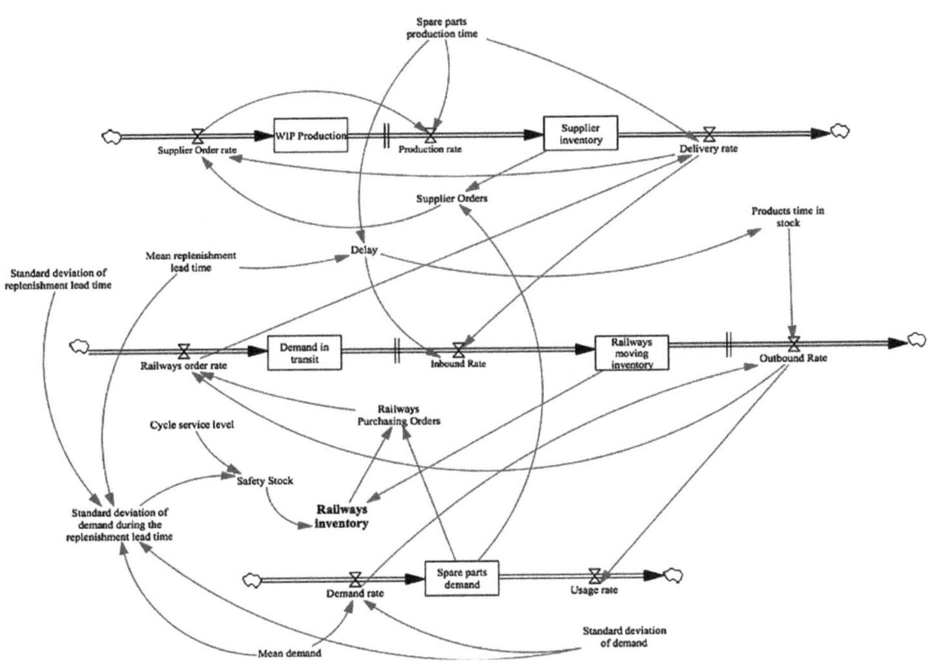

Since the objective was to provide a comparison based on the actual data and logistic configuration within the railways company, the causal relationships of the conventional and AM based systems are conceptually the same. The systems differ from each other, in terms of the spare parts production time. When injection moulding is used in the production of plugs, the required lead time from order to delivery is estimated to be 4 weeks, assuming the necessary tooling and other pre-processing operations have been already performed to fulfil previous orders. On the other hand, when SLS is employed, the order to delivery time is drastically reduced to 1-2 weeks, also assuming that the necessary STL files and related AM design specifications are already available for the supplier. This difference in terms of time has serious implications in the supply chain.

The SD model parameters are shown below:

```
Demand in transit = INTEG (railways company order rate -
Inbound Rate);
Demand rate =  INTEGER(RANDOM NORMAL(0, 100, Mean demand,
Standard deviation of demand, 0));
Delay =INTEGER (Spare parts production time + Mean
replenishment lead time);
Delivery rate = DELAY FIXED (railways company order rate, Spare
parts production time, 0);
Inbound Rate = DELAY FIXED (Delivery rate, Delay, 0);
Mean demand = constant = 2;
Mean replenishment lead time = constant = 1/7;
Outbound rate = DELAY FIXED (Demand rate, Products time in
stock, 0);
Production rate = DELAY FIXED (Supplier Order rate, Spare parts
production time, 0)
Products time in stock = 3 + Delay;
Railways inventory = Railways moving inventory + Safety stock;
Railways moving inventory =  INTEG (INTEGER (IF THEN ELSE
(railwayscompany  moving inventory >= Outbound Rate, Inbound
Rate - Outbound Rate,0));
Railways order rate = INTEGER (MAX (Outbound Rate + railways
company Purchasing Orders, 0));
Railways purchasing orders = Spare parts demand - Railways
inventory;
Safety stock = INTEGER (Cycle service level*Standard deviation
of demand during the replenishment lead time);
Spare parts demand = INTEG (Demand rate-Usage rate);
```

```
Spare parts production time = constant = 4;
Standard deviation of demand = constant = 1;
Standard deviation of demand during replenishment lead time
= SQRT (Mean replenishment lead time*Standard deviation of
demand^2 + Mean demand^2*Standard deviation of replenishment
lead time^2);
Standard deviation of replenishment lead time = constant = 0.3;
Supplier order rate = INTEGER (MAX (Delivery rate + Orders,
0));
Supplier orders = Spare parts demand - Supplier inventory;
Supplier inventory = INTEGER (IF THEN ELSE (supplier
inventory>=Delivery rate, Production rate-Delivery rate,0)),
initial value = 10;
Usage rate = Outbound rate;
WIP production = INTEG (Supplier Order rate - Production rate).
```

RESULTS AND DISCUSSION

Comparison of the Variable Costs

Several studies compare cost structures of different AM technologies with those of conventional manufacturing methods (Hopkinson and Dickens 2003; Ruffo et al., 2006). The cost categories considered in these studies include material cost, labour cost, machine cost and administrative cost (Li et al., 2017). There are also studies that directly consider the costs and benefits of using AM from a supply chain perspective (Holmström et al., 2010; Li et al., 2017). In these studies, inventory cost, processing cost, transportation cost and transaction costs are the main elements taken into account in a cost model for AM-based supply chains. These studies consider transportation cost, manufacturing cost, administrative cost and inventory cost are the variable costs considered.

In order to compare the costs of the two different supply chains based on the established SD model, the formulation of cost components and of the total variable costs are discussed below prior to the analysis of the results:

The inventory costs are calculated based on the weekly inventory level for each time step of the simulation. For the estimation of inventory costs, the company policy assumes this cost to be 8% of the price paid for the for the part. However, the price charged by the manufacturer varies according to the number and size of orders, placed by the railways company. Given the unavailability of the data necessary for an exact estimation of the price according to order sizes, inventory cost in the conventional configuration is calculated as a function of the price for the moulded part when a lot of 100 units is produced plus the total handling and setup costs over the 5 years divided by the total number of parts purchased. This estimation is easier for printed parts, where a flat price per unit (when 100 units are produced) is estimated to be 9 €/part. Thus, inventory cost per unit in the railways company in the case of injection moulding was 0,9656 €/unit*week, while with SLS was 0,72 €/unit*week. As no information is known of the inventory appreciation methods for the supplier, the aggregated value of inventory holding costs for the finished products is estimated to be 0,31 €/unit*week as found in Li et al., (2017).

Transportation costs are calculated as a function of the amount transported as a fixed rate, rather than considering distance and time as parameters. In particular, the transportation cost for raw materials is estimated according to the amount of raw materials transported, which, in turn, are calculated based on the amount of spare parts produced by the supplier. In the case of injection moulding, Olmstaed and Davis (2001) estimates up to 50% of material processing without value addition. When SLS is adopted, material yields range from 56 to 80% (Telenko et al., 2012). Thus, the indicative value 70% is used for this calculation. The weight of the product is 9,09 g,and is assumed to be the same in both cases. The weight of the part is estimated from the given dimensional specifications, multiplied by 1,01 g/cm- the value for the density of the utilized material polyamid 12 (Griehl & Ruestem, 1970). Thus, the amount of raw material required for the production of one plug is 18,18 g/part in the case of injection moulding and 15,45 g/part with SLS. The transportation cost of spare parts is calculated based on the amount of inbound parts are parts that enter inventory.

The administrative costs are estimated as a function of the number of orders. The number of orders is estimated ex-post the simulation counting the weeks in which orders have been issued to the supplier.

Variable costs of the two supply chains are compared on the basis of the same demand profile over the 260 weeks of the simulation. This is deemed necessary to ensure that both supply chains achieve the same service level, so that a clear and fair comparison can be performed.

The total cost function of the supply chain system is calculated in Excel, considering the output data provided by the Vensim software. The mathematical formulation can be expressed as:

$$Total\,Variable\,Cost = \left(\sum_{0}^{T} Production\,rate(t) \times Ct1 + \sum_{0}^{T} Inbound\,rate(t) \times Ct2\right)$$
$$+\left(\sum_{0}^{T} Production\,rate(t) \times (Cm1 + Cm2 + CL)\right)$$
$$+\left(\sum_{0}^{T} supplier\,inventory(t) \times CiS + \sum_{0}^{T} railways\,inventory(t) \times CiDB\right) + \sum_{0}^{T} Orders(t) \times CA$$

Only considering the railways company perspective, the formulation of the costs considered becomes:

$$Total\,Variable\,Costs\,railways = \sum_{0}^{T} Inbound\,rate(t) \times Ct2$$
$$+\sum_{0}^{T} railways\,inventory(t) \times Cirailways + \sum_{0}^{T} Orders(t) \times CA.$$

Given the data limitations faced in the research, manufacturing costs, and administrative costs are collected from previous research found in academic studies. The specific studies considered include Hopkinson and Dickens (2003), Ruffo et al., (2006) and Li et al., (2017), estimates the cost or a component of similar size, and weight. The estimated costs are summarized in Table 1 and the results of the simulations are shown in Table 2.

As shown in Table 2, the total variable cost of the conventional supply chain results much higher than the one adopting AM, proving the validity of previous qualitative studies such as Holmström et al., (2010). The AM-based supply chain can provide tangible economic benefits to the railways company, especially when considering the huge savings on inventory holding costs.

Further analysing the cost structure of the two systems, inventory expenditure appears as the dominant cost item in both configurations. Comparing the systems in terms of material and product transportation, the AM based system is, by a small extent, more economical than the conventional one. This is deemed justified by the reduced need of raw material allowed by SLS due to its higher material efficiency.

Table 1. Summary of cost components

Cost Components	Cost Data	
	Conventional Supply Chain (Euros)	**AM Based Supply Chain (Euros)**
Material transportation cost per part	2.02	2.02
Product transportation cost per part	2.85	2.85
Material cost per part	0.165	1.0725
Machine cost per part	1.17	1.235
Labour cost per part	0.065	0.4225
Supplier inventory cost per part	0.31	0.31
Railways company inventory cost per part	0.9656	0.72
Administrative cost per part	0.52	0.52

As mentioned, the only difference between the systems, was the longer production time required for the production of plugs using injection moulding. The effects of the difference in the longer lead time shows that the status of inventory over the period of the simulation in the alternative scenarios. With a longer lead time in a supply chain system, the manufacturer and the railways company are forced to hold additional inventory in order to avoid stockouts, consequently leading to higher inventory costs.

The results show that AM-based supply chain is characterized by higher manufacturing costs. This finding is deemed consistent with some of the existing studies indicating that material and manufacturing costs constitute a significant portion of an additive manufactured product (Thomas, 2015). Lastly, administrative costs as characterized in this research, are shown to be higher in the AM based configuration. This is also considered coherent, since the faster and more flexible production provided in the case by SLS, can allow the railways company to order less units more frequently to optimize inventory levels and, consequently, inventory costs.

To summarize, the results of the simulation show that the AM based supply chain is superior to the conventional one in terms of total variable costs from both the railways company and the whole supply chain perspectives, since the characteristics of AM enable a consistent reduction of the order to delivery time. The analysis also indicates that by exploiting the opportunities granted by AM, both the railways company and the supplier can take advantage of more cost-effective spare parts supply chains and, simultaneously, enjoy the benefits of increased manufacturing flexibility.

Table 2. Cost simulation results

Type of Cost	Category	Conventional Supply Chain (Euros)		AM Supply Chain (Euros)	
		Total	%	Total	%
Transportation	Material (supplier)	933.24	6	925.16	16
	Product (railways company)	1236.90	8	1177.05	20
Manufacturing cost (supplier)	Material	76.23	0.5	491.21	8
	Machine	540.54	4	565.63	10
	Labour	30.03	0.2	193.51	3
Inventory cost	Supplier	4158.96	27	871.10	15
	Railways company	8308.02	54	1503.36	26
Administrative cost	Supplier	32.76	0.2	60.32	1
	Railways company	32.76	0.2	60.32	1
Total variable cost	Supplier	5771.76		3106,93	
	Railways company	9577.68		2740.73	
	Supply chain	15349.44		5847.66	

Comparison of Carbon Emissions

The assessment of the carbon footprint of the supply chain systems is conducted based on the previously presented stock and flow diagram, and relies on the output provided by the Vensim software. Energy use, material consumption and waste management constitute the main sustainability indicators, which should be evaluated to provide a comprehensive sustainability assessment of the alternative scenarios. The model developed can allow the estimation of such indicators in the time horizon set for the simulation.

The aim of this section is to characterize the major sources of carbon emissions of the modelled supply chains and provide a quantitative comparison of the carbon emissions between the conventional and AM based supply chain systems. A carbon footprint is the measure of the exclusive total amount of carbon dioxide emissions, that is directly and indirectly caused by an activity or is accumulated throughout the lifecycle of a product. Spare parts supply chains involve several activities which possess a substantial carbon footprint, such as raw material production and processing, products' manufacturing and distribution.

Carbon emission data for spare parts supply chains are difficult to measure and account for and, specifically in the context of AM, a lack of detailed lifecycle inventory data and high data uncertainties are observed (Kellens et al., 2017). Moreover, a set of assumptions was provided to facilitate the estimation of carbon emissions considering the output of the simulation model:

- The products produced by both injection moulding and SLS are assumed be equivalent in terms of technical specification and quality.
- Given the unavailability of technical specifications of the conventionally manufactured plug, Polyamid 12 is assumed to be the constituting material of the part in both systems. The difference is found in the form in which the material is fed to the system: nylon granulate for the conventional system; nylon powder for SLS;
- To facilitate the calculations, an average fixed transportation distance is defined, and the carbon emissions generated during transportation are considered a function of the amount of products transported;
- As set for the estimation of the total costs, the emission profiles for the alternative systems are compared on the basis of the same demand profile over the 260 weeks of the simulation.

Table 3. Emission calculations

	Emission Data	
	Conventional Supply Chain	**AM Supply Chain**
Product weight (kg/part)	0.009	0.009
Total weight of primary material required in supply chain (kg/part)	0.018	0.005
Primary material production carbon emission (kg CO_2 equivalent /part)	0.035	0.029
Manufacturing carbon emission (kg CO_2 equivalent /part)	0.14	0.16
Material transportation carbon emission per part (kg CO_2 equivalent /part)	0.036	0.018
Product transportation carbon emission per part (kg CO_2 equivalent /part)	0.0225	0.0225

In order to compare the emission profile of the two different supply chains based on the developed SD model, the data used for the estimation of the relevant emission factors and the mathematical formulation for the calculation of the total emissions are discussed. Given the absence of primary data, the data presented in Table 4 were collected as follows:

- Since the production of the part takes place in Germany, the estimation of CO_2 emissions was based on the German energy mix of the year 2008 where 1 kWh = 3.6 MJ = 0,61 kg CO_2/kWh (Itten et al., 2014).
- Following the assumption that Polyamid 12 the constituting material of the plug in both systems, the weight of the final part is assumed to be 0,009 kg, and the required amount of primary material is estimated from the material yield of injection molding and SLS, respectively, as computed in the previous section.
- Given the lack of life cycle inventory data for AM powder production processes, both systems are assumed to use the same material and refining processes. The estimated specific energy consumption (SEC) for Polyamid 12 granulate was estimated to be 116 MJ/kg of polymer produced (Telenko et al., 2012). The value was converted in kg CO_2 equivalent based on the German energy mix, and the primary material production carbon emission further computed considering the different amount of primary material needed in the manufacturing of a part fabricated.
- The reported generic average SEC value for a hydraulic injection molding machine, including mold production and machining, was estimated to be 93,6 MJ/kg. As regards SLS, the value reported for the energy consumption was 130 MJ/kg (Thiriez, 2006). The manufacturing carbon emission value is calculated as in described in the previous point.
- Material and product transportation carbon emission per part were estimated as proportional to the values provided in Li et al., (2017).

The function characterising the total emissions of the supply chain system is calculated in Excel, considering the output data provided by the Vensim software. The mathematical formulation can be expressed as:

$$Total\,Emissions = \left(\sum_{0}^{T} Production\,rate(t) \times eT1 + \sum_{0}^{T} Inbound\,rate(t) \times eT2\right)$$
$$+ \sum_{0}^{T} Production\,rate(t) \times eM1 + \sum_{0}^{T} Production\,rate \times eM2$$

The results of the simulation related to the comparison in terms of carbon emissions are reported in Table 4 below.

The carbon emissions of the AM based supply chains are to a small extent lower than those of the conventional spare parts supply chain. For both conventional and AM based spare parts supply chains, the highest share of the carbon emissions is found in products manufacturing. Particularly, SLS is found to be more energy intensive than IM, as discussed in Chen et al., (2015). In fact, the difference in the total carbon emissions between the conventional and AM-based supply chains is caused by the carbon emissions related to raw materials production. The conventional supply chain requires a larger amount of raw materials, consequently leading to a significant increase in carbon emissions during materials transportation.

The results indicate that shorter lead times resulting in lower inventory costs and lower material consumption and hence lower need for materials transportation contribute to lower costs and carbon footprint for the AM produced spare parts, under consideration. It is important to note that labour and material costs in manufacturing and carbon emission in the parts manufacturing in AM are higher than those of conventional manufacturing. Unless, the other potential benefits exceed these disadvantages, AM may not be suitable for other parts using different AM technologies.

CONCLUSION AND MANAGERIAL IMPLICATIONS

The results of this simulation are coherent with the overall findings from the literature review, as the expected reduction of material consumption through decreased overproduction is clearly observed in the AM based system. Thanks to AM, raw material suppliers might be able to reduce the carbon emissions from the transportation of raw materials, as less feedstock is needed and most of the required

Table 4. Results of emission profile simulation

Sources of Carbon Emission	Conventional Supply Chain		AM Supply Chain	
	Total	%	Total	%
Raw material production	16.17	15	13.28	13
Manufacturing	64.68	60	73.28	70
Material transportation	16.63	16	8.24	8
Product transportation	9.77	9	9.29	9
Total CO_2 emissions (Kg CO_2 equivalent)	107.2	100	104.1	100

materials are in powder form. For manufacturers, the adoption of AM may not result in direct reduction of the carbon emissions from manufacturing processes. However, this might be achieved due to lower material requirements for manufacturing and the decreased need for raw materials processing.

In conclusion, the analysis shows that, for specific applications, the adoption of AM can reduce the carbon footprint of operations in the spare parts supply chain. It is expected that along with the development of AM technology, the carbon emissions related to its materials and manufacturing processes will be further reduced and AM can play a more important role in the sustainable development and more environmental benefits of the spare parts supply chain.

While considering to adopt AM for spare parts, companies should collect relevant data and try to conduct a thorough assessment of sustainability impact for the specific part using the suitable AM technology across the supply chain over the lifecycle of the usage of the part. It is important to note that using AM technologies may not always result in better economic and environmental performance compared to existing technologies and sometimes part redesign may be needed or the companies may need to wait for further advancement of the technologies.

REFERENCES

Angerhofer, B. J., & Angelides, M. C. (2000). System Dynamics Modelling in Supply Chain Management: Research Review. *Proceedings of the 2000 Winter Simulation Conference, 1*, 342–351. 10.1109/WSC.2000.899737

Cerdas, F., Juraschek, M., Thiede, S., & Herrmann, C. (2017). Life cycle assessment of 3D printed products in a distributed manufacturing system. *Journal of Industrial Ecology*, *21*(S1), S80–S93. doi:10.1111/jiec.12618

Chen, D., Heyer, S., Ibbotson, S., Salonitis, K., Steingrímsson, J. G., & Thiede, S. (2015). Direct digital manufacturing: Definition, evolution, and sustainability implications. *Journal of Cleaner Production*, *107*, 615–625. doi:10.1016/j.jclepro.2015.05.009

Ford, S., & Despeisse, M. (2016). Additive manufacturing and sustainability: An exploratory study of the advantages and challenges. *Journal of Cleaner Production*, *137*, 1573–1587. doi:10.1016/j.jclepro.2016.04.150

Gao, W., Zhang, Y., Ramanujan, D., Ramani, K., Chen, Y., Williams, C. B., ... Zavattieri, P. D. (2015). The status, challenges, and future of additive manufacturing in engineering. *Computer Aided Design*, *69*, 65–89. doi:10.1016/j.cad.2015.04.001

Gebler, M., Uiterkamp, A. J., & Visser, C. (2014). A global sustainability perspective on 3D printing technologies. *Energy Policy*, *74*, 158–167. doi:10.1016/j.enpol.2014.08.033

Ghadge, A., Karantoni, G., Chaudhuri, A., & Srinivasan, A. (2018). Impact of additive manufacturing on aircraft supply chain performance: A system dynamics approach. *Journal of Manufacturing Technology Management*, *29*(5), 846–865. doi:10.1108/JMTM-07-2017-0143

Gibson, I., Rosen, D., & Stucker, B. (2015). *Additive Manufacturing–3D Printing, Rapid Prototyping, and Direct Digital Manufacturing*. Springer.

Griehl, W., & Ruestem, D. (1970). Nylon-12-preparation, properties, and applications. *Industrial & Engineering Chemistry*, *62*(3), 16–22. doi:10.1021/ie50723a005

Holmström, J., & Gutowski, T. (2017). Additive manufacturing in operations and supply chain management: No sustainability benefit or virtuous knock-on opportunities? *Journal of Industrial Ecology*, *21*(1), S21–S24. doi:10.1111/jiec.12580

Hopkinson, N., & Dicknes, P. (2003). Analysis of rapid manufacturing—Using layer manufacturing processes for production. *Proceedings of the Institution of Mechanical Engineers. Part C, Journal of Mechanical Engineering Science*, *217*(1), 31–39. doi:10.1243/095440603762554596

Itten, R., Frischknecht, R., & Stucki, M. (2012). *Life Cycle Inventories of Electricity Mixes and Grid*. Treeze Ltd. Retrieved from: www.lc-inventories.ch

Kellens, K., Baumers, M., Gutowski, T. G., Flanagan, W., Lifset, R., & Duflou, J. R. (2017). Environmental Dimensions of Additive Manufacturing: Mapping Application Domains and Their Environmental Implications. *Journal of Industrial Ecology*, *21*(1). doi:10.1111/jiec.12629

Khajavi, S. H., Partanen, J., & Holmström, J. (2014). Additive Manufacturing in the Spare Parts Supply Chain. *Computers in Industry*, *65*(1), 50–63. doi:10.1016/j.compind.2013.07.008

Knofius, N., van der Heijden, M. C., & Zijm, W. H. M. (2016). Selecting parts for additive manufacturing in service logistics. *Journal of Manufacturing Technology Management*, *27*(7), 915–931.

Kohtala, C. (2015). Addressing sustainability in research on distributed production: An integrated literature review. *Journal of Cleaner Production*, *106*, 654–668. doi:10.1016/j.jclepro.2014.09.039

Li, G. J., Cheng, Y., & Hu, Y. (2017). Additive manufacturing technology in spare parts supply chain: A comparative study. *International Journal of Production Research, 55*(5), 1498–1515. doi:10.1080/00207543.2016.1231433

Liu, P., Huang, S. H., Mokasdar, A., Zhou, H., & Hou, L. (2014). The Impact of Additive Manufacturing in the Aircraft Spare Parts Supply Chain: Supply Chain Operation Reference (SCOR) Model Based Analysis. *Production Planning and Control, 25*(13–14), 1169–1181. doi:10.1080/09537287.2013.808835

Olmsted, B. A., & Davis, M. (Eds.). (2001). *Practical injection molding*. CRC Press. doi:10.1201/9781482294590

Ruffo, M., Tuck, C., & Hague, R. (2006). Cost Estimation for Rapid Manufacturing - Laser Sintering Production for Low to Medium Volumes. *Proceedings of the Institution of Mechanical Engineers. Part B, Journal of Engineering Manufacture, 220*(9), 1417–11427. doi:10.1243/09544054JEM517

Shamsuddoha, M. (2015). Integrated Supply Chain Model for Sustainable Manufacturing: A System Dynamics Approach. *Advances in Business Marketing and Purchasing Sustaining Competitive Advantage Via Business Intelligence, Knowledge Management, and System Dynamics*, 155-399.

Sterman J. D. (2000). *Business dynamics: systems thinking and modelling for a complex world*. McGraw-Hill Education.

Telenko, C., & Seepersad, C. C. (2012). A comparison of the energy efficiency of selective laser sintering and injection molding of nylon parts. *Rapid Prototyping Journal, 18*(6), 472–481. doi:10.1108/13552541211272018

Thiriez, A., & Gutowski, T. (2006). *An Environmental Analysis of Injection Molding*. IEEE International Symposium on Electronics and the Environment,·San Francisco, CA.

Chapter 8
Autonomous Vehicle in Industrial Logistics Application:
Case Study

Julius Fusic S.
Thiagarajar College of Engineering, India

Kanagaraj G.
Thiagarajar College of Engineering, India

Hariharan K.
Thiagarajar College of Engineering, India

ABSTRACT

Rapid technological advances have revolutionized the industrial sector. In the global market, it is necessary to consider the new paradigm of Industry 4.0 that presents a lot of features in the industrial logistics application. It has been seen through literature that innovation management practices enable companies to compete within the autonomous and connected vehicle market and is considered as an emerging and competitive differentiator towards the growth of the product and that of meeting customer demands within the changing markets. The first case study explores the integration of GPS and GLONASS signals in AGV for localization and navigation of customer destination and materials in the indoor and outdoor environment. The second case study implemented in obstacle environment that recognized the obstacle in front of the robot and also identified the dimension of the obstacle size, length, width, circumference, height, and distance from a robot. The strength and disadvantages of the system are discussed in the logistical application and future outlines are provided.

DOI: 10.4018/978-1-5225-9078-1.ch008

INTRODUCTION

Logistics is the lifeblood of the fourth industrial revolution and makes a key contribution to the Industry 4.0 concept. Automated Guided Vehicles (AGV) become an integral part of the solution portfolio for fully-automated logistics to transport loads to places where repetitive movements of material are required with little or no human intervention. In recent times, labour shortages and rising labour costs have forced manufacturers to adopt decisive solutions in order to remain competitive on the market. Shortage of labour and increasing labour cost in recent times has forced manufacturers to undertake decisive solutions in order to stay competitive in the market. AGV are unmanned automated systems that transport all kind of products within warehouse, distribution and logistics environment that reduce cost and increase profitability, reliability and flexibility in the system (Brent & Heard, 2018). The current challenges faced in all in-house and industrial need are to satisfy manufacturing and control system automation based on Internet of things technology. However, the main issue faced in current transportation is unavailable or shortage of drivers to transport goods. Thus, the research is going on to find alternate solution by replacing existing vehicles with AGV's in the Logistic application for public and commercial need. Sousa et.al., (2018) stated the loading and unloading operation of goods are arranged and carried out in the linear path inside the warehouse is a complicated task. Few researchers suggested that handling materials using AGV, without a change in path and carried out in a straight line using Scott-Russel Mechanism (SRM) was a proper solution. Logistics literature rigorously recommends that implementing automated vehicles in commercial indoor and outdoor applications and made a smart logistic system by adding intelligent over the system. The introduction of Industry internet of things (IIOT) in a logistic field made self-organization arrangement to solve all conditional tasks and problems. Currently, indoor and outdoor vehicle identification or navigation has been a major problem in the application of logistics warehouses. In this chapter we elaborate the navigation of autonomous vehicle and its obstacle avoidance system in detail.

BACKGROUND

Manufacturing companies are subject to permanently changing environments. The factories have to be continuously adapted to stay sustainable and competitive in the global market. Especially the smart driven logistic system facing huge problem relating to data processing and analyzing mechanism in navigation and control of the system. While the introduction of IIOT in logistic systems encounter solution for Automated Guided vehicle problems like planning path to reach destination, localize the material and localize the vehicle itself, navigate the customer location/ navigate the vehicle by the customer and sender itself, identifying the obstacle in the tracked path, recognize the obstacle and materials to pick and place operation with low budget by Mac et al., (2016). The laser and camera-based image processing-based vehicle positioning and obstacle avoid methodology were frequently followed (S. Lee & S. Lee 2013). For example, the PAN robot in advance where the laser mapping technology synchronized with stereo camera-based vision sensor to handle autonomous trouble-free loading and unloading in modern factory warehouses in European countries. Zhang et.al., (2018) pointed that the sensor data must be handle carefully and stored in repository in secured manner, by deploying fog computing cyber security system in industry 4.0. Thus, the data driven model based on data processing and analyzing data in a more secure manner to avoid the error-free system. These vehicles encounter problems like unable to predict alternate path when it identifies obstacles, when it deviates from a predefined path, the navigation and localization of vehicle are complicated.

The mobile robots are used to collect packages from vendor end to customer end in outdoor applications and as like the same in indoor application it carries spares, materials from warehouse. In the early stages, the Amazon tied up with Kiva robotics used nearly 30,000 robots for loading, sorting, lifting, pulling, shipping, transportation in the warehouse of Amazon. In the beginning, robots which predict the parcels to be transported from the warehouse to the dispatch section. Later the development of Amazon robotics made them into an automated total logistic section as both indoor and outdoor automated vehicle system by utilizing UGV, drones, and UAV. The transportation of a robot in the industry to move the rack stretches the fork, hold the packed goods, move back the robot and return to a destination point. The goods were selected based on QR code scanned by a camera attached to the robot head. Goran Vasiljevic et al., (2016), developed AGV along with forklift in a production plant, warehouse carrying system, assembly line and order picking system. The world logistic pioneer brands like Swiss log combined with DB Schenker Logistics, first introduce the Carry Pick technique in the logistics world. The Swiss log CarryPick solution made a revolution in click and picks an option in the logistics industry. This solution consists of 65 AGV's, 1500 mobile racking units, and 7 ergonomics workstation installed along with Swisslog control systems. In the present scenario customer requirements through robotic IOT system connected with the mobile robot and ERP inline in one network where the order placed, initially the robot check the stock and if it available then the robot move in a specific path to reach the ordered material using RFID tags, Barcodes, QR codes. Once the parcel is tracked and took materials from the rack by means of a hydraulic system or Industrial automation systems. Then the robot moved in a path to reach the packing and forwarding section as shown in Figure 4 (courtesy: exotec). Jasprabhjit

Mehami et al., (2018) described the introduction of mimicked AGV vehicles in industries may create smart manufacturing environment for implementation of industry 4.0. The mobile robots made the errors and inefficiency happen due to humans were reduced. Fernåndez et al., (2018) reviewed the basics of Industry 4.0 and smart labels, details the latest technologies used by them, their applications, the most relevant academic and commercial implementations, and their internal architecture and design requirements, providing researchers with the necessary foundations for developing the next generation of Industry 4.0 human-centered smart label applications. Recently, Tao et al., (2018) introduced big data concept to the companies to adopt data-driven strategies and proposed big data perspective for smart manufacturing. This good-to person technique made the AGV's are viable than human employees. In the future, the human power will replace by machines. The material handling systems are automated fully from individual goods to shelves to total warehouse control. Once the goods packed and forwarded from the logistics industry to customer end the carrying vehicle must possess special features like a vehicle to vehicle communication, Luís Conde et al., (2015). Synchronization with another automated vehicle, navigate path and reach a destination by carrying heavy loads. Yugi hosoda, et al., (2011) states that as like indoor system the outdoor automated vehicles suits for public roads and traffic control rules and regulation to follow.

The liability issues are not complex but vary between countries. The outdoor logistic applications like freight the goods from country to country, loading and unloading the goods in a harbor, drones and AGV's carrying goods to customer end, AGV truck carrying heavy loads from warehouse to another etc. The current trend and technology made the transportations made easier comfort and deliver faster. Bram Van Meldert and Liesje De Boeck, (2016) reviewed the famous AGV

Figure 1. Warehouse mobile robot operating system
Source: exotic warehouse system

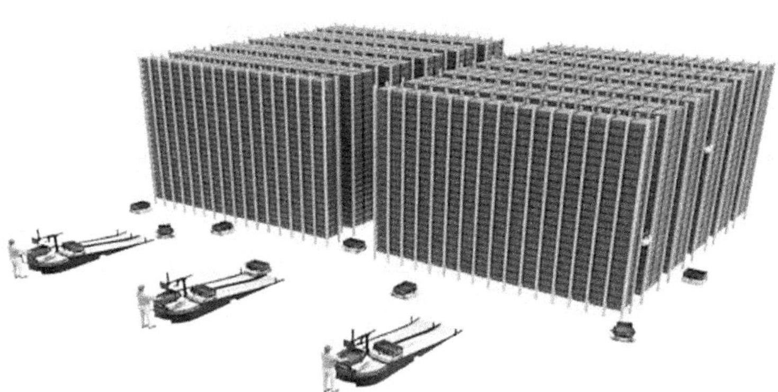

truck in Western Australia named RIO TINTO's AGV truck possessed advance technology like automatic navigation, Autonomous haulage system (AHS). The automatic navigation system used in these trucks is as like car navigation GPS technology system, also the laser scanners are also used for environment mapping and navigate the obstacle-free path. Because of this, the customer can easily track the truck and location of delivering goods more efficiently. Later this manufacturer introduced the first Automatic Drilling system attached to the truck and drilled nearly 2.6million meters (www.riotinto.com). Since the trucks are autonomous and the driving speed may be varied and the precision also varies with respect to environment and time. Since all the AGV's are time variant system, the same input may not produce accurate output as theoretical. The riot into trucks may have the speed range of 6 km/h and the precision range of 2 cm. The onboard sensors like vision sensors, Lidar, Laser mapping technology made the impossible activities to be likely possible. Later the robot-logistics made a great revolution in logistic business in which the robot logistic made complete planning, controlling, and testing of all internal flow of goods in the logistic platform. The study also confirms that the future food processing chain will not complete in the absence of autonomous trucks and vehicles. The autonomous vehicles used in this consumer process will consider the potential use of fuels, communication, and navigation to deliver food to customer home itself. In the review paper, Hector J. Carlo, et al., (2014) surveyed lists and types of heavy vehicles used in transportation. The classification scheme listed in the journal publication revealed the trend, metrics terminology and technology development in goods and container carrying vehicle in all applications. The future vehicles might fully be automated and autonomous which self-lift container and packages, improve accuracy, feasible, less time consumption, less quay crane work rate, self-stacking, self-navigate delivery end and so on. The role of self-driving vehicles is broadly utilized in Logistic, freight, and supply chain management system and courier applications to improve the system accuracy and develop the problem free customer services.

From the above literature reveals that the major problems faced in implementation of automated guided vehicles (AGV) for logistic application are localization, navigation in inside and outside environment and path planning to reach destination in trouble free path. The existing GPS satellite navigation system was not precise to identify consignment inside the warehouse or indoor application. Thus, the theft, misplacement of materials might increase in smart manufacturing system. To overcome this difficulty running of GPS-GLONASS satellite navigation system would consider L1, L2, L3, and L4 band such that the tracking of vehicle is possible even in indoor/ware house. Hence, in this chapter, two-wheeled autonomous mobile robot inbuilt with GPS and GLONASS signals system was developed and utilized for indoor and outdoor environment logistic application in a short span of time with high precision and adaptive abilities to manage the logistic cluster in cost effective manner. The performance, accuracy and position navigation have been confirmed with two case studies for both indoor and outdoor environment. The navigated robot will identify trouble free path to reach the destination based on image processing platform.

MAIN FOCUS OF THE CHAPTER

Technology Development in Logistic Vehicles

H. Song (2016) highlighted the various navigation techniques and sensor classification used for navigation itself made different types of AGV's. Guiding or steering system based on wires, magnetic tapes, lines are named as Tugger type vehicles in and guidance based on laser beams or lidar are classified as Laser guidance vehicle (LGV), guidance based on vision is Vision guidance vehicles (VGV) and so on. The important aspects of autonomous vehicles are:

- Locate the vehicle and the goods to pick and supply.
- Locate the bay, racks, and potential sections in warehouse.
- Identify the appropriate goods to transport using scanners.
- Control the forklifts or manipulators to pick and place.
- Plan the path to follow through AI techniques.
- Safe transportation and cognitive control based on environments.
- Traffic control and navigate according to time variant.

Navigation and Communication of AGV'S

During past decade, research went to develop human assisting machines for development of productivity, feasibility, and precision in industrial sectors. The robots were introduced in this field made rapid growth and repeating the natural task in a precise manner. The vast development made in industry 3.0 paved a way for mobile robot development in industrial sectors. The autonomy was giving to robots to decide and complete the industrial task based on the programmed priority. Initially the autonomous introduced to identify the obstacles in the prescribed path. Later the term navigation introduced to localize the robot, avoid the obstacle and identified the shortest path to reach customer or target end. Based on R.Siegwart et al., (2004), the primary classification of the navigation is Localization, Perception, Cognition, Path Planning, and Environment. Based on the environment the selection of perception system and cognition system will change on board to the robot (see Figure 2).

Perception

Normally the robot means combination of sensors and actuators. Farid Bounini (2016), the perception answers the generic question like what is around me. The "environmental awareness" is most important for the system to take decision by their own. In Automated system only the sensor and its measured values extract appropriate information from the environment. In specific the perception of AGV system is defined as data collected from the robot sensing devices and processed to predict the in and around environment. Generally, the perceptions are classified as Proprioceptive/exteroceptive and active/passive sensors.

Localization

The localization predicts the position of robot for every intermediate point moving in appropriate path. These predictions are made the robot or autonomous system to find where exactly the robot is positioned. Generally, the localization in AGV's are classified as feature-based localization and Map based localization. Below, in Figure 3, the reference position can be identified through Global positioning system (GPS) or other satellite position which are comparing with predicted position and locate the goods or vehicle. The localization accuracy depends on many factors like sensor noisy data, fusion of arrays of sensors and data analysis, end effectors/ actuators malfunction and mapping unwanted information from navigation.

Figure 2. General navigation system of AGV's

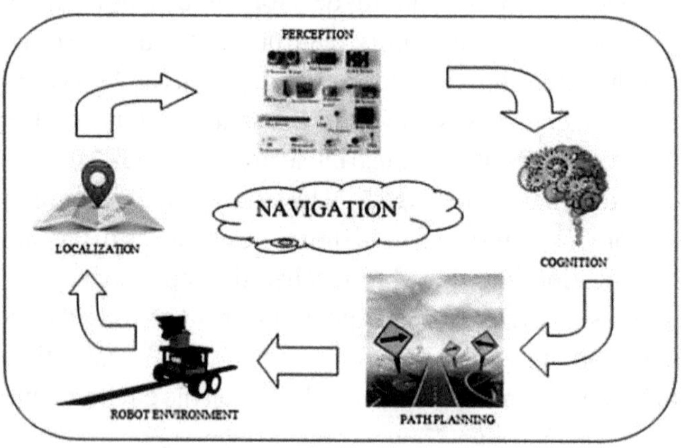

Figure 3. Localization system of AGV's

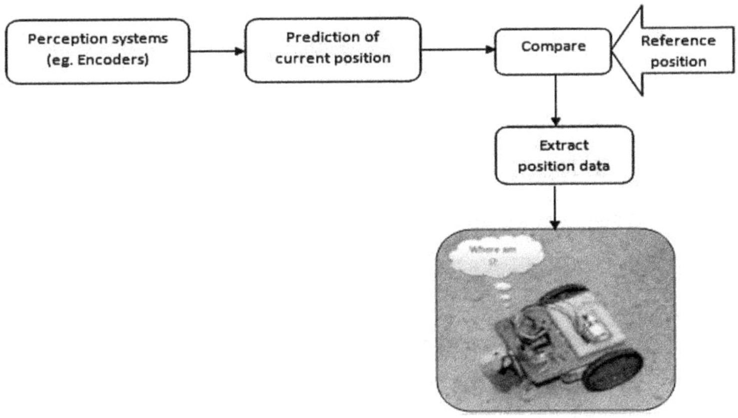

Cognition

The term cognition in mobile robotic is defined as acquire information and data from sensors to control kinematic operation in systems to complete task. As like human beings how the mental activities operate from the data acquired from human perception system. Initially the robots were designed to divide big task into few simple tasks and cluster the task based on features as per application environment. Based on human cognitive thinking and soft computing tools the path was predicted in moving robots and AGV's. Depend on mental states the levels of cognition classified as Long Term Memory level and Short Term Memory level.

Path Planning

Path planning in AGV's are defined as selection of optimal path to reach the destination within the environment. The path planning prediction may be future research topic, where the machine learning and deep learning plays a vital role in cognitive learning technique-based path or trajectory selection. Belkhouche (2009) introduced many classical and heuristic algorithms applied for planning the path based on selection of environment. Due to the selection of environment itself made the path planning techniques are classified as offline path planning and online path planning. If the selected environment contains static obstacles, then the selected path based on trajectories about obstacles itself navigate a path to reach destination (offline mode). If the obstacles are dynamic under non-holonomic condition then path planning will vary regarding intermediate sensor information to cognitive system. Some algorithms and methods like Genetic algorithm, Voronoi diagram, potential

Figure 4. Cognition levels and structure of AGV's

field method, fuzzy logic, neural network, particle swarm algorithm, Decision tree algorithms, quad tree, A* Algorithm, cell decomposition, sampling based, sub-goal networks, ANN, natural inspired algorithms and hybrid algorithms.

Machine Learning

In AGV's navigation system machine learning type cognitive system is widely used. Now the research and development move towards Deep Learning Algorithms for autonomous systems. Normally learning means any computerized system that improve its programming strategy to solve dissimilar task based on learning from experience is said to be machine learning. This learning methodology is widely used in autonomous driving vehicles, which learn by experience about different road surfaces in various environments (K. Zindler, et al., 2016). From that resultant data are incorporated and group into different clustering information and again train the system. Once training is completed and then give test data to check the trained hypothesis provide possible correct solution as output, as in Figure 5.

Figure 5. Machine learning structure for AGV's

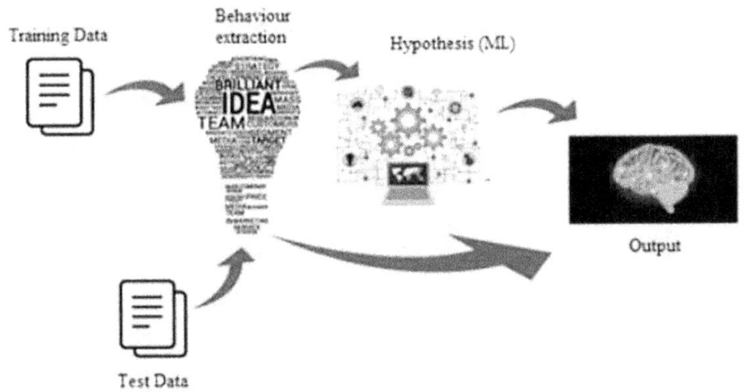

Indoor and Outdoor Navigation System

Based on environment, the navigation is classified as Indoor environment navigation and outdoor environment navigation system. The Satellite signals were used to localize the position of robot and made the cognitive instruction to follow the pre-defined path (Mohammed Z. Al-Faiz, et al., 2015). The positioning system of the satellite (such as GPS, GLONASS, IRNSS and LORAX) was recognized through satellite signals. The indoor navigation systems used Wireless technology, vision technology, and range techniques to locate the robot. The indoor navigation perception systems are classified as shown in Figure 6.

The navigation-based research was more useful to predict the actual position of robot and identify time taken to reach the destination from intermediate point. The importance of navigation in industry robotic application using techniques, optimization algorithms, coding and technologies have been implemented both in indoors and outdoors mobile application. More predicting the mobile robots in Logistic application have to navigate in both indoor and outdoor navigation techniques. Thus, several technologies were proposed and few were implemented to sort out the navigation problem in logistic applications. Few are GPS, GLONASS, Infrared sensors, Ultrasound, RADAR, SONAR, Magnetic, Vision sensor, Visible Light, Dead Reckoning, Inertial Navigation System (INS), Lidar, Kinect and Hybrid system used as mechanical perception systems. The range detection like Bluetooth, Ultra-wideband, Wireless local area network, Wireless Sensor Network (WSN), Radio-Frequency Identification tags, Near Field Communication (NFC). Sakpere.W, et.al (2017) stated that the range detection techniques used to identify robot or obstacle location through Time of Arrival (TOA) or Time of Flight (TOF), Angle of Arrival (AOA), Received Signal Strength Indication (RSSI), and Time Difference of Arrival

Figure 6. Indoor Navigation perception system classification

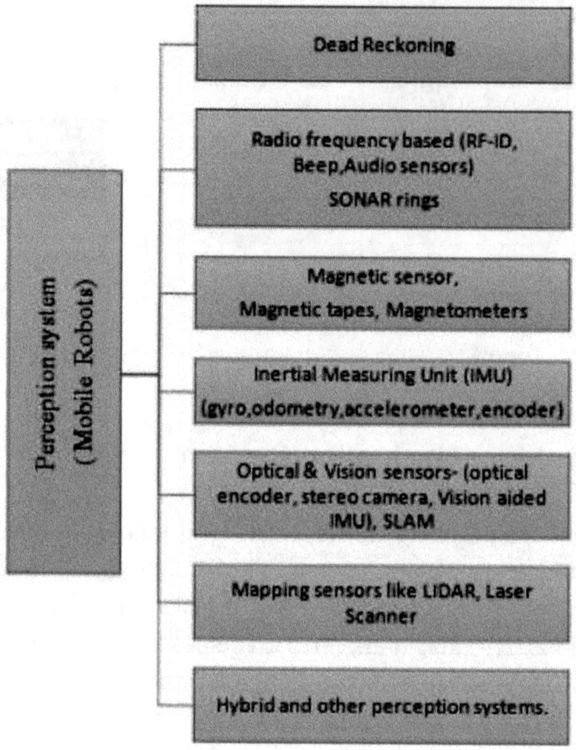

(TDOA) and the position algorithms like Triangulation, Trilateration, Proximity and Scene Analysis/ Fingerprinting, Gaussian motion model, and contour mapping. Once the perception system errors are reduced and the cognitive signals made the AGV's to follow the appropriate path (obstacle free path) to reach the destination.

Satellite Navigation System

The Navigation using satellites is bringing into play to identify autonomous geo-spatial positioning using satellites. This navigation system based on satellite which allows electronic antennas and electronic devices to identify latitude, longitude and altitude/ elevation with high precision based on signal transmitted along a line of sight by satellites. The signals received from satellites radio are provide high precision of object/vehicle position within few metre distances. The satellite navigation is used as satellite tracking system by providing position and orientation within the tracking receiver range. The GPS and GLONASS are current using global navigation satellite system (GNSS). As of 2016 the fully operated systems are Americas GPS, Russians

GLONASS, Europe union Galileo as GNSS in earth. In 2020 the China's BeiDou Navigation satellite system will expand as BeiDou-2. Also, the countries like India (IRNSS), France, Japan, are in process to develop navigation system as well.

GPS System

The Global Position System (GPS) is a radio signal communication based on space satellite signals. The satellite signal frequency is 1575.42MHz as L1 band and 1227.60MHz as L2 band radio frequency signal transmitted from the radio satellite. Currently the USA government provide the satellite and the data as open source than anyone can use to track and navigate the system at this frequency range receiver. The GPS system required minimum four satellites at three position coordinates to get prior and precise information about vehicle position. The current GPS tracked location can be identified in the form of (latitude, longitude, altitude) and current navigate time. The GPS receiver is an electronic device capable of receiving information from GPS satellite and used for geographical position of system. The GPS signal reception required maximum when four satellites at line of sight. The GPS band and application are listed as follows:

GLONASS System

GLONASS is the radio communication navigation system based on Russian space satellite signals. In much application this system acts as an alternate or second system for GPS. So, the GLONASS system enables with full code and Real time kinematic (RTK) positioning, also the ability to acquisition of raw GPS and GLONASS measurement. As like the GPS, the GLONASS required four satellites to receive

Table 1. GPS Band frequency range and application

Band	Frequency	Description
L1	1575.42 MHz	Coarse-acquisition (C/A) and encrypted precision (P(Y)) codes, plus the L1 civilian (L1C) and military (M) codes on future Block III satellites.
L2	1227.60 MHz	P(Y) code, plus the L2C and military codes on the Block IIR-M and newer satellites.
L3	1381.05 MHz	Used for nuclear detonation (NUDET) detection.
L4	1379.91 MHz	Being studied for additional ionosphere correction.
L5	1176.45 MHz	Proposed for use as a civilian safety-of-life (SoL) signal.

Source: Informatics Education Limited

three-dimensional position information. The GLONASS system consists of control segment, space segment and user segment to provide position, velocity and time data to the user. The combination of GPS and GLONASS system improves the RTK positioning significantly high.

Proposed GNSS Satellite Navigation System

GPS satellite navigation system was not precise to identify consignment inside the warehouse or indoor application. To overcome that the introduction of GPS-GLONASS satellite navigation system would consider L1, L2, L3, L4 band such that the tracking of vehicle is possible even in indoor/ware house. GNSS system-based satellite navigation combines GPS, GLONASS servers and receives signals without any interruption. The procedural steps for implementing GNSS satellite navigation system are illustrated in Figure 7.

A Micro Strip Patch Antenna (MSPA) was designed and manufactured. The MSPA antenna is a single layer antenna consists of patch, feeding part, ground plane and substrate. Once the frequency is identified regarding to the application the wavelength of the antenna can be calculated using the Equation (1):

$$\lambda = C / f \tag{1}$$

where λ – is the Wavelength of propagated signal, C- is Speed of Light (3×10^8 m/s), f – frequency of antenna signal range. The patch is a very thin metal strip that radiate signal as single strip or array of strip located on the substrate one sided layer, ground plane with metal same as strip material locate in other side of substrate. The patches were made by copper foil along with corrosive resistive metal like gold, tin or nickel. The designed antenna will receive the signal frequency range 1.575GHz and 1.602GHz respectively. The band width was $\mp$10MHz and $\mp$ 5MHz. Once the antenna was designed based on the prescribed format shown in Figure 8 and tested in anechoic chamber then it is ready to interface with IMU system in Mobile Robot.

Figure 8 shows the GPS – GLONASS antenna implemented with the existed GPS module to obtain more satellites signals from the satellites in indoor and outdoor

Table 2. GLONASS Band frequency Range

Band	Frequency	Bandwidth	Description
L1	1575.42 MHz	6.5 MHz Signal	FDMA
L2	1602 MHz	5 MHz	FDMA Signal

Figure 7. GNSS satellite navigation system

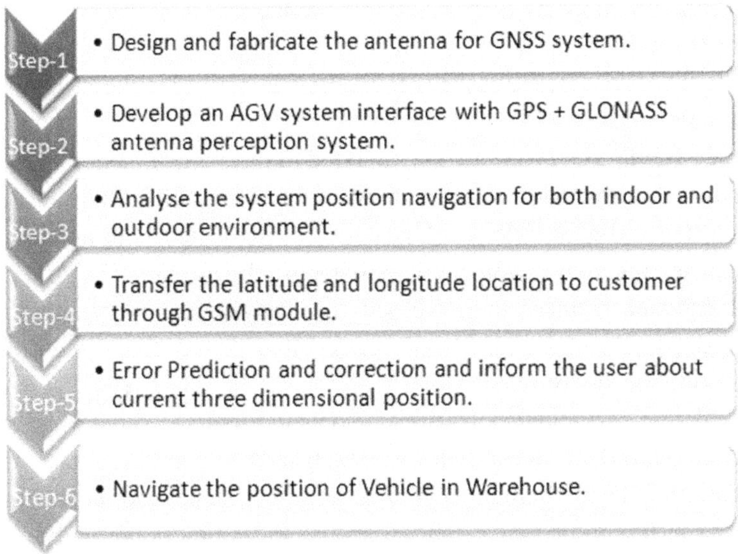

Figure 8. GPS – GLONASS Antenna

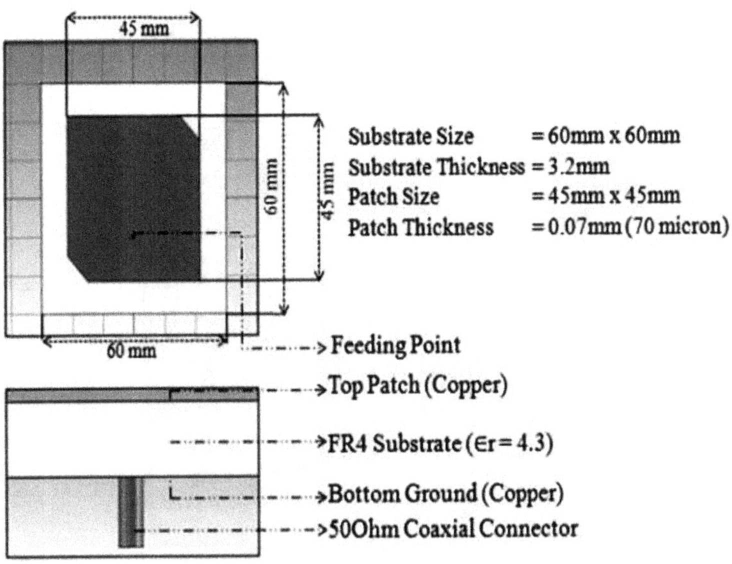

environments. The GPS module provides the support for connecting external antenna, where the module supports the GLONASS frequency also. The satellite signals received by the module are seen by using the Visual GPS View software as shown in the Fig 9. Thus, the total number of satellites in view is 19 and the number of satellite signals tracked is 18.

Once the Mobile robot is fixed and operated in the environment various sample data are acquired of about 40 data were shown in the chart with latitude and longitude information. Once the latitude and longitude are identified with antenna and the position will recognize by compare it with geographical map. The Antenna fixed along with encoder to calculate the velocity of the system at every interval.

The designed GPS – GLONASS antenna received the satellite signals in both the indoor and outdoor environment along with the satellite data received is recorded using the real time data logger software as shown in Figure 11. The Error estimation for the collected raw data from antenna was analyzed using least square fitting (LSF) method. The LSF is a mathematical approach to find the best possible fitting curve to the given set of latitude and longitude data in {x,y} function by minimizing the sum of squares of offset value from the curve. Assume i- number of iteration, xi- latitude values and yi- longitude values, and then the least square method based on linear regression can be calculated using the formula shown below, Equation 2:

Figure 9. Prototype model of Robot Design.

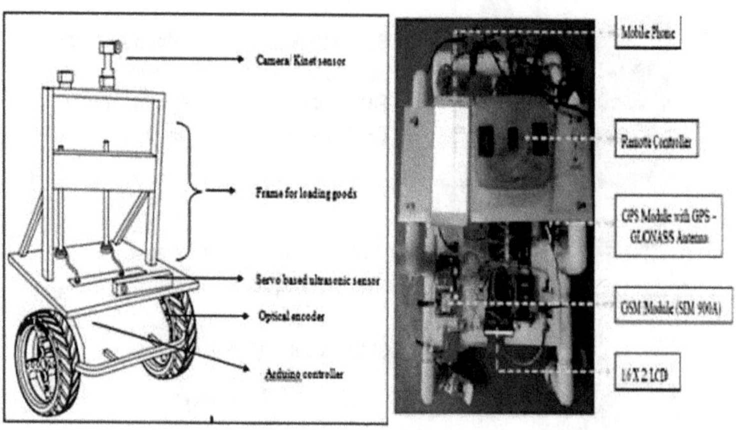

Figure 10. Number of samples with Latitude and Longitude from antenna

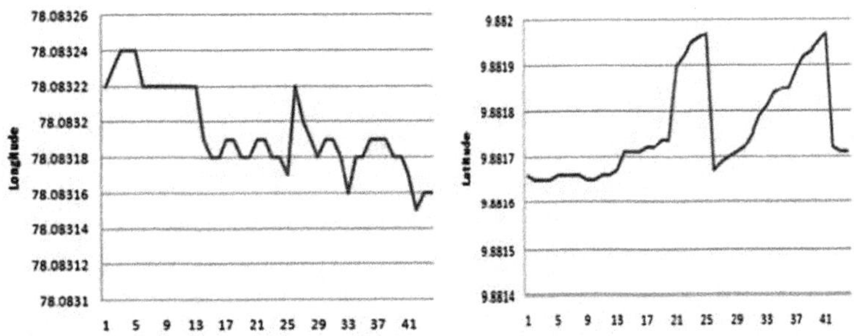

$$b_i = \frac{\sum_{i=1}^{n} x_i y_{i} - \frac{\left(\sum_{i=1}^{n} x_i\right)\left(\sum_{i=1}^{n} y_i\right)}{n}}{\sum_{i=1}^{n} x_i^2 - \left(\sum_{i=1}^{n} x_i\right)^2 \Big/ n} \qquad (2)$$

In this LSF approach, the sum of squares of offset value points is used instead of absolute values, for the reason that only allows the residual value treated as continuously differentiable quantity. The offset values used to outlay misappropriate points which effect on the fit as shown in Figure 12. The linear least square fitting method is simple and most commonly applied to error estimation problems. As the data points are being graphed to identify the cluster of data located near the straight line.

The Antenna setup along with inertial measurement unit (IMU) system is connected with either arduino controller or Raspberry Pi controller. After that the setup is then connected with GSM module and with power source. Once the system tracked then the information is delivered to the registered mobile numbers as message through GSM module. The flow chart for the CLRM system is shown in Figure 13.

The proposed GPNSS satellite navigation system is validated through two case studies. Case study 1 investigates the effectiveness of the navigation system in both indoor and outdoor environments. Case study 2 demonstrates the better accuracy of positioning and avoiding the obstacles due to the incorporation of vision sensor and computer vision-based system

Figure 11. GPS GLONASS satellite tracking system

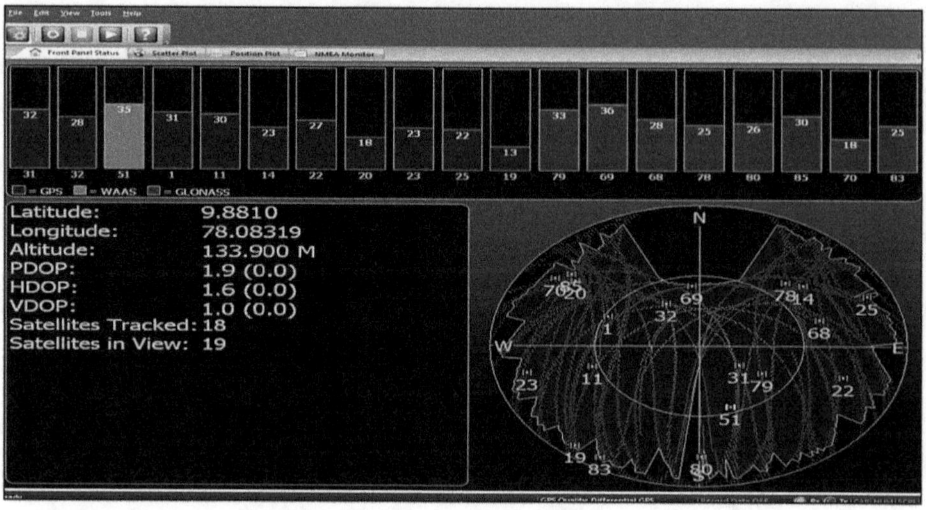

Table 3. Antenna positions received from both GPS & GLONASS

No. of Samples	Latitude	Longitude	No. of Samples	Latitude	Longitude
1	9.88166	78.08322	21	9.88195	78.08318
2	9.88165	78.08323	22	9.88196	78.08318
3	9.88165	78.08324	23	9.88197	78.08317
4	9.88165	78.08324	24	9.88167	78.08322
5	9.88166	78.08324	25	9.88169	78.0832
6	9.88166	78.08322	26	9.8817	78.08319
7	9.88166	78.08322	27	9.88171	78.08318
8	9.88166	78.08322	28	9.88172	78.08319
9	9.88165	78.08322	29	9.88175	78.08319
10	9.88165	78.08322	30	9.88179	78.08318
11	9.88166	78.08322	31	9.88181	78.08316
12	9.88166	78.08322	32	9.88184	78.08318
13	9.88167	78.08322	33	9.88185	78.08318
14	9.88171	78.08319	34	9.88185	78.08319
15	9.88171	78.08318	35	9.88189	78.08319
16	9.88171	78.08318	36	9.88192	78.08319
17	9.88172	78.08319	37	9.88193	78.08318
18	9.88172	78.08319	38	9.88195	78.08318
19	9.88174	78.08318	39	9.88197	78.08317
20	9.88174	78.08318	40	9.88172	78.08315

Figure 12. Accuracy estimation using linear least square fitting method

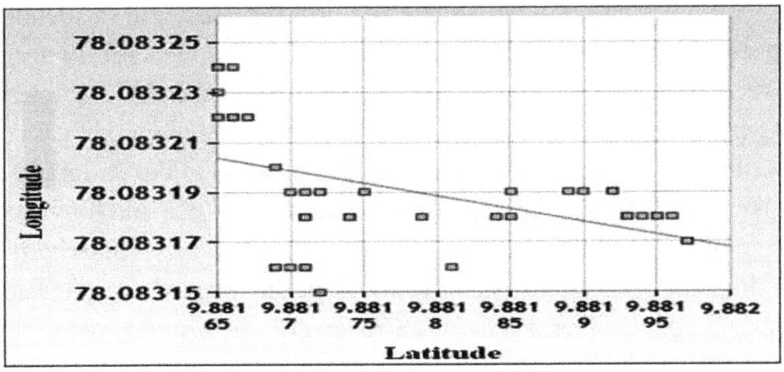

Figure 13. Flow control of vehicle with position data from antenna

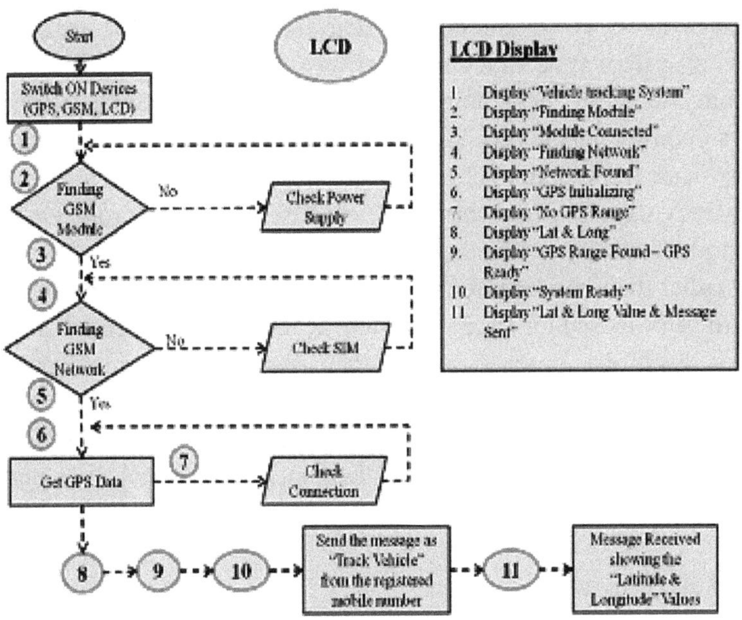

CASE STUDY 1

In this case study, the indoor system based AGV is tested in the indoor environment. It clearly troubleshoots the indoor system problems like navigating the vehicle and navigation accuracy improvement by introducing GNSS system-based navigation. The GNSS is the combination of GPS and GLONASS satellite signals are received and identify the latitude and longitude of the indoor vehicle and its operation. A prototype is designed by interfacing GNSS system along with Raspberry Pi as a controller. Normally a GPS receiver can perform the operation of identifying the location, but in NLOS areas it is difficult to receive satellite signals. Initially, the NEO - 7N GPS receiver is connected with Arduino Uno and tested alone to determine the satellite signals received in indoor and outdoor areas. The GPS module is connected with the Arduino UNO and Arduino coding is uploaded to the Arduino controller using the Arduino software and tested to display the latitude and longitude values at the current position. Initially, it is tested at the indoor location, where the GPS module is difficult to obtain the satellite signal in the indoor location. The number of satellites signals received by the GPS receiver is determined using the Visual GPS view software which is open source software. This method of navigating the vehicle inside the warehouse improves the system accuracy with low cost. The development of this low-cost high accuracy navigation system quite handy for all logistic application. The proposed navigation system was shown in Figure 14.

The localization of an autonomous car like mobile robot using available GPS tracking system at accuracy range of few meters (Macheng & Shen, 2016). Furthermore, the GPS technology is not applicable in indoor workspace for logistic warehouse applications. In that cases the modification or addition of new technology made the system to be reliable in indoor and outdoor environment. The radio frequency signals may absorb or reflect based on the environment obstacles. The object identification near Line of Sight can usually be dealt with using better antennas, rather than non-line of site are required alternating path propagation or new path to be proposed for improve system accuracy. To overcome the reduction in accuracy, this project proposes an idea that a system with a combination of both the GPS and GLONASS system, this makes the major part of GNSS tracking system made more satellites visible in Non-line of Sight areas and enabling positions to be fixed more quickly and accurately for localization. It provides real time position and

Figure 14. GNSS navigation system

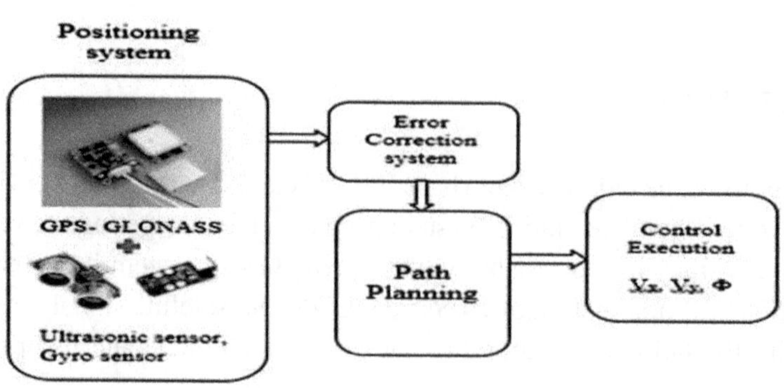

velocity determination quicker and more accurate. Recently, consumer devices like smart phones use the signal from both GPS and GLONASS satellites simultaneously to get accurate position and used for navigation:

- Based on this the radio frequency signals are combined to form GPS-GLONASS system to localize the robot in both environments.
- To improve the positioning and navigation of system accuracy by reducing the perception system errors.
- The tracked system latitude and longitude are transferred from the robot to the customer through GSM module.

CASE STUDY 2

Obstacle detection and avoidance happened in a secure manner in the path of the AGV in the real-time environment is monitored in this case study. The vehicle position and obstacle position determined by information provided by onboard sensors like wheel encoder, camera, and ultrasonic sensor. The CLMR with obstacle avoidance hardware is further developed to implement the GPS and GLONASS antenna for the localization of the robot, and identifies. The synchronization of range sensors and camera sensors will revolve future autonomous system. However, the high precision and accuracy need to use collected data like odometer information, range prediction of obstacle and vision prediction through images to be used in many heuristic algorithms for navigation and obstacle avoidance. The Mobile Robot Navigation with obstacle avoidance has some issues in accuracy which tends to a collision because the accurate measurement is needed for smooth navigation around the obstacles. Obstacle Avoidance based on Edge Detection is inaccurate, whereas the analytical based algorithms are more complex to determine the dimension of the obstacle. The improvisation of control system interfaced with the sensor is required to identify obstacles and avoid the obstacles. I-Hsum Lee et.al (2014), in advance the dimension of the obstacle to be measured and avoid the obstacle without stop the robot when it detects the obstacle through image processing technique based on the Morphological operation. Once the obstacle dimension is identified then the robot configuration whether the steering control servo is actuated or if the two-wheeled robots the driving wheel control is actuated. The control and interfacing are done through Raspberry pi interfaced with sensor and actuator. In the proposed system, the measurement of obstacle distance and dimension of obstacle using fusion of Our system uses a Logitech HD 720p webcam, Raspberry pi3 Model B, L293D Motor driver, LM317 Voltage Regulator, and an HC-SR04 ultrasonic transceiver device to achieve this. The range data is collected by the ultrasonic sensor for object detection with the video captured by the camera for object size measurement.

The vision sensor (camera) works as like our human eyes which grab the lateral range of area in front of eyes are covered. Based on that the camera is fitted to the robot at particular position to cover certain angle information. The covered information is recorded either as video or images based on accuracy level. Once the image captured by the camera and the resolution and pixels are same, but the size of the obstacle or object in front of robot is varying with respect to real life. The size of the object is not fixed, and it varies time to time depend upon the distance from the camera and angle of perception over the object. So, the irregularities can be identified by calculating the distance between the source and object and then the dimension of the object at static and dynamic condition. The geometry similarities achieved by the sensor fusion used to identify the dimension of object.

The abilities of this program are to detect and avoid obstacles autonomously in the real-time environment. In a challenging environment, it is necessary to make out the robot about the environment through sensor data. The localization of robot is more important at dynamic condition through fusion of position sensor and vision sensor. The sensors, like Inertial measurement unit, accelerometer, optical encoder, ultrasonic sensor, and infrared sensor are fused with vision sensor to locate the robot and detecting obstacle accurately.

Obstacle detection based on ranged-based sensors (like RFID, GPS, GNSS, Ultrasonic, RADAR, SONAR, Lidar etc.) are used to scan the environment to detect the obstacles distance precisely. Obstacle detection based on Appearance-based used to predict the physical dimension of the obstacle using various image processing techniques. Based on the combination of both ranged-based and appearance-based obstacle detection algorithm is used to measure object size and distance from the mobile robot are clearly explained in this work. In the proposed work the distance

Figure 15. Block diagram of Obstacle avoidance system

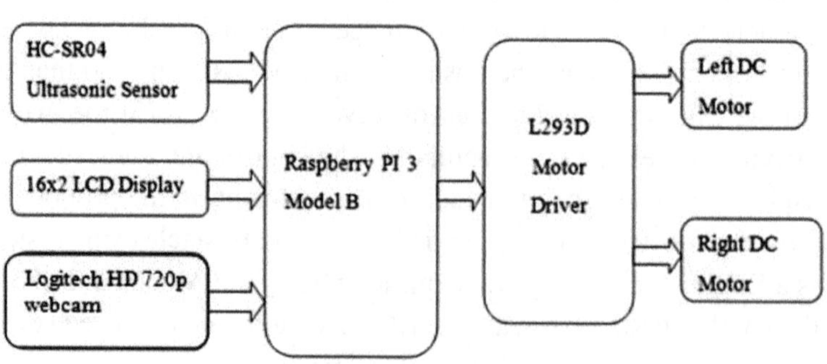

of obstacle can be easily identified using ultrasonic sensors as like bat principle. The ultrasonic sensor consist transmitter which emits sound signal range more than 20,000Db. If the sound signal collides with an obstacle in front of the robot, the wave will bounce back to the sensor receiver. Once the receiver receiving signal that provides the change in time by which the reflected wave from the obstacle. The time difference between the transmission of the ultrasonic wave and the reflection of the reflected wave is calculated and thus gives the distance between the robot and obstacle.

By doing so the obstacle dimensions are observed to protect the robot from physical damages. The two-wheeled robot along with ultrasonic sensor and camera are connected as hardware for obtaining experimental results. The obstacles are scanned by turning on the video mode or to take photos of the obstacle in order to determine the height of the obstacle & give a command to the raspberry pi to redirect the robot to move in the correct direction by actuating the motors which are interfaced to it through a motor driver.

The analysis of obstacle shapes and dimensions are acquired from the selected environment are compared based on image processing using morphing technique for non- holonomic condition. Introduction to obstacle avoiding system ensures the AGV will practically satisfy the automated requirement of freight and logistics application in real life. In continuation to that the sensors like vision, GPS and ultrasonic are interconnected in one network and collect data to localize and navigate the system by storing data in Cloud or Fog storage with betterment in security. Once the navigation and obstacle avoidance systems are ready then as like Figure 18, the prototype robot or vehicle is used to implement in warehouse and find the performance and accuracy of the system.

Figure 16. Static obstacle environment

(a) (b) (c) (d)

Figure 17. Dimensions of multiple obstacle detection

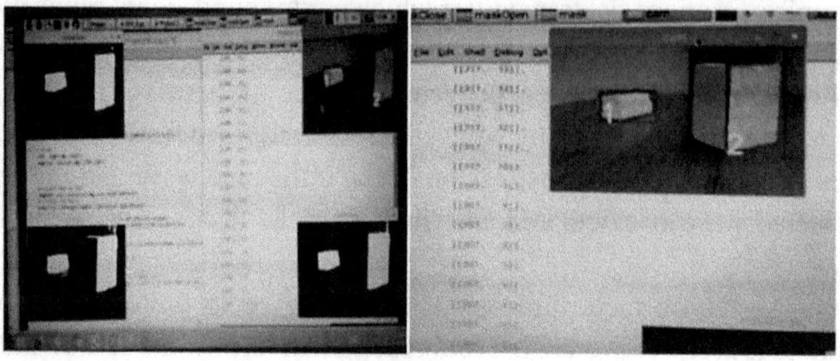

Figure 18. Application of mobile robot in logistics

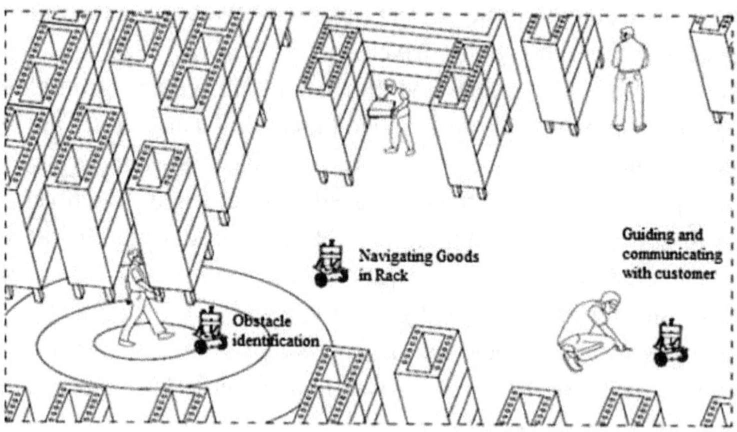

CONCLUSION

Two-wheeled autonomous mobile robot inbuilt with GPS and GLONASS signals system was developed and utilized for indoor and outdoor environment logistic application in a short span of time with high precision. Design and fabrication of Micro Strip Patch Antenna inbuilt with GNSS system to receive signals range from 1.575GHz and 1.602GHz. The performance, accuracy of GNSS navigation system confirmed with two case studies for both indoor and outdoor environment. The two case studies demonstrates that precise control and monitoring of a fleet management system using low cost GPS/GLONASS-based automatic vehicle to show the exact position of the desired vehicle on different maps and take detailed reports of the mission, obstacles present in the travelled path, and other necessary information according to the customers' requests. The completion of work relies on the performance of AI machines to perform an assigned scenario. Many control parameters are required to control the machines to work smartly and accurately. This chapter consider few control parameters like navigation of vehicles inside and outside environment are identified us GNSS system and the obstacle avoidance to predict trouble free path in warehouse environment are discussed using vision-based obstacle identification and dimension measurement to overcome the obstacle and reach the destination. The dimension measurement using spline or non-linear curves to predict exact shapes of the obstacle to be identified and the servo based steering mechanism to be implemented for the outdoor environment path planning system is another fruitful research area. Also, the introduction of smart traffic monitoring and controlling system in a warehouse or raw material handling system is welcome approach. The development of these systems is used to park the vehicle autonomously without any collision in the warehouse or to protect the vehicle from theft through the interconnection of sensor data in cloud storage.

REFERENCES

Al-Faiz & Mahameda. (2015). GPS-based Navigated Autonomous Robot. *International Journal of Emerging Trends in Engineering Research, 3*(4).

Bechtsis, D., Tsolakis, N., Vlachos, D., & Srai, J. S. (2018). Intelligent Autonomous Vehicles in digital supply chains: A framework for integrating innovations towards sustainable value networks. *Journal of Cleaner Production, 181*, 60–71. doi:10.1016/j.jclepro.2018.01.173

Belkhouche, F. (2009). Reactive path planning in a dynamic environment. *IEEE Transactions on Robotics, 25*(4), 902–911. doi:10.1109/TRO.2009.2022441

Bounini, F., Gingras, D., & Pollart, H. (2016). Real Time Cooperative Localization for Autonomous Vehicles. *IEEE 19th International Conference on Intelligent Transportation Systems (ITSC),* 1186 – 1191. 10.1109/ITSC.2016.7795707

Cho, J.-H., & Kim, Y.-T. (2017). Design of Automated Logistics Transportation Robot System with Fork-Type Lifter. *International Journal of Fuzzy Logic and Intelligent Systems, 17*(3), 177–186. doi:10.5391/IJFIS.2017.17.3.177

Conde, L., Chelim, R., & Nunes, U. (2015). Collaborative Vehicle Self- Localization Using Multi-GNSS Receivers and V2V/V2I Communications. *IEEE 18th International Conference on Intelligent Transportation Systems*, 2525 – 2532. 10.1109/ITSC.2015.406

Echelmeyer, W., Kirchheim, A., & Wellbrock, E. (2008). Robotics-logistics: Challenges for automation of logistic processes. *2008 IEEE International Conference on Automation and Logistics*, 2099-2103, 10.1109/ICAL.2008.4636510

Fernández-Caramés, T. M., & Fraga-Lamas, P. (2018). A Review on Human-Centered IoT-Connected Smart Labels for the Industry 4.0. *IEEE Access: Practical Innovations, Open Solutions*, *6*, 25939–25957. doi:10.1109/ACCESS.2018.2833501

Heard, B. R., Taiebat, M., Xu, M., & Miller, S. A. (2018). Sustainability implications of connected and autonomous vehicles for the food supply chain. *Resources, Conservation and Recycling*, *128*, 22–24. doi:10.1016/j.resconrec.2017.09.021

Héctor, Vis, & Roodbergen. (2014). Transport operations in container terminals: Literature overview, trends, research directions and classification scheme. *European Journal of Operational Research, 236*(1), 1-13. doi:10.1016/j.ejor.2013.11.023

Hosoda. (2011). Autonomous Moving Technology for Future Urban transport. *Hitachi Review, 60*(2).

Kamagaew, A., Stenzel, J., Nettsträter, A., & ten Hompel, M. (2011). Concept of Cellular Transport Systems in facility logistics. *The 5th International Conference on Automation, Robotics and Applications*, 40-45, 10.1109/ICARA.2011.6144853

Lee, S., & Lee, S. (2013). Embedded Visual SLAM: Applications for Low-Cost Consumer Robots. *IEEE Robotics & Automation Magazine*, *20*(4), 83–95. doi:10.1109/MRA.2013.2283642

Li, I.-H., Chen, M.-C., Wang, W.-Y., Su, S.-F., & Lai, T.-W. (2014). Mobile Robot Self-Localization System Using Single Webcam Distance Measurement Technology in Indoor Environments. *Sensors (Basel)*, *14*(2), 2089–2109. doi:10.3390140202089 PMID:24473282

Mac, T. T., Copot, C., Tran, D. T., & De Keyser, R. (2016). Heuristic approaches in robot path planning: A survey. *Robotics and Autonomous Systems*, *86*, 13–28. doi:10.1016/j.robot.2016.08.001

Mehami, J., Nawi, M., & Ray, Y. (2018). Smart automated guided vehicles for manufacturing in the context of Industry 4.0. *Procedia Manufacturing*, *26*, 1077–1086. doi:10.1016/j.promfg.2018.07.144

Sabattini, Aikio, M., Beinschob, P., Boehning, M., Cardarelli, E., Digani, V., ... Fuerstenberg, K. (2018). The PAN-Robots Project: Advanced Automated Guided Vehicle Systems for Industrial Logistics. *IEEE Robotics & Automation Magazine*, *25*(1), 55–64. doi:10.1109/MRA.2017.2700325

Sakpere, W., Adeyeye-Oshin, M., & Mlitwa, N. B. W. (2017). A state-of-the-art survey of indoor positioning and navigationsystems and technologies. *South African Computer Journal*, *29*(3), 145–197. doi:10.18489acj.v29i3.452

Shen, M., Zhao, D., & Sun, J. (2016). Enhancement of low-cost GNSS localization in connected vehicle networks using Rao-Blackwellized particle filters. *IEEE 19th International Conference on Intelligent Transportation Systems (ITSC),* 834 – 840.

Siegwart, R., & Nourbakhsh, I. R. (2004). *Introduction to Autonomous Mobile Robots*. MIT Press.

Song, H., Choi, W., & Kim, H. (2016). Robust Vision-Based Relative-Localization Approach Using an RGB-Depth Camera and LiDAR Sensor Fusion. *IEEE Transactions on Industrial Electronics*, *63*(6), 3725–3736. doi:10.1109/TIE.2016.2521346

Sousa, J., Galvão, J., Machado, J., Mendonça, J., Machado, T., & Silva, P. (2018). Modelling and Simulation of a Straight Line Motion Mechanism for Industrial Application. *5th International Conference on Control, Decision and Information Technologies (CoDIT)*, 1104-1109. 10.1109/CoDIT.2018.8394919

Tang, D., Zheng, K., Zhang, H., Sang, Z., Zhang, Z., Xu, C., ... Zechinelli-Martini, J.-L. (2016). Using Autonomous Intelligence to Build a Smart Shop Floor. *Procedia CIRP*, *56*, 354–359. doi:10.1016/j.procir.2016.10.039

Van Meldert & De Boeck. (2016). *Introducing autonomous vehicles in logistics: a review from a broad perspective.* Working Papers Department of Decision Sciences and Information Management.

Vasiljevic, G., Miklić, D., Draganjac, I., Kovačić, Z., & Lista, P. (2016). High-accuracy vehicle localization for autonomous warehousing. *Robotics and Computer-integrated Manufacturing*, *42*, 1–16. doi:10.1016/j.rcim.2016.05.001

Zhang, Y., Guo, Z., Lv, J., & Liu, Y. (2018). A Framework for Smart Production-Logistics Systems Based on CPS and Industrial IoT. *IEEE Transactions on Industrial Informatics*, *14*(9), 4019–4032. doi:10.1109/TII.2018.2845683

Zindler, K., & Geiß, N. (2016). Vehicle Ego-Localization in Autonomous Lane-Keeping Evasive aneuvers. *IFAC-Papers OnLine*, *49*(11), 160–167. doi:10.1016/j.ifacol.2016.08.025

KEY TERMS AND DEFINITIONS

AGV: Automated or autonomous guided vehicle is defined as vehicle or robot guided automatically in the pre-planned path without any human operation. These vehicles automatically navigate, plan obstacle free path to reach destination using artificial intelligence.

GNSS: The global navigation satellite system which receives position of the system using Receiver device to collect latitude and longitude data from satellites like GPS, GLONASS, Galileo, IRNSS, and so on. These data are used to predict automatic vehicle navigation, pedestrian navigation system, and tracking system in many applications.

Line of Sight: The line of sight (LoS) communication is defined as a type of propagation system where the transmitter and receiver which propagate and receive signal will be in view with each other without any obstacle block the signal. That improves the gain of signal whereas the NLoS (Non -line of sight) is a range in which physical interference across the signal propagate between transmitter and receiver.

Map-Based Approach: The navigation system of mobile robot in which the environment details are given in the form of graph and grid models. Using that models the system identify where the robot or AGV are located in the environment. In vision navigation early stages, the knowledge of environment is represent in 2D projection normally called occupancy map.

Mobile Kinematics: The process of understanding mobile motion based on operation of wheel constrains along with mechanical system behavior. The used mobile robots in case studies are two wheeled robot with generalized kinematics equation as, ε_R=R(∅)ε_I, where ε_R - Motion in local reference frame, R (∅) – Orthogonal rotational matrix and ε_I – Motion in the global reference frame.

Morphology Technique: The morphology technique is conversion of images into small structural elements and classifies the structural element based on binary values positioned at all places in image to identify the expected image pixel by comparing with neighborhood pixel values. This technique is used to identify the obstacle in pre-defined path.

Section 3
Industry 4.0

Chapter 9
Smart Make-to-Order Production in a Flow Shop Environment for Industry 4.0

Humyun Fuad Rahman
University of New South Wales, Australia

Mukund Nilakantan Janardhanan
University of Leicester, UK

Peter Nielsen
Aalborg University, Denmark

ABSTRACT

The permutation flow shop scheduling problem is one of the popular problems in operations research due to its complexity and also its practical applications in industries. With the fourth generation industrial revolution, decisional aspects in make to order flow shop environment needs to be decentralized and autonomous. One of the aspects is to consider a real-time or dynamic production environment where customers place orders into the system dynamically and the decision maker has to decide whether the order can be accepted considering the available production capacity and how to schedule the jobs of an accepted order. To answer these research questions, in this chapter, the authors introduce a new decision-making, real-time strategy intended to yield flexible and efficient flow shop production schedules with and without setup conditions, Numerical experiments based on realistic problem scenarios show the superiority of the proposed real-time approach over traditional right shifting approaches.

DOI: 10.4018/978-1-5225-9078-1.ch009

INTRODUCTION

The permutation flowshop scheduling problem (PFSP) is a popular scheduling problem seen in manufacturing environments such as the automobile industry, food, pharmaceuticals, steel making, sanitary ware, and furniture (Rahman et al., 2015; Rahman et al., 2013). It is a complex optimization problem, where the production system usually consists of a finite number of jobs or tasks that must be processed on limited machines or processors. Each machine performs a specific type of operation for each job within a certain amount of time, which is known as processing time. As a prime consideration of PFSPs, each machine has to process all jobs in the same processing sequence of jobs. On the other hand, each job has to follow the same sequence of all machines. Therefore, the work-flow is unidirectional. A schedule is accomplished by the completion of all jobs in all machines following a particular processing sequence of jobs. There are also specific constraints that must be satisfied. For example, a job cannot visit the same machine more than once, processing times cannot be negative, and a machine cannot handle more than one job at the same time. In addition, different objectives for solving PFSPs are makespan minimization, flow time minimization, completion time minimization, tardiness minimization, earliness maximization, and through output maximization. Makespan minimization is the most widely reported objective for solving PFSPs.

For everyday production processes, flow shop scheduling problems are encountered by manufacturers worldwide in different forms depending on the constraints and optimization criteria of the production environments. Thus, efficient planning and scheduling of flow shop production has always been principal criteria for the overall success of any manufacturing enterprise. The flow shop scheduling problem is a complex optimization problem and, therefore, finding efficient and effective techniques for solving production scheduling problems has attracted researchers and practitioners in the area of operations management and combinatorial optimization (Zobolas et al., 2009; Rahman et al., 2018). In literature, many different approaches have been developed for solving PFSPs under ideal conditions. Initially, the classical optimization techniques were developed for solving PFSPs for solving small-scale problems. For solving large-scale problems, many heuristics and meta-heuristics algorithms have been proposed. However, their performance varies significantly from problem to problem.

Although some characteristics of an ideal PFSP are exponential in nature and are complex to solve, practical production systems usually offer even more complex scenarios than the idealized problems addressed in current state. These include e.g. process interruptions, variability in product specifications, smoothness in product flow throughout the production floor, level of customer satisfaction, and competitive market scenarios. Therefore, based on the type of production, flow shop scheduling problems are classified as make to order (MTO) production systems and make to stock (MTS) production systems (Rahman et al., 2015). In MTO systems, manufacturers produce the products based on customer orders and, therefore, decision makers have to take order acceptance/rejection and scheduling decisions sequentially. Many modern manufacturing industries, academic researchers, have principally focused on static MTO flow shop scheduling problems, where a set of jobs of an order is scheduled on a set of machines in the flow shop without considering realistic settings. For example, Johnson (1954) proposed an optimal algorithm for two- and for a variant of three-machine static MTO PFSPs. For solving more than two-machine static MTO PFSPs, researchers proposed classical optimization techniques, such as the branch and bound (B&B) (Ignall & Schrage, 1965), and integer programming (Selen & Hott, 1986) algorithms. However, for more than two-machine MTO PFSPs classified as NP Hard (Garey et al., 1976), these techniques can find optimal solutions for small-sized problems only. To solve larger sized PFSPs, researchers have proposed heuristics (Nawaz et al., 1983) and meta-heuristic algorithms (Zobolas et al., 2009; Grabowski & Wodecki, 2004; Tasgetiren et al., 2007; Osman & Potts, 1989; Rajendran & Ziegler, 2004; Ruiz & Maroto, 2005; Ruiz et al., 2006; Rahmnan et al., 2013; Dasgupta & Das, 2015; Abdel-Basset, et al., 2018). Pinedo (2012) has argued that most of the theoretical studies in multiple machine scheduling problems considered static settings, i.e. scheduling an *n* set of jobs in an *m* set of machines only. However, with the emerging Industry 4.0 (I4.0) technologies (Ivanov et al., 2016; Zheng et al., 2018) practical MTO production systems become dynamic, whereby customers place orders to the manufacturers in real-time with delivery date or due date. Hence, the manufacturers do not have prior information about arrival time and due date of an order, number of jobs per order or order compositions, and processing time of each job in each machine. Therefore, the challenge managers face in practical MTO systems is that no prior information is available about the orders. Once an order is placed in the production, managers have to take decisions sequentially: (1) how is the acceptance or rejection decision of an incoming order taken? (2) how are the jobs of an accepted order scheduled on the machines?

Although the problem is practical and more relevant to the modern manufacturing industries in the context of Industry 4.0, it brings a number of challenges, such as: integrated decision making process of accepting/rejecting a newly arrived order, and scheduling the job set of an accepted order while considering both the orders already in process and also the accepted orders in the queue at that point in time. There is no doubt that these problems are more complex than the static MTO PFSPs. Although a handful of techniques were developed for solving similar problems in other shop floor problems (only in job shop), the real-time order acceptance and scheduling problem in the flow shop environment are still an unaddressed research problem. Therefore, it is difficult to implement existing methodologies available in the literature to solve real-time MTO production scheduling for the shop floor.

In order to address this problem, the major contribution of this chapter is to present a detailed implementation procedure for using metaheuristics to solve real-time MTO PFSP problems. Since the literature demonstrated that the focus on these aspects has been avoided till date. Two configurations of PFSPs (set time included in processing time and set time excluded from processing) from two previous studies (Rahman et al., 2015 (a); Rahman et al., 2015 (b)) will be discussed in detail. To solve the problem due to its complexity an evolutionary algorithm-based real-time (RT) approach will be used and experimental results obtained by using an RT approach is presented, and compared with respect to the right shifting (RS) approaches.

REAL-TIME MTO PFSP WITHOUT SETUP TIMES

In this section, a real-time multiple-order PFSP has been discussed, where customers randomly place orders in the production floor with a customer-specified due date (or delivery date). In this case, manufacturer does not have prior knowledge about the order details such as order arrival time and due dates and each individual order can be considered as similar to a static single-order PFSP, but without arrival time or due date. The schedule generated for an order can be implemented without any alteration, if the makespan (calculated considering single order PFSP) of the previously accepted orders is less than the inter-arrival time of orders. However, when processing times of orders overlap the schedule considering single order PFPS is not feasible. In this case, order acceptance/rejection decisions for a new order may not be feasible while satisfying machine availability to process the new order and due date constraints. The management has to make two decision strategies: (1) is it feasible to accept new order? (2) considering processing times, if the new order and other accepted orders overlap, then how the new order should be scheduled? The first decision can be made on basis of already accepted orders waiting in the queue or under process, already accepted orders waiting in the queue or under process, and availability of production capacity to process the order. The second research question can be dealt in two ways: (i) jobs of each order is scheduled as static single order PFSP and the order books the machines for time period equal to static makespan. Each accepted order starts to process immediately after finishing the orders currently under process. Alternatively, if the processing time of the new order is overlapping with the order under process, the new order needs to be shifted to right of the schedule. Therefore, we call this strategy is named as Right shifting (RS) strategy. (ii) In this strategy, a hypothetical schedule is generated for a new order considering the production capacity constraints (currently orders which are waiting in the queue and orders in process). The new order can start to process at the first or following machines when they are immediately available, and it may increase the chance of completing an order early as well as increases the chance of accepting the order. In order to take this advantage, jobs of a new order need to be scheduled/rescheduled while considering the capacity constraints. In this chapter, we have proposed a GA based memetic algorithm (MA) for scheduling an order while considering production constraints. This strategy is known as real-time (RT) strategy. MAs are successful in solving different types of complex scheduling problems (Hasan et al., 2009), and therefore it motivates us to developed RT strategy based on MA.

In order to study the problem in a systematic way, we have generated 40 problem instances of the 10-machine single order PFSPs from Taillard's benchmark (Taillard, 1993), since there is no standard benchmark instances available for real-time multiple-order PFSPs. We have generated a random inter-arrival time and due date for each order. To judge the quality of the generated solutions by RS and RT, we have compared the results with respect to the upper and lower bounds of possible solutions.

Related Literature on Real-Time MTO PFSP With Setup

Although real-time MTO PFSP is one the complex problems that occur in practical real-world industrial environments, limited literature is available for where order acceptance and scheduling decisions in flow shop and job shop environments. Nandi and Rogers (2004) proposed an order acceptance mechanism, where a new order is accepted by comparing the profit of the current orders. Rogers and Nandi (2007) also proposed an order acceptance/rejection rule that was based on total processing time of each job in all machines or the workload of a job in the busiest machines. Moreira and Alves (2009) developed an order acceptance/rejection mechanism based on negotiable due dates or workload. Once an order is accepted, orders are scheduled by simple dispatching rules, e.g.: earliest due dates (EDD) (Rogers & Nandi, 2007; Moreira & Alves, 2009), least slack per remaining operations (S/OPN) (Nandi & Rogers, 2004; Rogers & Nandi, 2007), and first come first serve rules (FCFS) (Rogers & Nandi, 2007; Moreira & Alves, 2009). All these studies consider the order acceptance decisions and the scheduling decisions separately and sequentially. However, in order to improve the overall performance of the system these decisions should be integrated. In the following section, these decisions are integrated.

Problem Definition and Assumptions for Real-Time MTO PFSP Without Setup Times

The problem under study focusses on real time MTO-PFSP without setup times and the following assumptions are made in defining real-time multiple-order PFSPs without setup times (Rahman, et al., 2015 (b)):

- Setup and transportation time are considered negligible or included in the processing time.
- Arrival time of each order is not known in advance.
- Each order contains multiple jobs (same as single-order PFSP).
- There is no prior knowledge of the composition of each order.
- Due date of each order is set by the customers.
- The manufacturer has the right to decide whether to accept or reject an order.
- An accepted order has a positive impact on productivity if it can be completed within the due date.
- There is no interruption while processing an order.
- An order accepted cannot be rejected later.
- There is no profit for complete before due date, but late completion has financial loss (tardiness).

Figure 1 presents the flow diagram for real-time multiple-order PFSPs, where orders are received by the manufacturer from customers. After receiving the new order, the production managers make the decision whether to accept or reject the order based on its completion time before due date. If a newly arrived order is accepted, it is scheduled based on available/unavailable production capacity at that point in time since some or all machines in the production line may process previously accepted orders. This means that some or all jobs of an order may need to be rescheduled based on the current production capacity. Therefore, an order may start to process if the first or the following machines are available. Alternatively, the accepted orders wait in the queue to be processed. After the completion, an order is immediately delivered to the customer. Otherwise, an order is rejected if it cannot be completed by its due date.

The mathematical model for real-time multiple-order PFSPs is presented as follows (Rahman, et al., 2015 (b)):

h – index for an order
i – index for a job in an order h
j – index for a particular machine
p^h_{ij} – processing time of job i (of an order h) on machine j

Figure 1. Flow diagram of processing orders

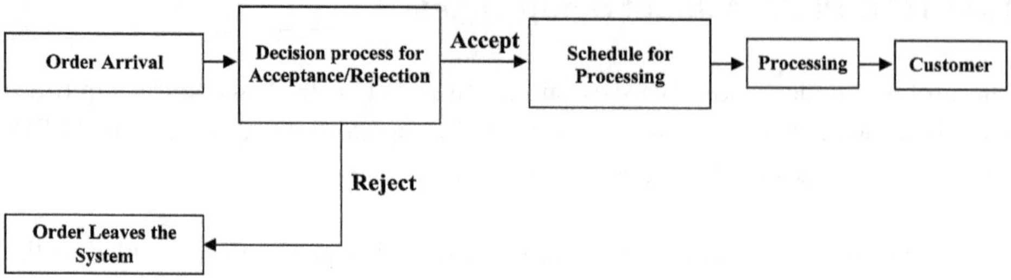

n_h – number of jobs in order h
m – number of machines in the flowshop
C^h_{max} – makespan for order h
C^h_{com} – completion time for order h
d_h – customer-specified due dates (or delivery time)
T_h – tardiness of order h
H – number of orders that arrive in one production shift
S^h_j – the starting time of the first job of order h on machine j
F^h_j – the finishing time of the last job of order h on machine j

The makespan, C^h_{max}, of an individual order l, can be determined as the makespan of an equivalent static single-order PSFP. The completion time for that order C^h_{com}, can be expressed as

$$C^h_{com} = S^h_1 + C^h_{max} \tag{1}$$

where the start time of the first job of the first order (start of production shift) is, $S^1_1=0$.

So the completion time for the first order is equal to its makespan,

$$C^1_{com} = C^1_{max} \tag{2}$$

The tardiness of an order is then:

$$T_h = C^h_{com} - d_h,\ h = 1,2,3,\ldots\ldots\ldots, H \text{ and } d_h \geq 0 \tag{3}$$

Order h is only accepted if $T_h \leq 0$, or else it is rejected (4)

In any machine, the $(h+1)^{th}$ order cannot be started until the hth order has been finished. In this case, the following condition must be satisfied.

$$S^{h+1}_{j} \geq F^{h}_{j} \forall h, j \tag{5}$$

The objective of the real-time multiple-order PSFP without setup time is to maximize the number of accepted orders.

Solution Approach for Real-Time MTO PFSP Without Setup Times

To solve real-time multiple order PFSPs problem, an order acceptance/rejection heuristic and a memetic algorithm (MA) (based on a genetic algorithm (GA)) based integrated approach have been proposed. This integrated process thus helps to reduce the completion time of each order (considering production capacity constraints and due dates) and selects a set of orders that would be feasible to process considering shop capacity and can be delivered within its due dates.

Acceptance/Rejection Decision

The order acceptance/rejection decision is made on the basis of a trial schedule that is generated while considering the start time, available capacity within the time window, completion time, and the due date of each order. An arriving order will only be accepted if it can be completed by its specified due date.

Heuristic for Order Acceptance/ Rejection Decision

The heuristic for the order acceptance/rejection decision can be described as follows. Assume that d_i is the due date and C^i_{com} is the completion time of ith order. Also assume that I is the total number of orders that arrive in a production shift, and T_i is the tardiness of that order. The completion time, C^i_{com}, is calculated by a scheduling algorithm that will be discussed in the next section.

```
[Start of algorithm]
1. Initialize Production Shift when the first order arrives in
the system.
    Set order number, i = 1
2. Repeat while (i = I)
A.        Calculate C^i_com for the i^th  order by MA.
B.        If T_i ≤ 0 (C^i_com <= d_i), Accept order i.
C.        Else Reject it.
D.        Set i = i+1.
              [End of the step 2 Loop].
```

```
            3. Save the total number of accepted orders of
the current production shift.
      [End of the algorithm].
```

Scheduling Decisions

Due to machine availability constraints, an order may not immediately be processed after its arrival to the system. If the first machine is free when the order arrives in the system, then the order may start to be processed without interrupting the processing of other orders. Alternatively, the order has to wait until the first or following machines are available. Based on the condition of production constraints at that time, the scheduling or rescheduling of some or all of the jobs of an accepted order will help to improve system utilization. To schedule or reschedule orders, two strategies referred to as Right Shifting strategy and Real-Time Multiple-Order strategy are described in the following section.

Right Shifting (RS) Strategy for Real-Time MTO PFSP Without Setup Times

In the RS approach, first each accepted order is considered as a single order PFSP and an initial schedule generated by sequencing all jobs in that order is generated by Memetic Algorithm with the objective of makespan minimization, assuming all machines are unconditionally available. If a job or some jobs can be processed immediately in the next machine, then the jobs can start immediately considering the initial schedule. However, if those jobs have to wait until the next machine is available and the start time of the jobs of that order is shifted to right of the schedule. That means the completion time after right shifting the jobs is more than its initial schedule. An illustration of the RS strategy without setup time with a three-machine flowshop with two orders is shown in Figure 2. A detailed description of the example is available in Rahman, et al., (2015 (b)).

Algorithm: RS Strategy for Real-Time Multiple Order PFSP Without Setup

The algorithm with the RS strategy is described as follows. Assume that the total number of orders that arrive in a production shift is N. Also assume the due date of the nth order is d_n and c is the current time of the clock. The algorithm first calculates the single-order makespan of the nth order $(C_{max})_n$ and, considering the start time of processing of that order, S_n, the completion time, $(C_{comp})_n$, and its tardiness, T_n is calculated Next, it also calculates the clock time when the first job of the ith order will start to be processed in the first machine t_i. Also assume that the number of accepted orders per production shift is N_{acpt}.

The algorithm initiate, when the first order is placed by the customers.

```
        [Start of algorithm]
I.         Set clock t_1=0, n = 1, and N_acpt =0
II.         Repeat while n = N (until the end of the current
production shift)
A.         Set the clock time, t_n = c, and S_n=c
B.         Calculate the single-order makespan for the current
order by MA.
C.         Recalculate the makespan (C_max)_n of the current order
considering the current system state:
a.         Right shift the n^th order, if the n^th order overlaps
with the (n-1)^th accepted order.
b.          Else, the makespan of the current order (C_max)_n is
unchanged.
c.          Calculate the completion time of the order, (C_comp)_n
=S_n+(C_max)_n
D.         Considering the completion time, (C_comp)_n
a.          Calculate tardiness as, T_n = (C_comp)_n - d_n
b.          Save the completion time, (C_comp)_n and tardiness, T_n
c.          If T_n  ≤ 0
i.         The order is Accepted.
ii.         N_acpt= N_acpt+1
d.          Else, the order is Rejected.
E.         Set n=n+1
[End of step II Loop]
III.          Save, the total number of accepted orders, N_acpt.
[End of the Algorithm]
```

Like the RS strategy, in this strategy, jobs of an accepted order start to process as soon as the first machine is free. However, instead of creating a best static PFSP and shifting the jobs to the right, jobs of an accepted order are scheduled considering the availability constraints of machines. Therefore, the scheduling in the strategy may differ from the static schedule of the order in this strategy and it may reduce the completion time of each order. An illustration of the RT strategy is given in Figure 3. Note that, in this example, with the RS strategy, the completion time of order N is bigger than it is for RT.

Figure 2. Right Shifting (RS) Strategy without setupReal-Time (RT) Strategy Without Setup

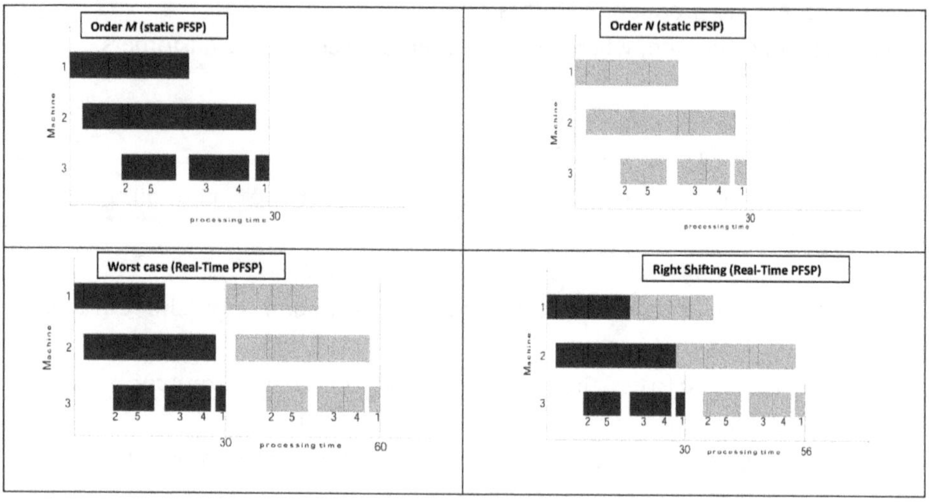

The algorithm for RT strategy without setup can be described as follows.

Assume that N is the total number of orders per production shift and is N_{acpt} is the number of accepted orders per production shift. Assume that the current time of the clock is c, the due date of the nth order is d_n, and the start time of that order is S_n. The algorithm first calculates the makespan of the nth order $(C_{max})_n$, the completion time $(C_{comp})_n$ and its tardiness, T_n. Finally, it has to calculate t_i, the clock time when the first machine starts to process the first job of the ith order.

The following algorithm starts when the first order arrives at the beginning of the Production Shift.

```
[Start of Algorithm]
I.          Set clock time, t1=0, n = 1, and Nacpt =0
II.          Repeat while (n = N) (until the end of the current
production shift).
A.          update Sn=c and the clock time, tn = c
B.          Use MA to generate a schedule for the order
considering current system state, i.e. machine availability or
unavailability constraints.
C.          Considering the completion time, (Ccomp)n, evaluate Tn:
a.          Evaluate the completion time of the order, (Ccomp)n
=Sn+(Cmax)n
b.          Calculate the tardiness, Tn = (Ccomp)n - dn
```

```
c.        Store the completion time, (C_comp)_n and tardiness, T_n
d.        if T_n ≤ 0.
i.        The order is Accepted.
ii.       N_acpt= N_acpt+1
e.        Else, Reject the order.
D.        Set n=n+1
[End of step II Loop]
III.      Store the total number of accepted order, N_acpt.
[End of the algorithm]
```

Experimental Study for Real-Time Multiple Order PFSP Without Setup

Extensive experiments are carried out to evaluate the performance of the RS strategy and RT strategy. To test the performance of the proposed approach, 46 realistic problem instances (40 instances for Scenario I, four instances for Scenario II and two instances for Scenario III) have been generated on basis of Taillard's Benchmark (Taillard, 1993), using information such as: (i) a randomly selected job from a pool of problem instances, (ii) a random arrival time of the order, and (iii) a random due date of the order. In Scenario I, only the same type of order is considered. For Scenario II, different types of orders are considered, where each order has an identical number of jobs. In Scenario III, different types (any number of jobs in each order) of order

Figure 3. Real-Time (RT) strategy without setup

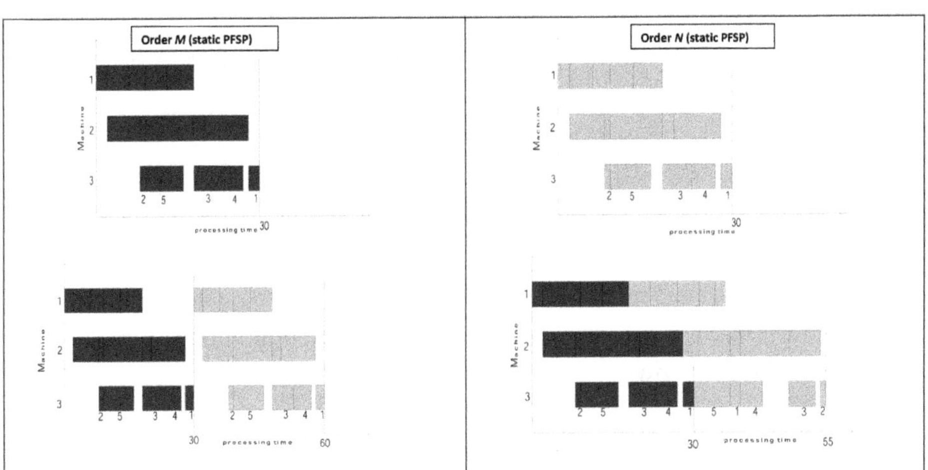

are considered. To compare the performance of each strategy, an upper bound (UB) and a lower bound (LB) for the accepted orders have been derived.

Upper and Lower Bounds

Suppose l^{th} order with the arrival time of At^l has been accepted by a manufacturer. Also, assume that Ct^l is the completion time of the lth order. If $(At^l - Ct^{l-1}) > 0$, between the completion time of order (l-1) and the arrival time of order l the machines will be idle for $(At^l - Ct^{l-1})$ time period. The positive time difference (TD^l) can be defined as follows:

$$TD^l = \begin{cases} At^l - Ct^{l-1}, & if\ (At^l - Ct^{l-1}) > 0 \\ 0 & Otherwise \end{cases} \text{for all } l \geq 2 \tag{6}$$

So the upper bound or the worst-case scenario in completing all accepted orders L can be expressed as follows.

$$UB = \sum_{l=1}^{L} Cmax^l + \sum_{l=2}^{L} TD^l \tag{7}$$

where $Cmax^l$ is the single order PFSP makespan of lth order.

If $(At^l - Ct^{l-1}) < 0$, machine-1 may start processing l^{th} order as soon as completing order l-1, and that will require $Cmax^l$ time for completing the order l. The best-case scenario for completing all of the accepted L orders is that if l^{th} order will arrive in the system just after completing order l-1 on machine-1. Further assume that the total processing duration required for l^{th} order on machine-1 is Pt^l_1 and the time of completing l^{th} order on machine-1 is CPt^l_1. The reduction in processing duration (RPT^l) is defined as follows:

$$RPT^l = \begin{cases} Cmax^l - Pt^l_1, & if\,(At^l - Ct^{l-1}) < 0\, and\, At^l = CPt^l_1 \\ 0 & Otherwise \end{cases} \text{for all } l \geq 2 \tag{8}$$

So the lower bound or best-case scenario of completing all L accepted orders can be expressed as follows.

$$LB = \sum_{l=1}^{L} Cmax^{l} - \sum_{l=2}^{L} RPT_{1}^{l-1} + \sum_{l=2}^{L} TD^{l} \quad (9)$$

RESULTS ANALYSIS AND DISCUSSION

The proposed algorithm was implemented in C++ and tested on a computer with a 4GB RAM and a core i7 processor. The parameters have been selected based on the previous study (Rahman et al., 2013 (b)). In the experimental study, we have considered three different scenarios. To demonstrate the applicability of the proposed algorithm, 10 orders for Scenario I and 25 orders for Scenario II and Scenario III are considered.

Table 1 shows the comparison between MA-based RS and RT strategies for 10 orders where each order contains 50 jobs (based on problem instance *Ta* 046) under Scenario I. Those orders arrive in point in time of production and each order has different due dates. Simulation results presented in the Table show that, out of 10 orders, six orders (1st to 5th and 9th order) can be accepted by the RS strategy and all orders are accepted by the RT strategy. The reason is, for overlapping orders, RS strategy blocks the machines more than the RT strategy and increases the completion time of those orders.

Table 1. Result for scenario I without setup

Order	Due Date, d	Arrival Time	Right Shifting (RS) Strategy			Real-Time Multiple-Order (RT) Strategy			Improvement
			Completion Time, $C_{comp}(RS)$	Tardiness= $C_{comp}(RS)$-d	Status	Completion Time, $C_{comp}(RT)$	Tardiness $= C_{comp}(RT)$-d	Status	
1	6297	0	3006	-3291	A	**3006**	-3291	A	0
2	5984	360	5740	-244	A	**5718**	-266	A	22
3	8573	4846	8436	-137	A	**8421**	-152	A	15
4	11983	8638	11644	-339	A	**11644**	-339	A	0
5	19126	13882	16888	-2238	A	**16888**	-2238	A	0
6	19606	15991	19624	18	R	**19594**	-12	**A**	30
7	22294	19112	22307	13	R	**22288**	-6	**A**	19
8	24987	20764	24997	10	R	**24982**	-5	**A**	15
9	30116	23051	27680	-2436	A	**27662**	-2454	A	18
10	30468	25101	33093	2625	R	**30356**	-112	**A**	2737

*Status of order: R-rejected, A-acceptedThe effectiveness of MA-based RS or RT strategies can be represented by the following equations:

$$\text{Average Deviation from the UB, } D_U = \left[\sum_{i=1}^{Ns}\left(\frac{UBi - Ri}{Ri}\right)\right] / N_S, \tag{10}$$

$$\text{Average Deviation from the LB, } D_L = \left[\sum_{i=1}^{Ns}\left(\frac{Ri - LBi}{LBi}\right)\right] / N_S \tag{11}$$

Where the makespan of the ith accepted order achieved by RS or RT is Ri and N_S is the total number of accepted orders by the particular method. This number may vary for both of the strategies.

Table 2 presents a summary of comparisons between the RS and RT strategies for Scenario I (random arrival of same types of orders). The average improvement from UB (worst case) for 10 different problem instances after applying RS and RT is shown in column D_U. The column titled D_L presents the improvement of RS and RT from LB, i.e. how quickly orders can start to process in the machines if processing times of the consecutive orders are overlapping. Note that, while calculating LB, capacity constraints are assumed to be violating, i.e. an order can immediately process even when the machines may be unavailable at that time. From all results, it can be seen that RT performs better than RS since bigger value of D_U and smaller value of D_L means better performance. In addition, it is clear that with the RT strategy, more orders can be accepted.

Tables 3 and 4 show the comparison between RS and RT strategies for Scenario II and Scenario III, where 25 orders randomly entered into the production shift with different due dates. These tables also show the better performance of RT over RS in terms of deviation from UB and LB, and also based on the number of orders accepted.

Table 2. Summary of comparisons for same order arrival

Number of Jobs	D_U		D_L		Number of Accepted Orders	
	RS	RT	RS	RT	RS	RT
20	1.566026	**2.045218**	0.796381	**0.501011**	64	**70**
50	0.3713685	**0.4072689**	0.24099229	**0.218244499**	73	**80**
100	0.16546857	**0.18658584**	0.17880491	**0.17248293**	59	**62**
200	0.11492327	**0.11501577**	0.06381179	**0.062574632**	59	**65**

Table 3. Mixed order with same order size

Number of Jobs	D_U		D_L		Number of Accepted Orders	
	RS	**RT**	**RS**	**RT**	**RS**	**RT**
20	0.145812	**0.185154**	0.054259	**0.038151**	14	**17**
50	0.070463	**0.0774516**	0.0435364	0.0444811	10	**12**
100	0.0204339	**0.0215277**	0.00742706	**0.00630862**	7	**7**
200	0.0274414	**0.0310825**	0.0049502	0.00606717	13	12

Table 4. Mixed order with different order sizes

Case	D_U		D_L		Number of Accepted Orders	
	RS	**RT**	**RS**	**RT**	**RS**	**RT**
1	0.292583	0.144235	0.0285443	**0.0231372**	19	**21**
2	0.335286	**0.335857**	0.00712116	**0.00665912**	25	**25**

Real-Time MTO PFSP With Setup

The aim of this section is to bridge the gap between the theory and actual practice of production scheduling by studying a problem in a real-life production environment. The case study presented refers to a MTO flow shop sanitary ware production system. Real-life sanitary ware production systems are complex and time-sensitive since they have to produce a wide range of variation in their products based on customer specification, real-time arrival of orders, dynamic batch adjustments, time for machine setup, and computationally expensive. Due to this complexity, many such companies run with sub-optimal solutions. To solve this problem, a MA-based real-time approach has been proposed in this section. Numerical experiments based on a practical production scenario are also presented in this section.

Real-Time MTO-Based Sanitaryware Production System

As described earlier, the case refers to a sanitaryware production plant capable of producing 10 different types of products, which are listed in Table 5. Each order consists of a product mix specified by customers, for example an order may consist of a total of 10 products, whereby three products are from wash basins, five are from galleries, and two are kitchen sinks.

Figure 4 presents a production flow diagram of a real-time multiple order PFSP with setup time. Like real-time multiple order PFSPs without setup systems, customers place orders with specific delivery times or due dates. Similar to real-time MTO PFSP without setup, an order acceptance and scheduling decision should be made when a new order arrives in the system. This decision depends on the arrival and start time of the order, the order composition (type of products in the order), available production capacity, batch size and number of setups required, and due dates. However, unlike the problem without setup time, setup time is critical in scheduling the accepted orders. If a machine processes a product after processing a product from the same family or category, setup time for that machine is zero, which means that there is to some extend sequence dependent setup times. Therefore, setup time or cost is minimum if each product family allows one setup. On the other hand, completion time or makespan of an order is improved if all products are scheduled together while allowing multiple setup of the same product category. However, setup time in each machine may increase the time for order completion. Hence, while scheduling an accepted order the number of setups for each product family should be adjusted on a real-time basis. The company will earn profit if an order can be completed within its due date. Otherwise, opportunity loss cost is generated by rejecting the order.

Table 5. Different product type

Product Number	Product Type
1	Foot rest
2	Water closet
3	Flower vase
4	Long pan
5	Wash basin
6	Ashtray
7	Cistern with lid
8	Gallery
9	Kitchen sink
10	Pedestal

Figure 4. Production flow of real-time multiple order PFSP with setup

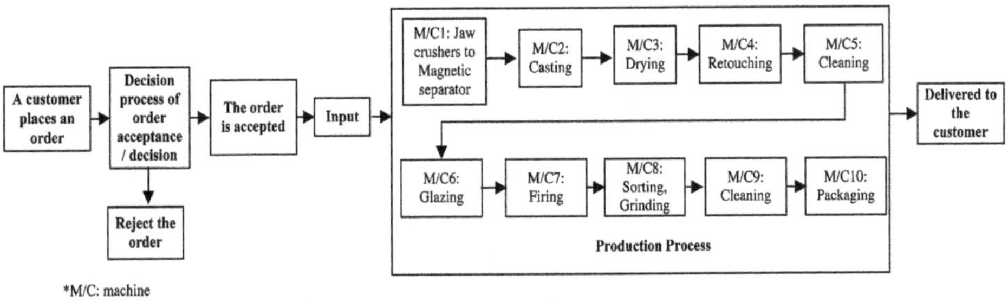

Related Literature on MTO PFSP With Setup

Even though, in reality, each machine may require a specific time to setup before it starts to process a job or manufacture a different product, only a few research contributions have addressed setup time for single order PFSPs. The following paragraph reviews those approaches. A brief review on this topic can be found in (Allahverdi et al., 2008).

Allahverdi (2000) proposed a branch and bound algorithm for a two-machine single order PFSP, where, in certain cases (up to 35 jobs), the proposed algorithm managed to find optimal solutions. For the same problem he proposed three heuristics and compared their results. Allahverdi and Al-Anzi (2006) studied a three-machine PFSP with minimization of total flow time as the objective. To solve the problem, they proposed a branch and bound algorithm able to find optimal solutions for up to 18 job problems. Moreover, for this problem they proposed an upper bound, a lower bound, and a dominance relation. Ng et al., (2007) developed a dominance relationship and several heuristics for a three-machine PFSP with the minimization of maximum lateness as the objective. Fondrevelle (2005) addressed an *m* machine PFPS with the objective of minimization of maximum lateness, where the case of negative time lags due to the overlapping of two or more jobs can be modelled as a sequence independent problem. They identified and studied some special polynomial cases. They developed a dominance relationship, an upper bound and a lower bound for the problem. Al-Anzi and Allahverdi (2005) developed a hybrid meta-heuristic for an *m* machine PFSP, with the objective of minimization of completion time variance. They compared their algorithm with other state-of- the-art algorithms and showed better performance. Levner et al., (1995). Kogan and Levner (1998) considered a scheduling problem of automated manufacturing cells integrated with computer-controlled transportation robots. They found that the problem is similar to a two-machine single order PFSP with setup times and found several polynomial cases. Moreover, they also found that, if makespan minimization is set as an objective of the problem, the problem can be solved in polynomial time by extending Johnson's algorithm (Johnson, 1954). Cheng and Sriskandarah (1999) addressed a two-machine PFSP with setup-independent release time. They studied the problem with both cases of separable and non-separable setup time. They showed that both cases have *NP-Hard* complexity and they proposed a heuristic algorithm to solve the problem. Glass et al., (2000) showed that the same problem is still *NP-Hard* without considering release time. Brucker, et al., (2005) made a different complexity analysis of *m* machine single order PFSP with makespan minimization. They found some polynomially solvable cases. They also studied the same problem with different objective functions, including minimization of total completion time, total tardiness, total weighted tardiness and total weighted completion times, and they found some polynomially solvable cases. Like two-machine PFSP, weighted tardiness minimization is polynomially solvable if the processing times of jobs are equal. From the above review, it can be concluded that all the previous studies have been limited to single order PFSP with setup times with small-scale problems (like two-machine PFSP) and all the problems considered a static permutation flowshop.

To authors' best knowledge, the only related work to PFSPs with setup in dynamic environments is that of Su and Chou (2000). They investigated a dynamic two-machine PFSP with the objective of minimization of the weighted sum of makespan and total flow time. A frozen strategy was used to convert the dynamic PFSPs into a static one and developed an integer programming model for the problem. To solve the problem in a small scale (only two machines, they proposed a heuristic algorithm with computational complexity of $O(n^3)$. Moreover, although jobs may arrive dynamically in the production floor, there are no restrictions to finish each job within a due date/delivery time.

Problem Formulation and Necessary Assumptions for Real-Time MTO PFSP With Setup

In this section, the problem for real-time PFSP is formulated and necessary assumptions are briefly presented. The studied problem has the following assumptions:

1. The production line consists of m-machines.
2. The sanitaryware production facility can produce only specific types of products.
3. Each product is completed after it is processed by maximum m-machines (m-operations). However, a product can skip a machine (operation) if it is not required.
4. Setup time is independent of sequence and it can be considered as a 'dummy job' in each machine.

5. Multiple setups are allowed for the products from the same order and from the same family (produced in multiple batches).
6. Inventory holding costs for work-in-process and finished goods are negligible. Transportation cost is also negligible.
7. Production process cannot be interrupted at any time.
8. Each machine processes a product or job only once.
9. Processing times and setup times are known in advance.
10. If a machine starts to process a product from an order, it should not process another product, i.e. no pre-emption is allowed in the process.
11. Each machine needs same time to process or setup for the products from the same family.
12. Production line maintain expected quality level to avoid any scrap or rework.
13. Finished orders are delivered instantly to the customers.

If each order consists of consists of $K \in \{h_1, h_2, ... h_f\}$ different product categories and each of the uth order consists of n products or jobs, and the order, then

$$n = \sum_{f=1}^{K} h_f \tag{12}$$

where f^{th} product family consists of h_f number of the product.

The setup time on the jth machine for the ith product can be defined as

$$S^K_j = \begin{cases} 0, & if\ the\ (i-1)^{th}\ and\ i^{th}\ product\ are\ from\ the\ same\ family \\ s^f_j, & otherwise \end{cases} \tag{13}$$

Note that all machines requires a setup before processing the first product from the first order. As described earlier, if makespan of the the uth order is C^u_{max}, then the completion time of that order, C^u_{com} can be defined as

$$C^u_{com} = S^u_j + C^u_{max} \tag{14}$$

Where the first product of first order start at, $S^u_1 = 0$ (15)

The earliness of the uth order is

$$E_u = d_u - C^u_{com},\ u=1, 2, 3, ..., U \tag{16}$$

The tardiness of the uth order is

$$T_l = C^u{}_{com} - d_u,\ u=1, 2, 3,........., U \tag{17}$$

where $d_u \geq 0$

The relation between earliness and tardiness of the uth order can be expressed as

$$E_u = - T_u \tag{18}$$

where,

T_u – u^{th} order's tardiness
E_u – u^{th} order's earliness
$S^u{}_j$ – starting time of the jth machine to process the first product of the uth order
$C^u{}_{max}$ – the makespan of the uth order under single-order PFPS conditions
$C^u{}_{com}$ – the completion time of the uth order
d_u –delivery time or due date of order u

The profit (or loss) of the uth order is

$$P^u = Sp^u + E_u \times C_e^u - T_u \times C_t^u - Op^u,\ u=1, 2, 3,........., U \tag{19}$$

Finally, the total profit in one production shift (which is our objective) can be expressed as,

$$P = \sum_{u=1}^{N} P^u \tag{20}$$

where,

U – total number of orders that is placed in a single production shift
P^u – profit from the uth order
Sp^u – revenue from the uth order
Op^u – opportunity loss cost of order u
C_t^u – tardiness costs/loss per unit time of order u
C_e^u – earliness bonus/benefit per unit time of order u

To maximize the total profit of in a production shift is the objective of the problem.

Solution Approach for Real-Time PFSP With Setup

In this section, the proposed order acceptance or rejection heuristic with setup is first described. Later, the MA-based real-time strategy for multiple order PFSPs with setup is described.

Order Acceptance Decision With Setup

When customers place an order in the system, an acceptance/rejection decision is made, based on the arrival time, start time, due date, makespan, and completion time of the order. The decision-making process is similar to the order acceptance decision described in Section 2.

Order Acceptance Decision for Multiple Order PFSP With Setup

[Start of the algorithm]

1. The production shift starts when first order, $u = 1$ arrives into the production plant.

2. Repeat while ($u = N$)

A. By MA, calculate C^{u}_{com} for order u.

B. If ($C^{u}_{com} <= d_{u}$), **Accept** order u.

C. Else order u is **Reject**ed.

D. Set, $u = u+1$.

E. Save the profit of the uth order as P^{u}

[End of Step 2 loop]

3. Save and return the total profit, P.

[End of Algorithm]

Scheduling the Orders

When an order arrives into the system, machines may not be available at that point in time due to the capacity constraints. However, products of a new order may be more efficiently scheduled by considering the machine status and also by starting the order immediately as soon as the first machine is free. To schedule the orders, we propose MA based on the RS strategy and RT strategy, which are similar to the strategies (without setup) discussed in the previous section.

RS Strategy With Setup

In this strategy, an accepted order is scheduled as a single order PFSP by MA while minimizing makespan as the objective. Products may be split into different batches with multiple setup, or be produced in a single batch with a single setup. The accepted order starts to process in the first machine immediately when it is available. However, products are shifted to right of their schedule if the following machines are unavailable at that time. Hence the makespan of the order may be greater than its single order makespan. A simple example of a three-machine flowshop problem with two orders is shown in Figure 5 and it demonstrates the overall process. A detailed description of the example is given in (Rahman et al., 2015 (a)).

Algorithm With RS Strategy With Setup

The procedure of the RS-based algorithm for the real-time MTO PFSP with setup problem can be described as follows. Assume that U is the total number of orders that arrive in a production shift. Also assume that the due date of the Uth order is d_u and the current clock time is ct. Let t_u be the clock time, when the first job of the uth order will start to be processed first and the number of accepted orders per production shift is U_{acpt}. The algorithm must determine the number of product categories available in that batch K, the setup time required per product machine on the jth machine s_j^K, and the start time of the uth order, St_u. The RS algorithm must also calculate the single order makespan of that order $(C_{max})_u$, the completion time $(C_{comp})_u$, and its earliness E_u, or its tardiness, T_u. Finally, the algorithm needs to calculate the profit/loss P^u, earned from the uth order and also the total profit earned from a production shift, P.

The algorithm starts when the first order arrives in the system.

```
[Start of the Algorithm]
1.          Set u = 1, U_acpt =0, and clock t_1=0
2.          Repeat until (u = U).
A.          Set the current clock time, t_u = ct, and St_u=ct
B.          Find out the total number of product types from
order u
C.          Calculate the required setup time of each product
for a given product sequence by equation (13)
D.          Calculate the single-order (static) makespan, (C_max)_u
of the u^th order by MA.
E.          Considering the current system state, re-calculate
the makespan (C_max)_u of the current order
a.           Right shift the products of the u^th order, if the u^th
```

and $(u\text{-}1)^{th}$ accepted orders are overlapping to each other.

b. Else, the single-order makespan $(C_{max})_u$ of the order u is unchanged.

c. Calculate the completion time of the current order, $(C_{comp})_u = St_u + (C_{max})_u$.

F. Considering the completion time, $(C_{comp})_u$

a. Calculate the earliness or tardiness, $T_u = (C_{comp})_u - d_u$

b. Store tardiness, T_u and the completion time, $(C_{comp})_u$

c. If $T_u \leq 0$

i. **Accept** the uth order.

ii. $U_{acpt} = U_{acpt} + 1$

d. Else, **Reject** the order.

G. Calculate the profit or loss, P^u, for the uth order .

H. Set $u=u+1$

[End of the step 2]

3. Save, U_{acpt} (total number of accepted orders in the current production shift).

4. Save and return the total profit, P.

[End of the Algorithm]

Figure 5. Right Shifting (RS) strategy with setup

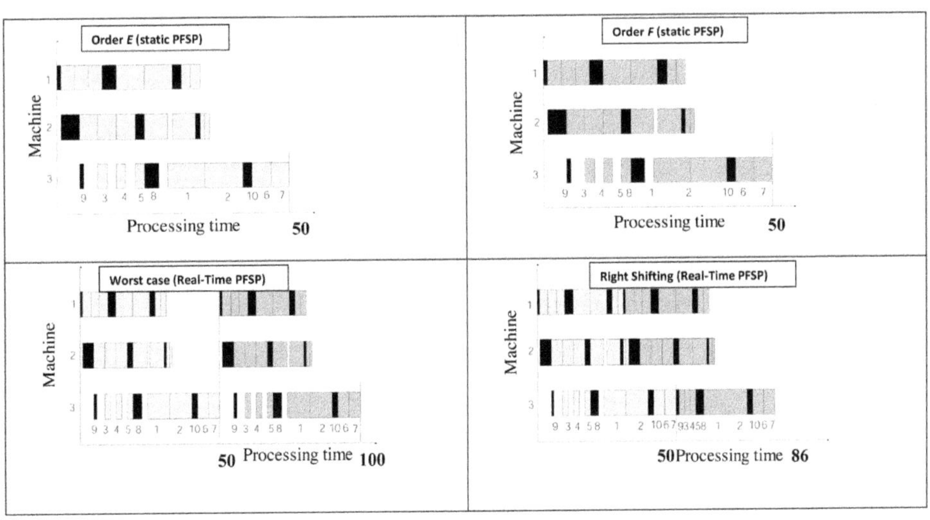

RT Strategy With Setup

In RT strategy, a new order is scheduled while considering the type of product category for which all machines are currently set up (dynamically adjust setup), the category of products currently under process, and machine availability/unavailability constraints. Similar to the RS strategy, an order is processed in the first machine as soon as it is free. On the basis of this hypothetical schedule, an order is accepted if it can be processed and completed within the due date, and brings profits to the company. Otherwise, due to order rejection, an opportunity loss cost is imposed. Figure 6 demonstrates a simple example of the RT strategy.

Algorithm for RT Strategy With Setup

Assume that U is the total number of orders that arrive in one production shift. The algorithm must then calculate the number of product categories available in that batch K, and the setup time required per product machine on the jth machine, s_j^K, and the start time of the uth order, St_u. The algorithm must also calculate the makespan of the uth order $(C_{max})_u$, the completion time, $(C_{comp})_u$, and its earliness, E_u or its tardiness, T_u, and the number of accepted orders per production shift, U_{acpt}. Finally, the algorithm determines the profit/loss gained from the uth order, P^u, and the total profit earned from a production shift, P.

The algorithm starts when customers place the first order in the system.

[Start of the Algorithm]

1. Set $u = 1$, $U_{acpt} = 0$, and clock time, $t_u = 0$

2. Continue until $(u < U)$

A. Set the current clock time, $t_u = ct$, and $St_u = ct$

B. Find out the total number of product categories in the current order u

C. Calculate the required setup time of each product for a given product sequence by equation (13).

D. Calculate a makespan of the current order $(C_{max})_u$ by MA while considering the machine setups in all machines and machine availability constraints, i.e. based on the current system state.

E. Calculate the completion time of the current order, $(C_{comp})_u = St_u + (C_{max})_u$

F. Considering the completion time, $(C_{comp})_u$

a. Calculate the earliness or tardiness, $T_u = (C_{comp})_u - d_u$

b. Store the completion time, $(C_{comp})_u$ and tardiness, T_u

c. If $T_u \leq 0$

iii. **Accept** the uth order.

iv. $U_{acpt} = U_{acpt} + 1$

d. Else, **Reject** the order.

G. Calculate the profit or loss earned form the uth order, P^u.

H. Set $u=u+1$

[End of step 2 Loop]

3. Store, U_{acpt} (total number of accepted order in one production shift).

4. Save the total profit, P gained from the production shift.

[End of the Algorithm]

Experimental Study for Real-Time Multiple Order PFSP With Setup

All the algorithms have been implemented in the C++ language and experimented on a computer with an Intel core i7, a 2.8 GHz processor and 4GB RAM. Parameters of the RS and RT strategies have been adjusted to the same value as used in (Rahman,

Figure 6. Real-Time Strategy (RT) strategy with setup

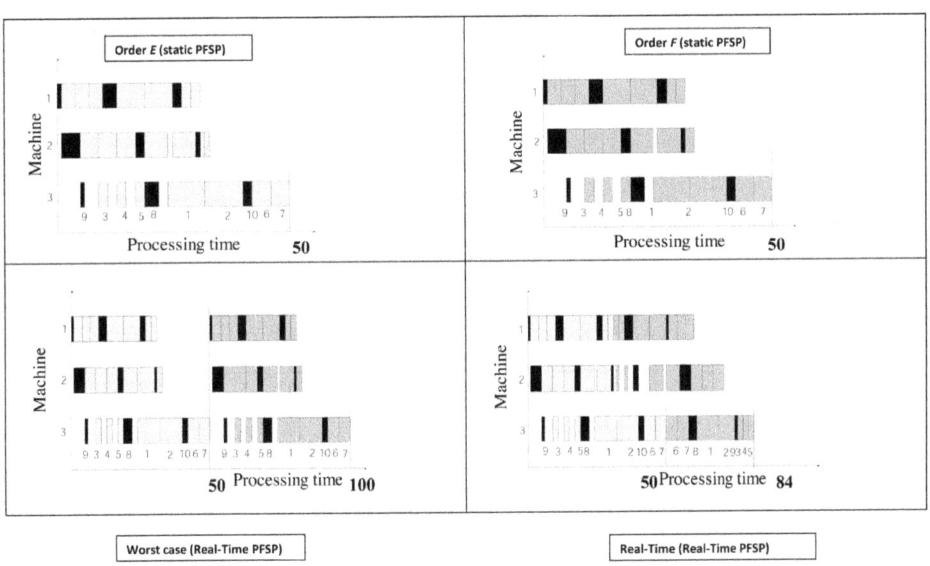

et al., 2013). To evaluate the performance of the algorithm, 40 instances from small- to large-scale practical instances (20 to 200 products per order) have been used. All production data, such as order arrival times, number of the product from a product category or family, processing times, setup times, and due dates are collected from the practical production floor. To test the performance, three different types of scenarios have been considered: 1) same combination and size of orders, 2) same combination but different size of orders, 3) different combination and size of orders. All solutions generated by RS and RT strategies have been compared against upper bound (UB) and lower bound (LB) to evaluate the quality and validity of the algorithms. The expressions of UB and LB can be seen in Equations 7 and 9 respectively.

Table 8 presents a comparison between the current algorithms for the 10 order problems under Scenario I. All products of an accepted order must be processed on 10 machines flow shop. Table 6 shows the composition of each order, and each order consists of 50 products. Table 7 represents all costs and selling prices that relate to production.

From Table 8, it can be seen that both RS and RT strategies accept five orders. However, since the accepted orders can be completed earlier by RT strategy over RS strategy, its helps to earn more profit for the accepted orders. So more profits can be earned by RT strategy. This also reflects in the total profits earned from both strategies (as shown in last column).

Table 9 illustrates a comparison between RS and RT for Scenario I, where each problem size contains 10 orders with random inter-arrival times and due dates. From the values in the column headed with D_U and D_L shows that the RT strategy performs better than the RS strategy (bigger values of D_U and smaller values of D_L). The following columns also show that considering the number of accepted orders and total profit the RT strategy outperforms the RS strategy.

Table 6. Order composition

Number of Types of Product in This Order	Product Types	Quantity
4	Foot rest	1
	Gallery	3
	Wash Basin	14
	Cistern with Lid	2
Total Jobs in this Order		20

Table 7. All sales revenue and costs

Order Number	Cost per Order for Tardiness, $	Bonus per Order for Earliness, $	Revenue Earned From sales,$	Opportunity Loss Cost, $
1	1.054	1.054	1348	393
2	1.524	1.524	2954	833
3	1.839	1.839	1595	257
4	2.702	2.702	1936	681
5	1.538	1.538	2598	645
6	1.430	1.430	2099	887
7	1.093	1.093	1007	692
8	1.041	1.041	2740	549
9	2.606	2.606	2325	391
10	1.001	1.001	1798	430

Tables 10 and 11 show a summary of comparison between RS and RT for Scenarios II and III respectively. The structure of these tables are the same as for Table 9. For each problem instance, 25 arrive randomly in each production shift. For all problem instances, RT performs better than RS strategy.

Table 8. Results for scenario I with setup

Order No.	Due Date, d_l	Order Arrival Time	Right Shifting Strategy (RS)				Real-Time Strategy (RT)			
			Completion Time, C_{comp}	Tardiness, T_l	Profit, $	Status	Completion Time, C_{comp}	Tardiness, T_l	Profit per Order, $	Status
1	11456	0	2115	-9299	2328.115	A	**2115**	-9299	2328.115	A
2	6289	494	3811	-2478	3331.647	A	**3806**	-2483	3332.409	A
3	12087	553	5445	-6642	2816.464	A	**5443**	-6644	2816.832	A
4	5671	561	7116	1445	-681	R	**7088**	1417	-681	R
5	6125	1036	7116	991	-645	R	**7088**	963	-645	R
6	3465	1677	7111	3646	-887	R	**7083**	3618	-887	R
7	7438	2173	7083	-327	1042.741	A	**7011**	-355	1045.802	A
8	1824	2602	8777	6953	-549	R	8717	6893	-549	R
9	5814	2784	8782	2968	-391	R	8717	2903	-391	R
10	20260	2815	8777	-11483	2947.448	A	**8717**	-11543	2953.454	A
Total profit from a production shift,$			9313.415				**9323.612**			

*Order acceptance status: A- Accepted, R-Rejected

Table 9. Summary of comparisons for scenario I

Problem Size Products	D_U		D_L		Number of Accepted Orders		Profit, $		
	RS	RT	RS	RT	RS	RT	RS	RT	Difference in Profit
20	0.98	**1.33**	0.45	**0.41**	36	**40**	45889.84	**46589.93**	700.49
50	0.742	**0.84**	0.68	**0.63**	33	**35**	41902.07	**43010.89**	1108.82
100	0.65	**0.81**	0.78	**0.61**	41	**42**	39953.26	**40575.96**	622.7
200	1.33	**1.78**	1.39	**1.31**	38	**38**	48864.87	**49574.49**	709.62

Table 10. Mixed order with same order size

Problem Size Products	D_U		D_L		Number of Accepted Orders		Profit, $		
	RS	RT	RS	RT	RS	RT	RS	RT	Difference in Profit
20	1.261	**1.331**	0.789	**0.745**	17	**19**	47947.16	**48916.89**	969.73
50	0.098	**0.092**	0.042	0.046	11	**13**	47665.72	**48262.06**	596.34
100	1.455	**1.687**	1.099	**1.078**	11	**12**	39307.49	**40026.66**	719.17
200	1.515	**1.651**	0.958	0.951	15	**17**	45146.86	**45242.21**	95.35

Table 11. Mixed order with different order sizes

Case	D_U		D_L		Number of Accepted Orders		Profit, $		
	RS	RT	RS	RS	RS	RT	RS	RT	Difference in Profit
1	1.673	1.668	0.211	**0.202**	22	**24**	50655.83	**50759.68**	103.85
2	1.122	**1.321**	0.516	**0.621**	21	**21**	53901.23	**53912.21**	10.98

SUMMARY

In this chapter, the particularities of scheduling procedures in Industry 4.0 flow shop production environments has been examined. To do this, first we have discussed the issues with real-time order acceptance and scheduling problems in a permutation flowshop environment without setup with the objective of maximizing the number of accepted orders. To deal with this problem, a real-time approach was presented, that involved an order acceptance/rejection heuristic and an MA based real-time algorithm. The order acceptance heuristics decided to accept or reject an arriving order on the basis of two decision criteria-completion time and due date. The MA based real-time algorithm introduced the ability to schedule the accepted order while considering current state of the capacity, that is the orders in process by machines, and the accepted orders on hand at that point in time. For the second problem (i.e. real-time PFSP without setup), we considered the dynamic order arrival and scheduling problem in a flowshop environment while assuming that setup time is included in processing time or it is negligible. However, in practice, the production system may need additional time to set up machines or equipment. In that case, setup becomes one of the influential criteria in order acceptance and scheduling decisions. To address this, we have extended our algorithms to deal with such types of problem. The case of a practical MTO sanitary ware factory was studied. The model used in section 2 was reformulated, in order to fit with the actual problem under study. To solve this problem, a real time approach was presented, which is similar to the order acceptance decision making process described in section 2. The order acceptance heuristics made the order acceptance or rejection decisions, and the MA based real-time strategy scheduled the different product mixes by dynamically adjusting the size of each batch.

Present study will certainly show road for examination and further improvement of real-world production systems and running Smart Scheduling for them. For example, introducing temporal constraints into the problem, like release dates, sequence dependent setup time, or unrelated parallel machines at each stage is a possibility. Further exploration of the effect and interactions of the other key technologies of Industry 4.0, like Big data analytics, cloud manufacturing, cyber physical systems, and internet of things in these problem settings under study seems also promising.

REFERENCES

Abdel-Basset, M., Manogaran, G., El-Shahat, D., & Mirjalili, S. (2018). A hybrid whale optimization algorithm based on local search strategy for the permutation flow shop scheduling problem. *Future Generation Computer Systems*, *85*, 129–145. doi:10.1016/j.future.2018.03.020

Al-Anzi, F. S., & Allahverdi, A. (2005). Using a hybrid evolutionary algorithm to minimize variance in response time for multimedia object requests. *Journal of Mathematical Modelling and Algorithms*, *4*(4), 435–453. doi:10.100710852-005-9027-9

Allahverdi, A. (2000). Minimizing mean flowtime in a two-machine flowshop with sequence-independent setup times. *Computers & Operations Research, 27*(2), 111–127. doi:10.1016/S0305-0548(99)00010-6

Allahverdi, A., & Al-Anzi, F. S. (2006). A branch-and-bound algorithm for three-machine flowshop scheduling problem to minimize total completion time with separate setup times. *European Journal of Operational Research, 169*(3), 767–780. doi:10.1016/j.ejor.2004.07.074

Allahverdi, A., Ng, C., Cheng, T. E., & Kovalyov, M. Y. (2008). A survey of scheduling problems with setup times or costs. *European Journal of Operational Research, 187*(3), 985–1032. doi:10.1016/j.ejor.2006.06.060

Brucker, P., Knust, S., & Wang, G. (2005). Complexity results for flow-shop problems with a single server. *European Journal of Operational Research, 165*(2), 398–407. doi:10.1016/j.ejor.2004.04.010

Cheng, T., Wang, G., & Sriskandarajah, C. (1999). One-operator–two-machine flowshop scheduling with setup and dismounting times. *Computers & Operations Research, 26*(7), 715–730. doi:10.1016/S0305-0548(98)00087-2

Dasgupta & Das. (2015). A Discrete Inter-Species Cuckoo Search for flowshop scheduling problems. *Computers & Operations Research, 60*, 111-120.

Fondrevelle, J., Allahverdi, A., Oulamara, A., & Portmann, M. C. (2005). Permutation flowshops with exact time lags to minimize maximum lateness. Nancy, France: Academic Press.

Garey, M. R., Johnson, D. S., & Sethi, R. (1976). The complexity of flowshop and jobshop scheduling. *Mathematics of Operations Research, 1*(2), 117–129. doi:10.1287/moor.1.2.117

Glass, C. A., Shafransky, Y. M., & Strusevich, V. A. (2000). Scheduling for parallel dedicated machines with a single server. *Naval Research Logistics, 47*(4), 304–328. doi:10.1002/(SICI)1520-6750(200006)47:4<304::AID-NAV3>3.0.CO;2-1

Grabowski, J., & Wodecki, M. (2004, September). A very fast tabu search algorithm for the permutation flow shop problem with makespan criterion. *Computers & Operations Research, 31*(11), 1891–1909. doi:10.1016/S0305-0548(03)00145-X

Hasan, S. K., Sarker, R., Essam, D., & Cornforth, D. (2009). Memetic algorithms for solving job-shop scheduling problems. *Memetic Computing, 1*(1), 69–83. doi:10.100712293-008-0004-5

Ignall, E., & Schrage, L. (1965). Application of the branch and bound technique to some flow-shop scheduling problems. *Operations Research, 13*(3), 400–412. doi:10.1287/opre.13.3.400

Ivanov, Dolgui, Sokolov, Werner, & Ivanova. (2016). *A dynamic model and an algorithm for short-term supply chain scheduling in the smart factory industry 4.0*. Academic Press.

Johnson, S. M. (1954). Optimal two-and three-stage production schedules with setup times included. *Naval Research Logistics, 1*(1), 61–68. doi:10.1002/nav.3800010110

Johnson, S. M. (1954). Optimal two- and three-stage production schedules with setup times included. *Naval Research Logistics Quarterly, 1*(1), 61–68. doi:10.1002/nav.3800010110

Kogan, K., & Levner, E. (1998). A polynomial algorithm for scheduling small-scale manufacturing cells served by multiple robots. *Computers & Operations Research, 25*(1), 53–62. doi:10.1016/S0305-0548(98)80008-7

Levner, E., Kogan, K., & Maimon, O. (1995). Flowshop scheduling of robotic cells with job-dependent transportation and set-up effects. *The Journal of the Operational Research Society, 46*(12), 1447–1455. doi:10.1057/jors.1995.205

Moreira, M. R. A., & Alves, R. A. F. S. (2009, May). A methodology for planning and controlling workload in a job-shop: A four-way decision-making problem. *International Journal of Production Research, 47*(10), 2805–2821. doi:10.1080/00207540701725083

Nandi, A., & Rogers, P. (2004, March). Using simulation to make order acceptance/ rejection decisions. *Simulation, 80*(3), 131–142. doi:10.1177/0037549704045046

Nawaz, M., Enscore, E. E. Jr, & Ham, I. (1983). A Heuristic Algorithm for the M-Machine, N-Job Flowshop Sequencing Problem. *Omega-International Journal of Management Science, 11*(1), 91–95. doi:10.1016/0305-0483(83)90088-9

Osman, I. H., & Potts, C. N. (1989). Simulated Annealing for Permutation Flowshop Scheduling. *Omega-International Journal of Management Science, 17*(6), 551–557. doi:10.1016/0305-0483(89)90059-5

Pinedo, M. (2012). *Scheduling: theory, algorithms, and systems* (4th ed.). New York: Springer. doi:10.1007/978-1-4614-2361-4

Rahman, Sarker, & Essam. (2015). A real-time order acceptance and scheduling approach for permutation flow shop problems. *European Journal of Operational Research, 247*, 488-503.

Rahman, H., Sarker, R., & Essam, D. (2013). Permutation Flow Shop Scheduling with dynamic job order arrival. *Cybernetics and Intelligent Systems (CIS), IEEE Conference on*, 30-35. 10.1109/ICCIS.2013.6751574

Rahman, H. F., Sarker, R., & Essam, D. (2018). Multiple-order Permutation Flow Shop Scheduling under Process Interruptions. *International Journal of Advanced Manufacturing Technology*, *97*(5-8), 2781–2808. doi:10.100700170-018-2146-z

Rahman, H. F., Sarker, R., & Essam, D. (2015). A genetic algorithm for permutation flow shop scheduling under make to stock production system. Computers & Industrial Engineering, 90, 12-24. doi:10.1016/j.cie.2015.08.006

Rahman, H. F., Sarker, R. A., & Essam, D. L. (2013). A Memetic Algorithm for Permutation Flow Shop Problems. *2013 IEEE Congress on Evolutionary Computation (CEC)*, 1618-1625. 10.1109/CEC.2013.6557755

Rahman, H. F., Sarker, R. A., & Essam, D. L. (2013). A Memetic Algorithm for Permutation Flow Shop Problems. *IEEE Congress on Evolutionary Computation*, 1618 - 1625. 10.1109/CEC.2013.6557755

Rajendran, C., & Ziegler, H. (2004, June 1). Ant-colony algorithms for permutation flowshop scheduling to minimize makespan/total flowtime of jobs. *European Journal of Operational Research*, *155*(2), 426–438. doi:10.1016/S0377-2217(02)00908-6

Rogers, P., & Nandi, A. (2007, October). Judicious order acceptance and order release in make-to-order manufacturing systems. *Production Planning and Control*, *18*(7), 610–625. doi:10.1080/09537280701582422

Ruiz, R., & Maroto, C. (2005). A comprehensive review and evaluation of permutation flowshop heuristics. *European Journal of Operational Research*, *165*(2), 479–494. doi:10.1016/j.ejor.2004.04.017

Ruiz, R., Maroto, C., & Alcaraz, J. (2006, October). Two new robust genetic algorithms for the flowshop scheduling problem. *Omega-International Journal of Management Science*, *34*(5), 461–476. doi:10.1016/j.omega.2004.12.006

Selen, W. J., & Hott, D. D. (1986, December). A mixed-integer goal-programming formulation of the standard flow-shop scheduling problem. *The Journal of the Operational Research Society*, *37*(12), 1121–1128. doi:10.1057/jors.1986.197

Su, L.-H., & Chou, F.-D. (2000). Heuristic for scheduling in a two-machine bicriteria dynamic flowshop with setup and processing times separated. *Production Planning and Control*, *11*(8), 806–819. doi:10.1080/095372800750038418

Taillard, E. (1993, January 22). Benchmarks for Basic Scheduling Problems. *European Journal of Operational Research, 64*(2), 278–285. doi:10.1016/0377-2217(93)90182-M

Tasgetiren, M. F., Liang, Y. C., Sevkli, M., & Gencyilmaz, G. (2007, March 16). A particle swarm optimization algorithm for makespan and total flowtime minimization in the permutation flowshop sequencing problem. *European Journal of Operational Research, 177*(3), 1930–1947. doi:10.1016/j.ejor.2005.12.024

Zheng, P., Sang, Z., Zhong, R. Y., Liu, Y., Liu, C., & Mubarok, K. (2018). Smart manufacturing systems for Industry 4.0: Conceptual framework, scenarios, and future perspectives. Academic Press.

Zobolas, G. I., Tarantilis, C. D., & Ioannou, G. (2009, April). Minimizing makespan in permutation flow shop scheduling problems using a hybrid metaheuristic algorithm. *Computers & Operations Research, 36*(4), 1249–1267. doi:10.1016/j.cor.2008.01.007

Chapter 10
Evaluation of Influence of Principles Involved in Industry 4.0 Over Coal Industries Using TISM

Bathrinath Sankaranarayanan
https://orcid.org/0000-0002-5502-6203
Kalasalingam University, India

Rahul K.
Rajalakshmi Engineering College, India

Pradeep J.
Rajalakshmi Engineering College, India

S. G. Ponnambalam
https://orcid.org/0000-0003-4973-733X
University Malaysia Pahang, Malaysia

Saravanasankar S.
Kalasalingam University, India

ABSTRACT

Coal is the major source of energy in the world. But, the process of extraction and use of coal has adverse effects on the environment. In this chapter, the authors try to reduce these effects by considering the principles and technologies involved in Industry 4.0, also known as the Fourth Industrial Revolution. From a few expert reviews and research works, eight crucial factors were taken into account and were analyzed. The eight factors are consumer, water resources, smart transportation, smart factory, smart grid, smart mining, smart home, and renewable energy. The analysis has been made using the total interpretive structural modeling (TISM) method. The model distinctly demonstrates the influence of the principles of Industry 4.0 over coal industries. This chapter also aims to pave the way for future research and tries to contribute towards the sustainable extraction and usage of coal in energy industries. Consumer plays the most influential role in this regard.

DOI: 10.4018/978-1-5225-9078-1.ch010

INTRODUCTION

Coal, being the major energy source, is also the cause of most of the carbon gas release in the world. The extraction process of coal and its by-products leads to environmental damage, health problems, climate change, pollution of water bodies, depletion of ground water and acid rain thus affecting the day-to-day life of human beings as well as other living beings. Many alternatives have been found, yet, the effects are still too high to be ignored. Switching to natural gases, renewable resources and hydrogen fuel can match neither the energy requirements nor the production output of coal industries. Hence, researchers and experts suggest the principles of Industry 4.0 as a means to make the coal production process more sustainable and less environment affecting. The main objective of our paper is to evaluate the factors influencing coal industries by using Total Interpretive Structural Modeling (TISM) for lessen the impacts of environment with the help of Industry 4.0 principles.

The process of industrialization has gone through three phases, namely, The First, Second and Third Industrial Revolutions. Each revolution has its own unique characteristics that enabled them to protect the environment and avoid material wastage. For the first time in 2011, during the Hannover Fair event, a proposal for the development of German economic policy based on higher technological strategies was made, symbolizing the beginning of the Fourth Industrial Revolution (Lee 2013 & Masconi 2014). Global industrialization and the development of technologies such as automation, cloud computing and deep learning have led to Industry 4.0. Production processes have become more sustainable, effective and persistent due to the aforementioned technological developments (Wahlster 2012). Industry 4.0 also deals with Internet Of Things (IOT) and effective Information and Communication Technology (ICT) infrastructure. The effective communication network between the related sectors formed through these technologies provide better real-time communication capabilities, thus helping with the reduction of cost, wastage, CO_2 emission and lead time of coal production. Thus, the above- mentioned influential factors are proposed through intensive literature research and TISM method. The factors related to Industry 4.0 are studied by many researchers (Lee et al., 2014; Wollschlaeger et al., 2017; Wan et al., 2019; Zheng et al., 2018).

In the Section 2 literature review is presented, the problem description is discussed in section 3, the proposed methodology is discussed in section 4, in section 5 results and discussion is presented, the managerial implications and conclusions are presented in section 6 and 7 respectively.

LITERATURE REVIEW

Literature review is classified into eight sections namely (i) Smart Transportation (ii) Smart Factory (iii) Smart Mining (iv) Consumer satisfaction (v) Water Resource (vi) Smart Grid (vii) Smart home (viii) Renewable Energy . They are detailed as follows:

Smart Transportation

Managing an industry's transportation system using various information and communication technologies in an effective manner is called smart transportation (Stefannson & Lumsden 2008). In the case of electric vehicles, based on the service requirements, charging stations are set up and programmed for the effective recharge of electric vehicles. The intelligent system also minimizes waiting period and cost (Kim et al., 2010).

Smart Factory

A factory that is run using the technologies that are being developed in Industry 4.0 such as big data analysis, machine learning, IOT and so on is called a smart factory. The deployment of these technologies in factories helps in increasing the production of the factory while making the production process sustainable too (Wang et al., 2016). The development of flexible production systems with integrated IOT in factories would enable sustainable production (Shrouf et al., 2014). An effective supply chain management system that enable continuous flow of processes also plays a major role in sustainable production by helping to save time and energy (Ivanov et al., 2016). Smart factories may also feature emerging advanced technologies such as Augmented Reality (AR) that can be used to train the workers for effective production (Paelke 2014).

Smart Mining

Smart mining can be referred to the process of mining raw materials in a sustainable manner by making use of the principles and technologies of Industry 4.0. Sustainability must be considered in both the micro and macro perspectives of Industry 4.0 such as mining, human resource, eco-friendly products, product life cycle and so on. (Stock & Seliger 2016).

Consumer

Consumers do not get the time to search and buy eco-friendly products due to their current busy lifestyle. But, the existing green consumer's buying pattern must be studied and technologies supporting those kind of purchases must be developed (Young et al., 2010).

Water Resource

Using Information and Communication Technology (ICT), high production industries such as coal industries must ensure the creation and adoption of smart usage procedure of water resources (Mutchek & Williams 2014). Properly treated water can also be viewed as a promising solution for the water problems and ensure sustainable usage of water (Lee et al., 2015).

Smart Grid

Integration of renewable energy, such as solar, wind and hydro energies, into the grid, which results in a smart grid, would help in lowering the dependence on coal energy, thus reducing the production of coal, in turn reducing the harmful emissions (Liserre et al., 2010). Efficient battery systems can be used to store electricity and charge electric vehicles while the batteries themselves are charged using renewable energy sources (Rahimi-Eichi et al., 2013).

Smart Home

Smart home is formed by integrating all sorts of smart technologies into a traditional home. Thus, a number of smart homes present over some geographical area together form smart city. Components of Industry 4.0 such as IOT, ICT and smart grid also form parts of smart city (Mohanty et al., 2016). Smart city also paves way for the intelligent conservation and provision of electrical energy that can be controlled either on the provider side or the consumer side (Corno & Razzak 2012). Collection of electricity usage data and manipulating the technological components of the smart home accordingly is a great way to conserve energy (Han et al., 2014).

Renewable Energy

Smart grid with renewable energy integration and smart transportation may go hand in hand as the concept of using renewable energy to power electric vehicles seems feasible, considering the current research works and their promising results

(Mwasilu et al., 2014). Solar energy being the largest form of renewable energy poses a great opportunity to install eco-friendly technological components at smart homes (Panwar et al., 2011). Solar energy systems can also be utilised by almost all the types of industries such as agriculture and manufacturing industries to make their production processes more sustainable (Mekhilef et al., 2011).

PROBLEM DESCRIPTION

A case evaluation of the proposed model has been conducted in a coal industry. The industry extracts and produces coal products and mainly electricity. They have captured the entire world by suppling power source. The industry desires to assess their risk profiles and to expand their risk management practices. Since environment is being polluted, having flexibility is the only way to survive in the world and to increase their eco-friendly products. Before developing their management system, the industry needs to know whether they are technologically and financially capable of adopting and expanding practices for industry 4.0. This research was conducted to assess the macro perspective of Industry 4.0 and to identify the most influential factors affecting industry 4.0 of coal industry. The results of the proposed model could enable and improve managerial decision making through identifying, analyzing and improving macro perspective of industry 4.0 of the coal industry for adopting sustainability in the industries.

METHODOLOGY

A Total Interpretive Structural Modeling (TISM) is employed to evaluate the important influential relations amongst the macro perspectives of industry 4.0 in coal industry. Interpretive structural modeling (ISM) is a widely used tool for establishing the prominent relations among factors (Rajesh 2017). Interpretive Structural Modeling (ISM) is an effective methodology for dealing with complex issues. It has been used for over twenty-five years by specially trained organization managers to help their clients understand complex situations and evaluate solutions to unusual problems and implementing new developments to their works. It is a computer assisted learning process that allows individuals or groups to develop an overview chart of the complex relationships between the many factors involved in a complex situation. It establishes the interrelationship among the factors and discusses the managerial implication of the research. The link interpretations are comparatively weak in ISM. Along with that, dealing with qualitative criteria is a serious task having obscurities and vagueness. In order to overcome this, ISM is

Table 1. Factors influencing Industry 4.0

S.No	Factors	Factors Code	References
1	Smart Transportation	F1	Kim et.al.,(2010); Stefansson & Lumsden(2008); Sherly & Somasundareswari (2015);
2	Smart Factory	F2	Shrouf et.al.,(2014); Wang et.al.,(2016); Ivanov et.al.,(2016); Paelke (2014, September);
3	Smart Mining	F3	Stock & Seliger (2016); Minggao (2010);
4	Consumer	F4	Young(2010); Grau et.al.,(2014);
5	Water resource	F5	Mutchek & Williams (2014);lee et.al.,(2015);
6	Smart grid	F6	Liserre et.al.,(2012); Kim et.al.,(2010); Rahimi-Eichi et.al.,(2013);
7	Smart home	F7	Mohanty, Choppali, & Kougianos (2016); Han et.al.,(2014); Corno & Razzak(2012); Zanella et.al.,(2014).
8	Renewable energy	F8	Mwasilu et.al.,(2014); Panwar et.al.,(2011); Mekhilef et.al.,(2011);

further modified to TISM. In TISM the inter-linkage of factors are identified by following these steps (Agarwal 2015; Rajesh 2017).

Fuzzy Total Interpretive Structural Modeling

This paper proposes Fuzzy-TISM, approach for industry 4.0 in coal industry. Fuzzy-TISM is multi-criteria decision making technique. TISM changes the fuzzy sets into a organized set. This method provides flexibility to understand about the influence of one factor over another factor. The user can number the factor in binary codes (0, 1) numbers. 0 represents no influence and 1 mentions influence between factors. It does not care about the level of influence.

Step 1: Identify the Factors and Define

Industry 4.0 contributes to the resilience for coal industry to achieve sustainability. The factors need to be identified through intensive analysis of literature and expert reviews is shown in able 1.

Table 2. Applications of TISM

S.No	Applications	References
1	Modelling and measuring code smells in enterprise applications	Gupta et.al., (2016).
2	Modeling of critical success factors of smartphone manufacturing ecosystem in India.	Jena et.al., (2016).
3	Modelling the enablers of agile performance in healthcare organization	Patri & Suresh (2017).
4	Modeling structural behavior of inhibitors of cloud computing	Radhika & Pramod (2014).
5	Technological capabilities and supply chain resilience of firms	Rajesh (2017).
6	A framework to enhance agile manufacturing system	Sindhwani & Malhotra (2017)
7	Green supply chain management drivers analysis	Soda et.al.,(2018).
8	Modeling post-disaster challenges of humanitarian supply chains	Yadav & Barve (2016).
9	Strategic performance management for Indian telecom service providers.	Yadav (2014).
10	Modeling enablers of TQM to improve airline performance	Singh and Sushil (2013).
11	Building theory of green supply chain management	Dubey et.al.,(2015).
12	Bionic model of organizational excellence.	Agarwal & Vrat (2015).
13	The enablers of a flexible control system for industry. Global Journal of Flexible Systems Management	Jayalakshmi & Pramod (2015)

Step 2: Relevant Relationship Definition

Relevant connections should be built up among the factors of intrigue. Contextual relationship between different factors is analyzed. Factor 1 is compared with remaining factors and studied how factor 1 influence/enhance factor 2 and how factor 1 influence/enhance factor 3 and so on. To evaluate and analyze the contextual relationship among factor intensive literature research and experts opinion is solicited.

Step 3: Direct Reachability Matrix

If the factors influence each other or have inter-link applications those relations are mentioned by Y (yes) and the non-influential relations are marked as N (no) in the matrix table. Later, this matrix is converted into binary matrix (0 or 1). The binary matrix represents the direct reachability matrix. Table 3 shows the direct reachability matrix.

Step 4: The Final Reachability Matrix

The influence relation between the factors are identified and represented in the direct reachability matrix to develop the final reachability matrix. If there are possible relations where a factor A influences factor B and the influenced factor B influences factor C; then the Factor A and Factor C are called as transitive relation. Final reachability matrix was shown in Table 4.

Step 5: Identifying the Level of Factors in the Final Reachability Matrix

Level partitions are required to place the factors in level-wise. The table consists of reachability set, antecedent set, intersection set. Reachability set denotes the influence factors (denoted by 1) in the row-wise; Antecedent set denotes the influence factors in the column-wise. The Intersection set denotes the influence factors which present in both sets; if the intersection set represents the reachability set then it tops the level. The top level is removed and the reachability and antecedent sets are determined again. This is process is repeated again until factors at all levels are determined. The partitioning of the reachability matrix is shown in Table 5.

Step 6: Flowchart

The protuberant relations in the final reachability matrix are represented in flowchart. The factors are arranged in the level-wise and the most influential factors are plotted in flowchart. Final level of elements in TISM is presented in Table 6.

Step 7: Construct of TISM Model

The flowchart is translated into a binary interaction matrix form depicting all the interactions by 1 in cells. Remaining cell entry is 0. TISM model is depicted in Figure 1.

Validation of TISM

The TISM method is the efficient methodology because each logical relationship among the factors is defined by the specialists. It is difficult for the organization to consider each and every factor; TISM helps to identify the influential factor. Pictograph portrayal (TISM) of connections is given in Figure 1.

Table 3. Factors, contextual relationship and interpretation

Factors	Smart Transportation	Smart Factory	Smart Mining	Consumer	Water Resource	Smart Grid	Smart Home	Renewable Resource
Smart transportation		Transportation improves the manufacturing process	Transportation maintains the mining schedule on time	Output products must be delivered in time				
Smart factory					Factory needs water resource	Power supply is vital for production		
Smart mining		Mining is inter-linked to the factory production			Mining raw materials requires water for dumping waste.	Smart grid improves mining process		
Consumer								
Water resource		Without water resource factory cannot work	Water resource plays the major part in mining process				Water is the basic need for humans to sustain	
Smart grid		Electricity is the power source for factories to operate	Mining equipment requires electricity.				Smart grid and smart homes are connected through (IOT)	
Smart home								
Renewable resource		Renewable energy provides the energy source to smart grid. Thus the energy is transferred to factories	Likewise, in mining too renewable energy is the energy provider			Smart grid is developed with the help of renewable energy		

Table 4. Initial Reachability Matrix

Factors	F1	F2	F3	F4	F5	F6	F7	F8
F1	1	1	1	1	0	0	0	0
F2	0	1	0	0	1	1	0	0
F3	0	1	1	0	1	1	0	0
F4	0	0	0	1	0	0	0	0
F5	0	1	1	0	1	0	1	0
F6	0	1	1	0	0	1	1	0
F7	0	0	0	0	0	0	1	0
F8	0	1	1	0	0	1	1	1

Table 5. Final Reachability Matrix

Factors	F1	F2	F3	F4	F5	F6	F7	F8
F1	1	1	1	1	1*	1*	0	0
F2	0	1	1*	0	1	1	1*	0
F3	0	1	1	0	1	1	1*	0
F4	0	0	0	1	0	0	0	0
F5	0	1	1	0	1	1*	1	0
F6	0	1	1	0	1*	1	1	0
F7	0	0	0	0	0	0	1	0
F8	0	1	1	0	1*	1	1	1

RESULTS AND DISCUSSION

Based on the relation developed between industry 4.0 and coal industries eight factors were analyzed using TISM method and these eight factors were partitioned into four levels of influence thus resulting in the following interpretations:

1. Consumer satisfaction is the ultimate goal to be achieved by the industries. They need to develop the industry to eco-friendly and economical products. Coal products release harmful gases that affect the health of all living beings and also cause climate change. These effects make the consumer think about their future and act to reduce them. So initially to manage the consumer satisfaction in Industry 4.0, the consumers are allowed to associate and visualize the process using intelligent manufacturing system. Also, industry 4.0 makes a major impact towards betterment of environment. Industry 4.0 plays a major

Table 6. Partitioning of the Reachability Matrix

Iterations	Factors	Reachability set	Antecedent set	Intersection	Level
ITERATION 1	F1	1,2,3,4,5,6	1	1	
	F2	2,3,5,6,7	1,2,3,5,6,8	2,3,5,6	
	F3	2,3,5,6,7	1,2,3,5,6,8	2,3,5,6	
	F4	4	1,4	4	I
	F5	2,3,5,6,7	1,2,3,5,6,8	2,3,5,6	
	F6	2,3,5,6,7	1,2,3,6,7,8	2,3,6,7	
	F7	7	2,3,5,6,7,8	7	I
	F8	2,3,5,6,7,8	8	8	
ITERATION 2	F1	1,2,3,5,6	1	1	
	F2	2,3,5,6	1,2,3,5,6,8	2,3,5,6	II
	F3	2,3,5,6	1,2,3,5,6,8	2,3,5,6	II
	F5	2,3,5,6	1,2,3,5,6,8	2,3,5,6	II
	F6	2,3,5,6	1,2,3,6,8	2,3,6	
	F8	2,3,5,6,8	8	8	
ITERATION 3	F1	1,6	1	1	
	F6	6	1,6,8	6	III
	F8	6,8	8	8	
ITERATION 4	F1	1	1	1	IV
	F8	8	8	8	IV

Table 7. Final level of elements in TISM

Factors Codes	Factors	Levels
F1	Smart Transportation	IV
F2	Smart Factory	II
F3	Smart Mining	II
F4	Consumers	I
F5	Water Resource	II
F6	Smart Grid	III
F7	Smart Home	I
F8	Renewable Resource	IV

Figure 1. TISM Connections

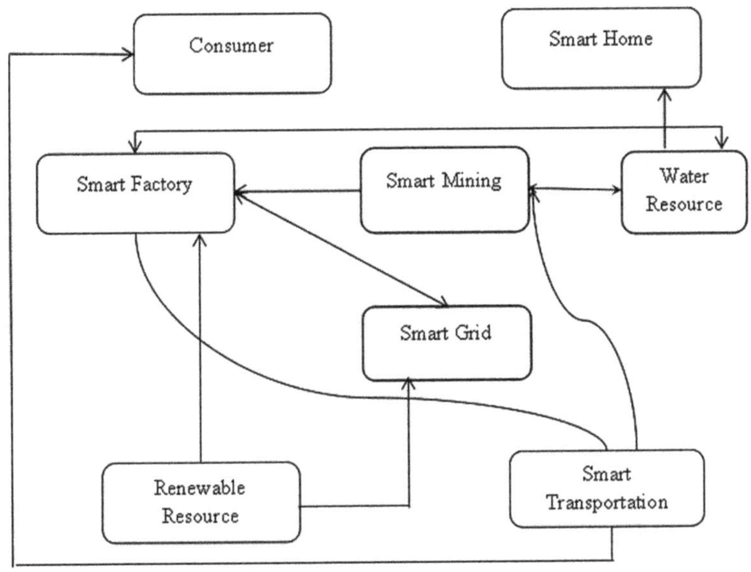

role in developing smart homes. These homes are fully controlled by means of cloud computing and IOT. These developments reduce the carbon footprint from houses. Smart homes are powered by smart grids with renewable energy as their main energy source. More significantly, the smart home isn't unified with a group of IOT empowered and related kitchen appliances.

2. Smart Factory is manufacturing organization that use lean standards and the most recent advances to boost efficiency. It is an all-encompassing way to deal with the outline and advancement of process-driven plants to convey facilities that are productive, responsive and sustainable. Smart factories and smart mining can be combined to make the processes of extraction of raw materials and manufacturing the product eco-friendly and release zero percent carbon gases. Also, smart mining is acted as a vital role in significance with the phases of human growth. Factories and mining sectors require huge quantity of water. Generally, straight impact of coal mining on water resources is the obliteration of hydrologic and geologic condition. It will affect the water contamination originated from the mine pit and mine shaft water amid digging process. The concerned organization must take the required steps to avoid pollution of water and invest in new technology, so that less water is consumed. Burning of coal and dumping of the wastages into water bodies must be avoided.

3. Smart Grid introduces a 2-way path where electricity and information can be exchanged between the utility and the customer at the same time. It is a developing network of communication, control systems, automation and other new technological tools those work together to make the power grid more reliable, secure, efficient and eco-friendly. This smart grid enables newer technologies such as wind, solar energy production and plug in electric vehicle charger to be integrated. Smart grid will replace the ancient infrastructure of today's grid and utilities can communicate better to help manage the electricity needs of both the industry and average consumers. Coal industry consumes large amount of electricity resulting in people's suffering from power shutdown and hike in electricity cost. Smart grid will reduce these problems and make the life of the people more peaceful.
4. Renewable energy is the next generation power source but it is currently not as efficient as the present day thermal power plants that use coal products as their primary energy source. Renewable energy sources such as solar and wind energy are sustainable and growing sources of electric power, yet, they are dependent on nature and adds complexity to the normal grid operations. Smart grid thus provides the data and automation needed to enable solar and wind farms to be set up and integrated into the grid. Logistics of products is made easier and efficient by controlling the concerned processes via cloud computing. Solar vehicles and plug-in electric vehicles are the next step towards transportation as they reduce carbon emission. The issues in smart transportation such as traffic blockage, road security, mishap recognition, automatic fare collection and constrained vehicle parking facilities can be solved by IOT with help of industry 4.0 principles.

MANAGERIAL IMPLICATIONS

The TISM model can be used as a influential tool for analyzing the success and failure of any project. In our present study we have identified the influencing factors of industry 4.0 for coal industry. The TISM model further provides a vision that, rather than aiming on all factors, it is important to understand the levels of influence of the factors. The TISM model helps managers to identify the factors which influence other factors, and can further be used to develop strategy through brainstorming exercises to achieve the linkage between any two factors. The theoretical framework developed in this study can be used as a guideline by the organizations to implement a new approach which incorporates sustainability into their strategic prioritization with full confidence based on their current developments.

CONCLUSION

This research paper has attempted to provide better vision concerning the factors behind industry 4.0. Subsequently, these factors were subjected to confirmation by reviewing their relevance in current development and future plans as well as by taking expert and research scholar opinion. Total Interpretive Structural Modeling (TISM) technique has been used to link the proposed factors of Industry 4.0 with coal industries. The model resulted from TISM methodology is named as EVALUATION OF INFLUENCE OF PRINCIPLES INVOLVED IN INDUSTRY 4.0 OVER COAL INDUSTRIES USING TISM. The factors involved in the TISM model have a vision for the betterment of organization development. These factors are based on theoretical assumptions and it requires enhancement. The factors are analyzed by few research scholars and experts. It necessitates vital endorsement for wider acceptance. It is hoped that in some ways this study will help in developing cleaner and sustainable environment. This in turn may help organizations in achieving set goals and objectives, which in turn will pave way for their growth and expansion. There are several limitations with this pragmatic study. Firstly, this study, in which the objective is to analyze and evaluate the influential factor of industry 4.0 in coal industry, is relatively new and in general, the influence values of these factors can change dynamically. Secondly, despite the fact that the consequences of this research study give a decent estimation of the proposed model, different factors might be added to enhance the expectation ability of the proposed model such as the business model, organization and finance. Lastly, it is also worthy to conduct a comparative analysis of sustainable process for development of coal industry in future work.

REFERENCES

Agarwal, A., & Vrat, P. (2015). A TISM based bionic model of organizational excellence. *Global Journal of Flexible Systems Management, 16*(4), 361–376. doi:10.100740171-015-0107-z

Carvalho, N., Chaim, O., Cazarini, E., & Gerolamo, M. (2018). Manufacturing in the fourth industrial revolution: A positive prospect in Sustainable Manufacturing. *Procedia Manufacturing, 21*, 671–678. doi:10.1016/j.promfg.2018.02.170

Coalition, S. M. L. (2011, June). *Implementing 21st century smart manufacturing*. Workshop summary report.

Corno, F., & Razzak, F. (2012). Intelligent energy optimization for user intelligible goals in smart home environments. *IEEE Transactions on Smart Grid*, *3*(4), 2128–2135. doi:10.1109/TSG.2012.2214407

Dubey, R., Gunasekaran, A., Wamba, S. F., & Bag, S. (2015). Building theory of green supply chain management using Total Interpretive Structural Modeling (TISM). *IFAC-PapersOnLine*, *48*(3), 1688–1694. doi:10.1016/j.ifacol.2015.06.329

European Commission. (2015). *Factories of the Future*. Retrieved from http://ec.europa.eu/research/industrial_technologies/factories-of-thefuture_en.html

Frey, C. B., & Osborne, M. A. (2017). The future of employment: How susceptible are jobs to computerisation? *Technological Forecasting and Social Change*, *114*, 254–280. doi:10.1016/j.techfore.2016.08.019

Grau, G., Kitsomboonloha, R., Swisher, S. L., Kang, H., & Subramanian, V. (2014). Printed transistors on paper: Towards smart consumer product packaging. *Advanced Functional Materials*, *24*(32), 5067–5074. doi:10.1002/adfm.201400129

Gupta, V., Kapur, P. K., & Kumar, D. (2016). Modelling and measuring code smells in enterprise applications using TISM and two-way assessment. *International Journal of System Assurance Engineering and Management*, *7*(3), 332–340. doi:10.100713198-016-0460-0

Han, J., Choi, C. S., Park, W. K., Lee, I., & Kim, S. H. (2014). Smart home energy management system including renewable energy based on ZigBee and PLC. *IEEE Transactions on Consumer Electronics*, *60*(2), 198–202. doi:10.1109/TCE.2014.6851994

Ivanov, D., Dolgui, A., Sokolov, B., Werner, F., & Ivanova, M. (2016). A dynamic model and an algorithm for short-term supply chain scheduling in the smart factory industry 4.0. *International Journal of Production Research*, *54*(2), 386–402. doi:10.1080/00207543.2014.999958

Jayalakshmi, B., & Pramod, V. R. (2015). Total interpretive structural modeling (TISM) of the enablers of a flexible control system for industry. *Global Journal of Flexible Systems Management*, *16*(1), 63–85. doi:10.100740171-014-0080-y

Jena, J., Fulzele, V., Gupta, R., Sherwani, F., Shankar, R., & Sidharth, S. (2016). A TISM modeling of critical success factors of smartphone manufacturing ecosystem in India. *Journal of Advances in Management Research*, *13*(2), 203–224. doi:10.1108/JAMR-12-2015-0088

Kang, H. S., Lee, J. Y., Choi, S., Kim, H., Park, J. H., Son, J. Y., ... Do Noh, S. (2016). Smart manufacturing: Past research, present findings, and future directions. *International Journal of Precision Engineering and Manufacturing-Green Technology*, *3*(1), 111–128. doi:10.100740684-016-0015-5

Kim, H. J., Lee, J., Park, G. L., Kang, M. J., & Kang, M. (2010). An efficient scheduling scheme on charging stations for smart transportation. In *Security-Enriched Urban Computing and Smart Grid* (pp. 274–278). Berlin, Germany: Springer. doi:10.1007/978-3-642-16444-6_35

Kumar, H., Singh, M. K., Gupta, M. P., & Madaan, J. (2017). Smart neighbourhood: A TISM approach to reduce urban polarization for the sustainable development of smart cities. *Journal of Science and Technology Policy Management*.

Lee, J. (2013). *Industry 4.0 in Big Data Environment*. German Harting Magazine.

Lee, J., Kao, H. A., & Yang, S. (2014). Service innovation and smart analytics for industry 4.0 and big data environment. *Procedia Cirp*, *16*, 3–8. doi:10.1016/j.procir.2014.02.001

Lee, S. W., Sarp, S., Jeon, D. J., & Kim, J. H. (2015). Smart water grid: The future water management platform. *Desalination and Water Treatment*, *55*(2), 339–346. doi:10.1080/19443994.2014.917887

Liserre, M., Sauter, T., & Hung, J. Y. (2010). Future energy systems: Integrating renewable energy sources into the smart power grid through industrial electronics. *IEEE Industrial Electronics Magazine*, *4*(1), 18–37. doi:10.1109/MIE.2010.935861

Mekhilef, S., Saidur, R., & Safari, A. (2011). A review on solar energy use in industries. *Renewable & Sustainable Energy Reviews*, *15*(4), 1777–1790. doi:10.1016/j.rser.2010.12.018

Minggao, Q. (2010). On sustainable coal mining in China. *Journal of China Coal Society*, *35*(4), 529–534.

Mohanty, S. P., Choppali, U., & Kougianos, E. (2016). Everything you wanted to know about smart cities: The internet of things is the backbone. *IEEE Consumer Electronics Magazine*, *5*(3), 60–70. doi:10.1109/MCE.2016.2556879

Mosconi, F. (2015). *The new European industrial policy: Global competitiveness and the manufacturing renaissance*. London, UK: Routledge. doi:10.4324/9781315761756

Mutchek, M., & Williams, E. (2014). Moving towards sustainable and resilient smart water grids. *Challenges*, *5*(1), 123–137. doi:10.3390/challe5010123

Mwasilu, F., Justo, J. J., Kim, E. K., Do, T. D., & Jung, J. W. (2014). Electric vehicles and smart grid interaction: A review on vehicle to grid and renewable energy sources integration. *Renewable & Sustainable Energy Reviews*, *34*, 501–516. doi:10.1016/j.rser.2014.03.031

Ntuli, N., & Abu-Mahfouz, A. M. (2016). *Simple security architecture for smart water management system*. Academic Press.

Paelke, V. (2014, September). Augmented reality in the smart factory: Supporting workers in an industry 4.0. Environment. In Emerging Technology and Factory Automation (ETFA), 2014 IEEE (pp. 1-4). IEEE.

Panwar, N. L., Kaushik, S. C., & Kothari, S. (2011). Role of renewable energy sources in environmental protection: A review. *Renewable & Sustainable Energy Reviews*, *15*(3), 1513–1524. doi:10.1016/j.rser.2010.11.037

Patri, R., & Suresh, M. (2017). Modelling the enablers of agile performance in healthcare organization: A TISM approach. *Global Journal of Flexible Systems Management*, *18*(3), 251–272. doi:10.100740171-017-0160-x

Qi, W., Liu, J., & Christofides, P. D. (2011). A distributed control framework for smart grid development: Energy/water system optimal operation and electric grid integration. *Journal of Process Control*, *21*(10), 1504–1516. doi:10.1016/j.jprocont.2011.05.010

Radhika, N., & Pramod, V. R. (2014). Modeling structural behaviour of inhibitors of cloud computing: A TISM approach. *Transactions on Networks and Communications*, *2*(5), 60–74.

Rahimi-Eichi, H., Ojha, U., Baronti, F., & Chow, M. Y. (2013). Battery management system: An overview of its application in the smart grid and electric vehicles. *IEEE Industrial Electronics Magazine*, *7*(2), 4–16. doi:10.1109/MIE.2013.2250351

Rajesh, R. (2017). Technological capabilities and supply chain resilience of firms: A relational analysis using Total Interpretive Structural Modeling (TISM). *Technological Forecasting and Social Change*, *118*, 161–169. doi:10.1016/j.techfore.2017.02.017

Schlick, J., Stephan, P., Loskyll, M., & Lappe, D. (2014). Industrie 4.0 in der praktischen Anwendung. In Bauernhansl, Industrie 4.0 in Produktion, Automatisierung und Logistik. Anwendung, Technologien und Migration (pp. 57–84). Academic Press.

Sherly, J., & Somasundareswari, D. (2015). Internet of things based smart transportation systems. *International Research Journal of Engineering and Technology*, *2*(7), 1207–1210.

Shrouf, F., Ordieres, J., & Miragliotta, G. (2014, December). Smart factories in Industry 4.0: A review of the concept and of energy management approached in production based on the Internet of Things paradigm. In *Industrial Engineering and Engineering Management (IEEM), 2014 IEEE International Conference on* (pp. 697-701). IEEE.

Shrouf, F., Ordieres, J., & Miragliotta, G. (2014, December). Smart factories in Industry 4.0: A review of the concept and of energy management approached in production based on the Internet of Things paradigm. In *Industrial Engineering and Engineering Management (IEEM), 2014 IEEE International Conference on* (pp. 697-701). IEEE.

Sindhwani, R., & Malhotra, V. (2017). A framework to enhance agile manufacturing system: A total interpretive structural modelling (TISM) approach. Benchmarking. *International Journal (Toronto, Ont.)*, *24*(2), 467–487.

Singh, A. K., & Sushil. (2013). Modeling enablers of TQM to improve airline performance. *International Journal of Productivity and Performance Management*, *62*(3), 250–275. doi:10.1108/17410401311309177

Soda, S., Sachdeva, A., & Garg, R. K. (2018). Green supply chain management drivers analysis using TISM. In *Flexibility in Resource Management* (pp. 113–135). Singapore: Springer. doi:10.1007/978-981-10-4888-3_8

Stefansson, G., & Lumsden, K. (2008). Performance issues of smart transportation management systems. *International Journal of Productivity and Performance Management*, *58*(1), 55–70. doi:10.1108/17410400910921083

Stock, T., & Seliger, G. (2016). Opportunities of sustainable manufacturing in industry 4.0. *Procedia Cirp*, *40*, 536–541. doi:10.1016/j.procir.2016.01.129

Wahlster, W. (2012). From Industry 1.0 to Industry 4.0: Towards the 4th Industrial Revolution. *Forum Business Meets Research.*

Wan, J., Tang, S., Li, D., Imran, M., Zhang, C., Liu, C., & Pang, Z. (2019). Reconfigurable smart factory for drug packing in healthcare industry 4.0. *IEEE Transactions on Industrial Informatics*, *15*(1), 507–516. doi:10.1109/TII.2018.2843811

Wang, S., Wan, J., Zhang, D., Li, D., & Zhang, C. (2016). Towards smart factory for industry 4.0: A self-organized multi-agent system with big data based feedback and coordination. *Computer Networks*, *101*, 158–168. doi:10.1016/j.comnet.2015.12.017

Wollschlaeger, M., Sauter, T., & Jasperneite, J. (2017). The future of industrial communication: Automation networks in the era of the internet of things and industry 4.0. *IEEE Industrial Electronics Magazine*, *11*(1), 17–27. doi:10.1109/MIE.2017.2649104

Yadav, D. K., & Barve, A. (2016). Modeling post-disaster challenges of humanitarian supply chains: A TISM approach. *Global Journal of Flexible Systems Management*, *17*(3), 321–340. doi:10.100740171-016-0134-4

Yadav, N. (2014). Total interpretive structural modelling (TISM) of strategic performance management for Indian telecom service providers. *International Journal of Productivity and Performance Management*, *63*(4), 421–445. doi:10.1108/IJPPM-04-2013-0081

Young, W., Hwang, K., McDonald, S., & Oates, C. J. (2010). Sustainable consumption: Green consumer behaviour when purchasing products. *Sustainable Development*, *18*(1), 20–31.

Zanella, A., Bui, N., Castellani, A., Vangelista, L., & Zorzi, M. (2014). Internet of things for smart cities. *IEEE Internet of Things Journal, 1*(1), 22-32.

Zheng, P., Sang, Z., Zhong, R. Y., Liu, Y., Liu, C., Mubarok, K., ... Xu, X. (2018). Smart manufacturing systems for Industry 4.0: Conceptual framework, scenarios, and future perspectives. *Frontiers of Mechanical Engineering*, 1–14.

Chapter 11
Supply and Demand Management During Industrial Evolutions:
Present and Future Outlook

Ponnusamy Venkumar
University of Sussex, UK & Kalasalingam Academy of Research and Education, India

ABSTRACT

The aim of this chapter is to understand the status quo of academic research on demand and supply management in terms of cause and mitigation strategies during the third industrial revolution and estimate the scope in the fourth industrial revolution. The chapter uses a systematic literature review approach to classify the past studies published in the International Journal of Production Research during the third industrial revolution based on cause and mitigation strategic framework. Similarly, the study estimates the scope in future by brainstorming academicians and practitioners using Q-methodology. This analysis reveals that dependence on technology will simplify tracking of transit inventory and real-time sharing transparency and continuous updating will simplify demand forecasting.

DOI: 10.4018/978-1-5225-9078-1.ch011

INTRODUCTION

In a company, demand forecasting is one of the initial and important activities. Because in this initial stage, if it is forecasted properly, then it is based on capacity planning, raw material procurement, manufacturing, distribution, retailer and finally sell the product to the customers. If the demand is not forecasted properly then the organization cannot sell whatever it has produced. So this demand forecasting is the survival process of any industry.

The next foremost process in any industry is the inventory management. Inventory is how the stock is maintained at various levels of the organization. Of course, company could maintain the high level stock at various levels of the SC so that we could satisfy the customers because it is easily available in the market. But the inventory involves the cost. So how the industry's financial position is? Based on financial position only the organization will survive and grow further. At the same time, company cannot maintain the stock level very low in all the levels of SC because when the customer wants to buy the product, it should be available; otherwise, the product will not be there in the market.

Of course, demand forecasting and inventory management may look like different activities in an industry; but it is very close activity. Because based on the demand forecasting only the procurement process, manufacturing process, finished products, distribution process and retailer sales process will be activated. In each and every process, the inventory level is maintained like inbound inventory, in-process inventory, finished goods inventory and outbound inventory. Based on the inventory level only the transshipment and replenishment in the entire supply chain.

So, the two critical and important activities are well handled in the industrial era and it will be handled in future. Matching supply and demand were found to be critical until now and the major disruption for the mismatch is caused by vital supply chain drivers such as Inventory and forecasting information. At times, these drivers helped companies to mitigate the cause and micro matched supply with demand. Demand forecasting is one of the critical activities and the errors occurred during forecasting may build up issues starting from the procurement till it reaches the customer end (Dellino et al., 2017). To deal with customer requirements, focal company's supply chain operations need to forecast demand accurately (Zhang et al. 2018).The other driver on the supply side, namely inventory, is quite often used as a mitigation strategy to match supply and demand is also with other supply chain drivers such as transportation and customer service level (Taleizadeh et al., 2015).

It is vital to understand how supply and demand management have been done in the past along sides of evolution of technologies. That too industry 4.0 can widely support to micro match supply and demand. Hence it is essential to revisit and learn supply demand management over the evolution of industry 1.0 to 4.0.

Hence, to the best of my knowledge I did not find any study that typically analyze the development in matching supply and demand during industrial evolutions. In particular, my aim is to trace back and look at the studies surrounding inventory and forecasting in a traditional production research journal. Typically, this study focuses on International Journal of Production Research (IJPR) which is one among the leading very first journal that publishes manufacturing, production and operations management research for more than fifty years (since 1961) with industrial applications. In addition, the number of publications related to managing supply and demand in IJPR is increasing the trend as shown in the Figure 1.

The IJPR usually publishes papers on decision support for product design, selection of manufacturing technology and production resources, solving problems of analysis and control that arise in combining these resources within the design of production systems. The reputation of IJPR was based on a strong link with industrial applications and persuasive scientific results with clear real life applications. Also, there has been continuous improvement in IJPR IF in the year of 2017 is 2.623. The IJPR Impact Factor (IF) and Journal Citation Reports (ICR) in the years are shown in Figure 2.

This research chapter deals with demand and supply evolutions in industrial revolutions through an extensive literature review. The review is based on the important causes and mitigation framework of demand and supply in concept note mentioned by Pfohi et al. (2015). In addition, to understand the perspective of academic researchers and practitioners in the era of Industry 4.0 this study uses Q-methodology to rate their views. The detailed literature review, Q-methodology

Figure 1. The evolution of forecasting and inventory with supply chain in IJPR

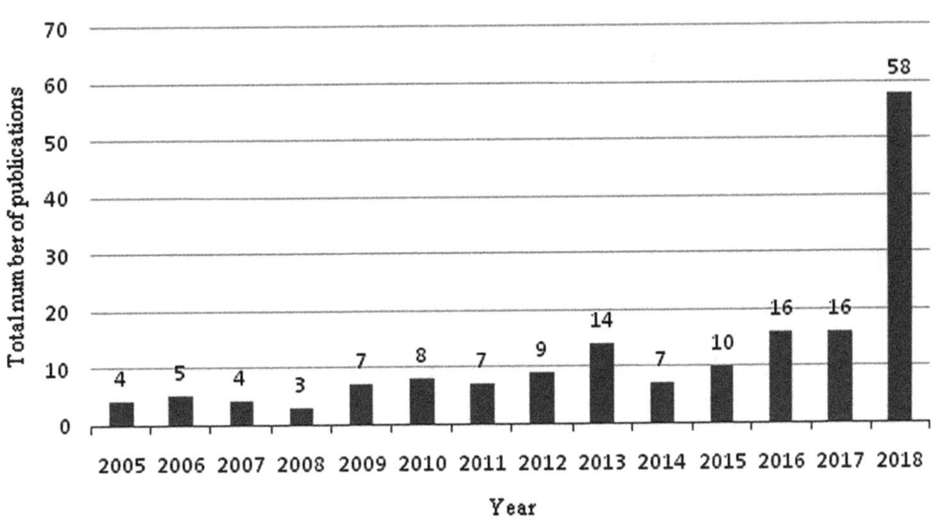

Figure 2. The IJPR IF and JCR years

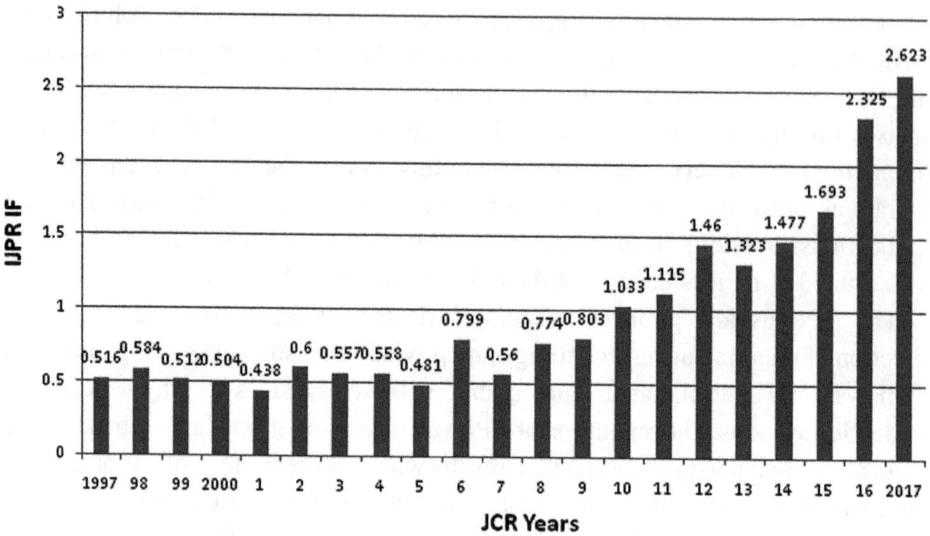

and identified areas of future advanced technologies to be implemented in Industry 4.0 on supply and demand aspects are presented in this chapter.

This chapter is organized into eight sections. The second section deals with the industrial revolution. The third section deals with the review methodology. The fourth and fifth sections are literature review based on causes and mitigation of supply and demand. The sixth section represents future causes and mitigation of supply and demand. The seventh section is Q-methodology survey on the future Industry 4.0 and the eighth section presents the conclusion and future scope.

INDUSTRIAL REVOLUTIONS

Several factors affected managing supply chain from Industry 1.0 to till now. For example, customer's instant need and variety is predominant now, whereas in the past companies focused on volume and price. Later in 1980's quality become an unforgettable mantra to companies other than volume and price. During 18th century the companies typically followed push process that totally relied on mechanized mass production where Companies are not able to satisfy the customer demand. The pioneer work during this period is from legendaries such as Adam Smith's known as a great scholar in economics regarding wealth nations (Yin et al., 2018, Xu et al. 2018).

Figure 3. Industrial revolutions with technology, demand and customer expectations

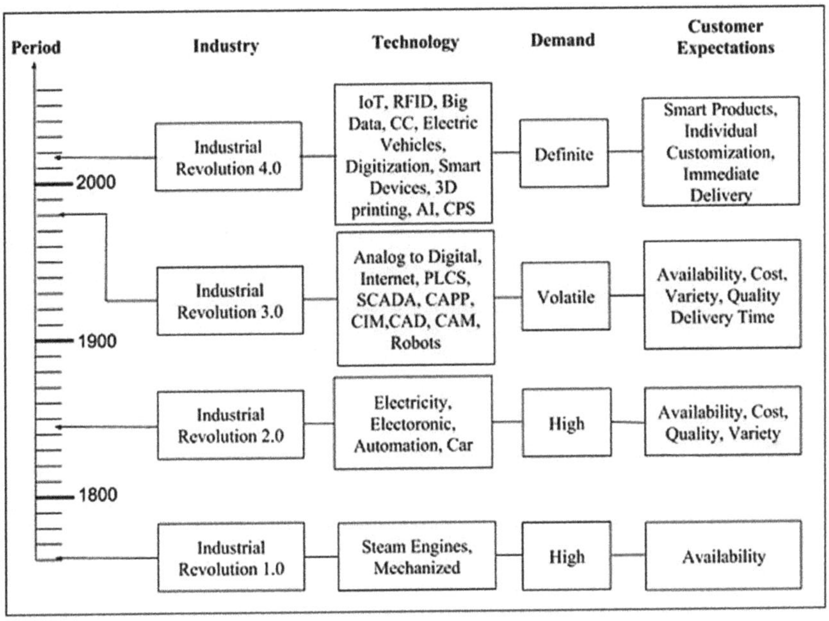

From the end of nineteenth century till 1980s, the customers' expectations were the product with variety. Industrial evolution during this period is known as Industry 2.0 where customer requirements are dealt with stable markets' demand through automated production. Seminal works that appeared during this period were from Frederick Taylor's (Father of Management), Henry Ford's and Taiichi Ohno's scientific principles were used to meet the customer demand with variety of products (Yin et al. 2018).

From the 1980s to till date the volume of demand and variety is very high in addition to the customer expectations in terms of frequent purchase and immediate delivery. To meet all these customer requirements the technological innovations are introduced such as from analog to digital, Internet-connected PCs, Programmable Logic Controllers (PLCs), mainframes, networks, sensors, Supervisory Control and Data Acquisitions (SCADA) systems, Computer Aided Process Planning (CAPP), Computer Integrated Manufacturing (CIM), Computer Aided Design (CAD), Computer Aided Manufacturing (CAM) and robots. With these advancements, the demand for products during this period is increased to three dimensions. They are volume, variety and delivery time. So, today's market is a volatile market and dramatic reductions in the average product lifecycle, to meet that Flexible Manufacturing System (FMS), serus, Supply Chain (SC) concepts are introduced (Yin et al., 2018, Xu et al., 2018).

Industry 4.0 began to implement in European and Chinese companies with the technology innovations like cyber-physical systems (CPS), Internet of Things (IoT), Radio-frequency identification (RFID), big data, electric vehicles, digitization, 3D printing, cloud computing (CC), artificial intelligence and smart devices (Xu et al. 2018). Many aspects of in this Industry 4.0 are unknown and uncertain, such as the demand dimensions of customers and the future product architecture of electric vehicles. The manufacturing execution systems, shop floor control and product life cycle management were farsighted concepts that lacked the technology needed to make their complete implementation for the Industry 4.0 (Yin et al., 2018).

Figure 3 shows the industrial revolutions with corresponding period, technology used, and the demand rate and customer expectations.

REVIEW METHODOLOGY

This study considers two primary issues such as demand and supply management. Demand is captured in terms of forecasting and for supply management I looked at inventory management policies and procedures carried out in the focal companies. Hence, the major search strings are forecasting and inventory in the IJPR publications.

In the third revolution period, the customers' expectations were huge and the market was volatile and so the variety of the products came as a major role. The SC became popular during this period. So, I decided to start the review process from this period to till date.

Initially, it was decided to review the papers containing search strings inventory and forecasting anywhere in the IJPR articles during the period corresponding to Industrial 3.0. The search yielded 1425 articles in forecasting and 4122 articles in inventory management. Then we decided to select the keywords either forecast or inventory; by this we ended up 304 and 62 papers respectively. Then in during the period between Industry 3.0 and till now, we have number of publications in the forecasting and inventory with Industry 3.0 as potential keywords. Our final search strings are Industry 3.0 keywords along with forecasting or inventory. In that search, we found 96 papers under inventory with SC and 13 papers forecasting with SC category. In terms of Industry 4.0, the search revealed 5 papers. Altogether, the sample for review contains 114 papers. Out of which three papers are review papers and three overlapping papers counts were removed. Hence the final number of papers to be reviewed is108 papers as per the procedure shown in Figure 4. All the 108 papers were critically reviewed.

Figure 4. The general frame work for review

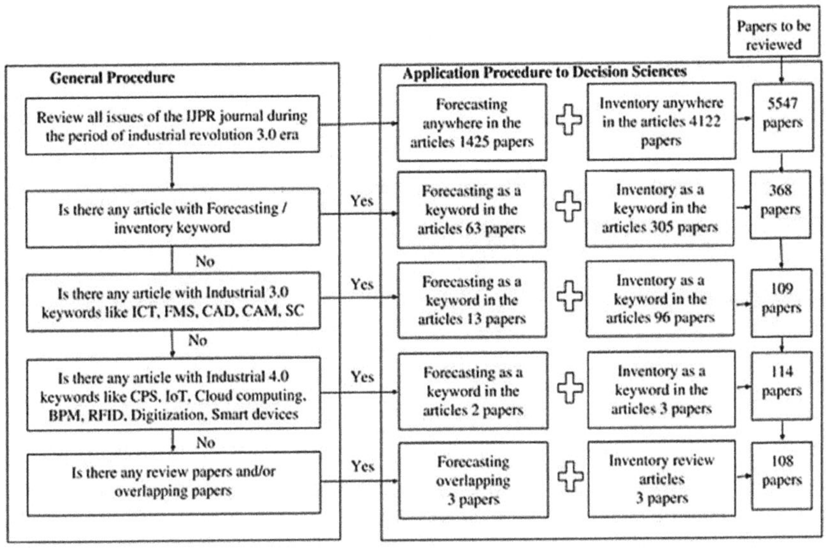

The entire review is based on the causes and mitigation of inventory which are identified through the literature (Gunasekaran et al. 2018, Gunasekaran et al. 2015) as shown in Figure 3. Pfohi et al. (2015) analyzed 49 Industry 4.0 technologies and concepts and indentified 11 most relevant technologies and concepts through conceptual analysis. In this review, I have categorized the causes and mitigation of supply and demand based on Pfohi et al. (2015).

The relative supply and demand terms are selected as the causes and mitigation of supply and demand, which is shown in Figure 5. The demand forecasting is conventionally based on the previous sales data. So, we are failing to predict the actual demand. The forecasting is dominant in the causes for supply. Similarly, supply is inventory based. So, we considered supply as dominant for mitigation.

Figure 5. The causes and mitigation of demand and supply

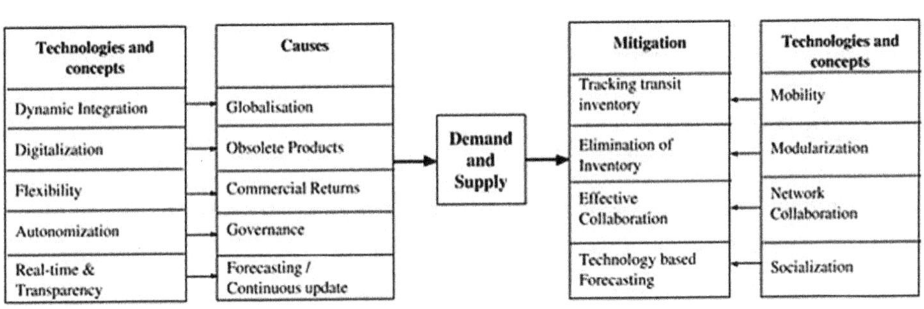

CAUSES OF DEMAND AND SUPPLY

The causes of demand and supply in the companies are continuously changing its trend. To meet the customers' expectations the industrial evolutions are evolved. For the demand and supply, the numbers of causes are involved in every industrial evolution. In this literature review some of the important causes of demand and supply are considered. They are globalization, obsolete products, commercial returns, governance and forecasting and continuous update.

Globalization

Globalization causes number of issues in demand and supply. Now each and every industry in the world is facing global competition. The industries should consider the various costs involved in demand and supply of globalization. A continuous review intercontinental and domestic outsourcing with hidden direct and indirect costs creates longer SC and their impact is substantially affecting profitability (Co et al. 2012). In the global market the industries are facing uncertain demand. To reduce the demand uncertainty, a combination of dual-channel SC that is the manufacturer have direct channel to sell the product through an e-store, and if the customer is not nearby then forward the orders to the nearest retailer to supply the product, is proposed to solve the issues like delivery lead time and customer hands on experience of the product, etc. (Zhao et al. 2016, Hennet 2009, Sirias and Mehra 2005).

Recently customers are expecting their needs to be met instantaneously. In that context, a guarantee quoted model is developed by combining long-term decisions such as facility and supplier selection with the midterm decisions such as inventory placement, replenishment and delivery lead time. In addition, multi-stage hybrid model is also developed to analyze a SC network collapse recovery possibility (Vahdani et al. 2011).The actual application of industry is based on better product availability for centralized production-inventory models in decentralized decision-making settings is also presented (Tayur 2013, Fattahi et al. 2015).

There are still a few open unanswered questions in demand and supply such as current demand rate of the particular variety, frequent purchase possibilities and locating distributors and retailers with effective collaboration, effective distribution and competition is even more aggressive.

Obsolete Products

Identifying and removing of obsolete products from the stock is one of the major issues in the inventory management system. In that context, single vendor and single buyer analytical model is developed to take into account the effects of obsolescence in a SC managed with a consignment policy (Persona et al. 2005).

The perishable products become obsolete by time delay in SC. The production, inventory and distribution policy for the perishable items in a multi-stage SC with optimized policy were solved by using GA (Dolgui et al. 2018). But the GA solutions are not necessarily optimal solutions and it is a viable approximate solution and requires a contextualized tuning.

One of the major SC strategic issues is fresh food SC management. In that context, a Decision Support System (DSS) is developed which is the combination of sales forecasting and order planning for the SC of packaged fresh food (Dellino et al. 2017). The DSS gives only acceptable and satisfactory performance, because the decision makers depend on computer based decision making.

Coordination between the SC players is necessary to tackle the obsolete product issues. In that aspect, a VMI for deteriorating goods designed for multi-period channel coordination (Chen and Wei, 2012).

The preservation and shipment are also important in the perishable/fresh/deteriorate items. In that aspect the optimal production and shipment under preservation policy for a single-vendor and single-buyer for deteriorating items are also developed (Dellino et al. 2017, Tsai 2011). Distribution of Perishable goods SC networks is also important. In that aspect the locating Distribution Centres (DCs), assigning retail stores to DCs, joint replenishment cycle time and investing in preservation technology are also addressed Tsao (2016). The supplier-buyer integrated production-inventory model with random yield is considered with and without a considerable inspection time (Liu and Cetinkaya 2011).

But so far, identifying the obsolete items from the inventory or before due date, it should be transferred to the required place which are not addressed by the researchers.

Commercial Returns

The demand and supply for remanufacturing/rework/buyback are also one of the important issues in SC. In that context, it is interesting to note that numbers of industries are involving in end-of-use (i.e. cores) operations such as remanufacturing and recycling. It is evident that these operations require specific purchase policy for purchasing cores. The distributed lag model is proposed for the purchasing policy for manufacturing or recycling cores in a finite planning horizon (Clottey2016). In the manufacturing and remanufacturing of products, there is a chance of bullwhip effect. That is also addressed in a hybrid manufacturing and remanufacturing with different control policies to analyze the cost performance and bullwhip effect (Zanoni et al. 2006). The demand rate is different for the newly manufactured or remanufactured products. In that aspect, a closed loop decision making model is designed to satisfy the retailer by newly manufactured products or by remanufactured with same function and quality products (Yuan and Gao 2010, Yuan et al. 2015, Subtamanian et al. 2012).

In the production process, there is a chance of producing good and defective items. The defective item may be identified by the retailer or customer and the manufacturer should face the buyback. In that context, the pricing and ordering decision in SC with imperfect quality items and inspection/buyback policies presented (Taleizadeh et al. 2016) and shortcoming identified and addressed as note (Brauer and Bushcher2017).

The companies can find the product service level with the customers by using the upcoming technology. Throughout the life span the product can be monitored by the company and provided preventive service without customers' complaint by using upcoming technology.

Governance

The inventory governance is also one of the important concepts in SC. The planning of order is important in inventory governance. In that aspect, the planned orders are obtained by simulating the future replenishment of each item based on the information about the delivery schedules, the inventory levels, the order policies and the point-estimate of the customer demand (Sillanpas and Liesio 2018). The coordination of order quantities amongst the players is also addressed in four-level SC (Jaber and Goyal 2009).

The production and distributions are also important in the inventory governance. The stochastic replenishment in multi-echelon SC by considered both production and distribution functions (Suwanruji and Enns 2006). The proportional, proportional-integral and proportional-derivative control schemes are developed to manage the inventory replenishment process in a SC under different forecasting situations (Sourirajan et al. 2008).

The lead time and throughput are also included in the inventory governance. The complex SC is evaluated with the impact of demand and lead time assumptions under the various SC configurations (Hwarng et al. 2005). The order timing preferences within a strategic framework involving a supplier and one of its retailers in a multi-retailer system is also presented (Bakal and Geunes 2010). The lead time is reduced through multiple sourcing which has been addressed elaborately (Glock and Ries 2013). By using Theory of constraints (TOC) throughput inventory and operating expenses were calculated to monitor SC and to ensure on-line delivery (Gupta and Anderson 2012, Gupta and Anderson 2018). TOC based replenishment is also successfully implemented (Wu et al. 2012).

There is a chance of stock-out in one node which may lead to other node of the SC networks. The existence and magnitude of stock-out propagation and stock-out amplification are investigated in the context of SC inventory systems (Chatfield 2013). The operational planning is also important in the inventory governance. An algorithm has developed a delivery-sequencing/inventory-allocation which is used to control a SC in the operational planning level (Monthatipkul and Kawtummachai 2007). The service level and market selection are also considered in the multi-product selective newsvendor model (Abel-Aal et al. 2017) and newsvendor with multiple unreliable supplier problems is also addressed (Merzifonluoglu and Feng 2014). The availability of the products influences the customers' purchasing decisions have also been addressed (Kurata 2014).

The safety stock is also to be considered in the inventory governance. In that context an advanced planning systems for safety stock policies with variation in production and distribution were also considered (Villegas and Smith 2006). The use of form postponement is based on the positioning of the differentiation point and stocking policy is also presented (Wang et al. 2009). The inventory performance on the impact of impulsive demand disturbance on the inventory control policies is also addressed (Wadhwa et al. 2009).

The governance of inventory and replenishing are not addressed by the researchers. By using the upcoming technology live monitoring and replenishing are also possible.

Forecasting/Continuous Update

The demand forecast updating is very important in SC. But in SC, the forecasting fails in real-time sharing, transparency and continuous updating. A novel reservation contracts has been addressed in which the demand forecast is partially updated and also found that demand forecast updating between the SC members will improve their performance (Yang et al. 2015).

To meet the uncertainties in demand, the forecast accuracy is a prime factor. The effects of demand uncertainties and forecast inaccuracies are partially incorporated and fully integrated the sales and operational planning (Feng et al. 2010, Kaku et al. 2009). The predictive control strategy is developed for biogas inventory control in sugarcane industries (Alvarez et al. 2017)

The information sharing will increase the accuracy of the demand forecast. In VMI also the independent decision making with full information sharing model is developed with service level constraints (Choudhary and Shankar (2015), Salzarulo and Jacobs (2014)).

Instead of using previous sales data, real-time sharing, transparency and continuous updating of information will improve the accuracy of demand and eliminate the bullwhip effect. That is possible by using the upcoming technologies.

MITIGATION OF DEMAND AND SUPPLY

The mitigation of demand and supply has continuously changed in every industrial revolution. In this literature review how the mitigation of demand and supply varies over the time, is analyzed. The tracking of transit inventory, elimination of inventory, effective communication and technology-based forecasting are considered as the important points of mitigation of demand and supply.

Tracking of Transit Inventory

In the transit of inventory, the location network design is an important factor. The location–inventory network design problem is designed to optimize the warehouse location, retailer assignments, inventory replenishment and the safety stock-level decisions over an infinite planning horizon (Shu et al. 2015).

The optimized distribution is also necessary in transit inventory. The inventory distribution optimization for multi-echelon multiple products decision model is designed with an objective of inventory, transportation and location decisions (Manatkar et al. 2016).

The third party logistics (3 PL) is one of the important concepts in transshipment of inventory. In that context, location-inventory decisions is presented with 3 PL in closed loop SC (Li et al. 2016,, Arabzad et al. 2015).The transshipment between the retailers is inevitable to increase the customer service level. A decentralized SC model is presented for one manufacturer and two retailers who could transit their excess inventory between each other (Li and Li 2017).

Tracking of transit will increase the inventory reliability. By using the interconnected logistics services with physical internet, the SC issue in replenishment of inventory is reduced (Yang et al. 2017).

Misplacement of inventory and transit inventory reliability can be improved by using the upcoming technologies.

Elimination of Inventory

The excess inventory elimination is the main motive of the SC networks. But in the SC, network is difficult to tackle the bullwhip effect. Analytical expressions are used for the two-level SC in which demand is price sensitive and the price follows auto regression with an objective of the study is reduced the net variance of amplification (i.e. bullwhip effect) (Ma et al. 2013). One of the major issues in demand forecasting errors is the bullwhip effect. That effect is addressed by number of researchers (Zanoni et al. 2006, Sourirajan et al. 2008, Cannella and Ciancimino 2010, Ma et al. 2013, Lin et al. 2014, Vicente et al. 2017, Li and Disney 2017). The main reason behind the bullwhip effect is SC players are not sharing their real-time data. If the real-time data is shared between the players, then the bullwhip effect will be eradicated.

While eliminating inventory, we have to consider the level of customer satisfaction also. A pragmatic solution to inventory control is developed with practical operational constrains. The proposed schemes are feasible and outperformed when compared to the conventional policy in reduction of operating cost, mitigation of bullwhip effect and increased customer satisfaction level (Fu et al. 2016, Yan et al. 2008). The optimal markdown price for remaining inventory level is to maximize the profit during and after the holiday is also presented (Wang et al. 2018).

By using the upcoming technologies (like IoT, 3D printing etc.) with updated information, there is a scope for elimination of inventory also.

Effective Collaboration

Collaboration is essential in SC to mitigate the demand and supply issues. The Information and Communication Technology (ICT) mechanisms are used for collaborative planning of forecasting and replenishment (Danese, 2006 and Danese, 2011). It is proved that the ICT coordination is necessary and it should update with the current data. Special type of contract is designed to enhance the value of forecast data issues, has also been solved by a game analysis and found that how the contract parameters affect the SC performance (Lackes, 2016).

For identifying the true source of shift in multivariate process, the neural-network-based algorithms are used and the performances are also more stable when compared with the statistical control charts (Brian Hwang and Wang, 2010). More number of parameters can be controlled by using neural-network-based models. Training the neural-network and generalized model in neural-networks are the complex processes.

The Vendor Managed Inventory (VMI) is one of the good concepts in SC collaboration. The vendor should monitor the retailers' inventory level and replenish whenever it reached the reorder level. Numbers of research papers were published in this VMI, particularly coordination mechanism model (Chakraborty, 2015) and approximation model (Berling and Markland, 2014). The VMI with Consignment Stock (CS) developed coordinated model vendor uses buyers warehouse (Hemmati et al. (2017), Battini et al. 2010). A VMI and CS policy is applied to the integrated vendor-buyer SC, where the demand at the buyer side is stock dependent (Zanoni and Jaber 2015, Yi and Sarker 2013). The analytical model is used to solve CS and VMI policy for a single-vendor and multiple buyers SC with known demand in which agreement is more beneficial (Ben-Daya et al. 2013). The effect of inspection errors on the costs occurred in VMI and CS issues are solved by mathematical model in which the vendor is not in perfect quality (Alfares and Attia 2017). The collaborative

SC adoption of VMI issues are solved by Analytical Hierarchical Process method (Borade et al. 2013). The Information Technology driven VMI two-echelon SC comprising m-vendor and n-buyers non-linear integer programming problem are also solved by using knowledge based system and using Genetic Algorithm (GA) to find the optimal quantity and optimal channel profit (Nachiappan et al. 2007).

The benefit of VMI to which party either vendor or retailer, is studied by analytical method to solve three cases (Bookbinder et al. 2010). The VMI implementation issues and to support corporate practice methodology is analyzed with a questionnaire and ten case-studies (Niranjan et al. 2012).

The various VMI based collaboration issues discussed like the collaboration in vertical, horizontal and lateral with conceptual model (Chan and Prakash (2012)), three-echelon SC inventory system in the beverage sector with design of experiments were presented (De Sensi et al. (2008)), flow of material between two vendors and a buyer and develops delivery structure (Glock 2012), emission control coordination (Jaber et al. 2013), coordination of order quantities amongst the players in four level SC with centralized decision process (Jaber and Goyal 2009), information exchange mechanism in a four-echelon SC (Viswanathan et al. (2007)), integrated inventory model fuzzy annual demand and/or production rate (Pan and Yang 2008), inventory management for a buyer by a multi period adjustment contract allows buyer to adjust his inventory level upwards or downwards during order period (Li and Ryan 2012), multiple period replenishment policies one retailer and one manufacturer with uncertain demand (Fattah et al. 2015), connection between SC performance and company financial performance (Martin and Patterson 2009), decision making in multi-echelon serial SC management in the presence of imprecision or uncertainty arising from human reasoning, emphasizing the computational resolution (Chen and Cheng 2014), approximation model for tandem manufacturing systems (Ruifeng and Subramanian 2011), integrated approach to SC risk management using operational methods and financial instruments (Bandaly et al. 2014), product and process design policies in the extended enterprise by studying dynamics of SC demand propagation (Meixell and Wu 2005) and DC synchronize inbound and outbound flows for cross-docking by maintaining inventory (Kellar et al. 2016).

A decentralized decision-making approach is developed for a coal multi-party coordination SC, with three independent parties – multiple mines, a rail operator and a terminal (Thomas et al. 2016). The different coordinating strategies are presented to enhance the SC (Taleizadeh et al. 2016). The two stage EOQ/EPQ model is revisited with inelastic demand with decentralization and coordination (Geunes 2018).

By using the latest technologies, effective collaboration is possible without relationship. The effective collaboration will solve most of the important issues in the demand and supply.

Technology Based Forecasting

The technology based forecasting will simplify planning and reduces complexities in production capacities, improve service level, product management, marketing and product design. A production-distribution and inventory planning model are developed for multi-product SC (Taxakis and Papadopoulos 2016). The sequencing, scheduling and lot-sizing decisions for multiple-products manufactured through multi-firms in a serial-type SC model is developed to minimize the costs of setting up and holding for customers' demand (Huang and Yao 2013). The dynamic multi-plant production planning and distribution model is developed for simultaneous optimization of production, inventory control, demand allocation and distribution decisions (Darvish et al. 2016).

FUTURE THE CAUSES AND MITIGATION OF SUPPLY AND DEMAND

Industrial experts state IoT technologies with parts, machines and products will be inextricably linked to the information, providing huge boost to lean manufacturing and the cash-conversion cycle (Pasquali 2015).

In demand planning, the mass customization requires a connection of production capabilities with the SC and translation of fluctuating demand patterns into targeted production units (Staff 2015).

Hofmann and Rüsch (2017) explained the opportunities and consequences of Industry 4.0 with the following points:

1. Through the auto-identification technology, the material and products traceability will increase and manufacturing planning will be more accurate with respect to forecast demand. The real-time material flows can be tracked precisely. If a product is sold at a particular point in the delivery chain and regulate their production in line with the current demand. The bullwhip effect might be prevented or mitigated through all players and would be able to act at the real-time statistics.
2. Real-time consumption of products can be automatically and continuously triggered the supply orders about the intake. The production orders are aligned with actual consumption. Based on that, delivery order is released. The transparency throughout the SC will improve the production planning.
3. By using RFID tags with less manual interaction more precise demand assessment and inventory levels are measured automatically by autonomous systems.

4. In the digitization of material flows, delivery processes are simulated with respect to adjacent processes and delivers the right product at the right time through fully integrated and transparent optimized process and more efficient delivery of material.
5. Through auto-ID, technologies automatically identify the incoming of the material and control it with regard to the respective order.

Luthra and Mangla (2018) proposed the potential challenges in adopting Industry 4.0 and it is useful for the Indian manufacturing industry practitioners, policy makers, regulatory bodies and managers for in-depth understanding.

SC should respond to demand with information technology tools with the operations of customers that may have deeper foresight to handle the demand (Sheffi 2009).

RFID can be successfully implemented in tactical SC management, operational SC management, procurement, distribution, logistics management and inventory management (Maxwell and Lal 2013, Nabelsi and Gagnon 2017).

The upcoming technologies are started implementing demand forecasting and inventory management. RFID is successfully adopted in newsvendor model in which the price and buy-back contracts with a manufacturer and retailer and the output shows the increases in SC efficiency, reduction in forecast error by shortening lead-time, the minimized human intervention and the misplacement also minimized (Zhang et al. 2018).

RFID may well collaborate with work simplification when multiple readings are implied such as containers or platforms in a warehouse, or for items directed at customers and then they return (Sheffi 2009).

The SC replenishment policies and customer demand forecasting methods are analyzed by using RFID enabled with ARIMA simulation experiment (Wang et al. (2010)). They compared the experiments with non-RFID and Taquchi experiment, found that RFID enabled system reduce the total inventory cost and inventory turnover increase. This research shows the importance of RFID in forecasting. The shrinkage errors that cause a difference between physical and information system inventory level is identified and eliminated by using RFID technology (Rekik and Sahin 2012).

The forecast accuracy is measured in the demand planning process by using agent-based modeling and simulation modeling (Hanke et al. 2017). In addition to that agent-based simulation optimization methodology is developed for the online auction policy to control the inventory in agriculture SC (Huang and Song 2017).

Upcoming technology oriented forecast and inventory will simplify the planning and reduces complexities in production capacities, improve service level, product management, marketing and product design.

Q-METHODOLOGY

The Q-methodology "Correlates persons instead of tests" (Stephenson 1935). Q-methodology is a combination of qualitative and quantitative analysis which is used to analyze the subjective data and the individual group according to their response. The perspective of each person is collected through statements. The opinion about the statements of each person is grouped and analyzed based on mathematical-statistical technique that enables the researcher to understand and examine the structure of the complex issues.

What is in the people's minds about something new? That could be easily identified and analyzed by using Q-methodology. If we want to know the opinion of the Industry 4.0 there is possibilities of immediate implementation, then one of the methods available to discover is Q-methodology. This methodology combines both qualitative and quantitative methods. Q-method discovers the correlations between participants.

Q-methodology permits the participants to give their opinion that reflects their subjectivity. Only in this method, capture what is actually in the participants' mind about a topic from all the participant responses, and at the same time identifying fine differences between some of the participants. We could do the Q-study online and the number of web support programs which are also available to analyze the Q-study.

The Q-methodology is used in this review to find the causes and mitigation based on Pfohi et al. (2015) in Industry 4.0 era. In most relevant technologies, nine statements were formulated and collected the survey from the practitioners and the academicians. The Q-survey statements are representing all the major causes and mitigation of the demand and supply in Industry 4.0. The various strategies on global sourcing resilience (Gunasekaran et al. 2015) and twenty-first-century sustainable manufacturing competitiveness (Gunasekaran et al. 2018) were addressed elaborately and based on that we formulated the causes and mitigation of the demand and supply.

The statements are uploaded in LinkedIn and invited the suitable relevant professional to participate that ended up in sixteen responses.

The Q-survey is conducted through LinkedIn on-line survey. Initially the respondents were requested to rate the statements as per the cell allocation. They are requested to separate three agreed statements, disagreed statements and the remaining three are neutral statements. The participants were requested to fill the Q-sort pattern which is quasi-normal distribution pattern. The strongly agree/disagree statement numbers are placed in the extreme columns of the Q-sort pattern. The next set of columns, closer to the centre, which they have less strong opinions. The collected Q-sort survey is calculated to by-person statement analysis and the same opinion group is identified and correlated. The statements with final factor scores are shown in Table 1. The factor-wise results are given below:

Table 1. Q- Statements for forecasting and inventory and the final factor scores

S. No.	Q-Statements	Factor Arrays	
		1	2
1	In future advanced technologies will eliminate inventory and enable companies to meet customers' demand in real time.	-1	1
2	Future advance tracking facilities will enhance in transit inventory reliability.	2	1
3	Effective collaboration is possible without relationship during IoT era to deal with uncertainties and complexities in managing supply and demand.	1	-2
4	In future, by using IoT technology the obsolete products will be removed from the inventory or before the due date or it will be transferred to the required place.	-1	0
5	Globalization challenges such as demand uncertainty, high competition, high variety, high inventory levels and delays are manageable in the future using IoT technologies.	-2	0
6	In future, demand and supply of remanufacturing/rework/buyback of defective items will be easily manageable with the support of futuristic technologies.	0	-1
7	It is very important to establish governance mechanism to monitor the inventory system by using upcoming technologies (like IoT, RFID, etc) in future.	0	0
8	Real-time sharing, transparency and continuous updating of information will simplify the forecasting in Industry 4.0 era.	0	0
9	In future, IoT based demand forecasting will simplify planning and reduces complexities in production capacities, improve service level, product management, marketing and product design.	0	2

Factor 1: These respondents are more than twenty year experience in both industry and academic put together and they are opined that future advance tracking facilities will enhance in transit inventory reliability (refer Table 1 loading score for statement 2). They tend to view the positive statements ranked higher in Factor 1 than in other factor array (refer Table 1 loading score for statements 9, 3, 7, and 6). They have positive opinion on implementing the upcoming technology in tracking IoT based forecasting, effective collaboration, inventory governance and commercial returns. The negative statements ranked in Factor 1 array than in other factor array (refer Table 5 loading score for the statements 7, 8, 4 and 1). They have negative opinion on implementing upcoming technology in inventory governance, forecasting/continuous update, obsolete products, and eliminate inventory. The lowest rank statement is globalization (refer Table 5 loading score for the statement 8).

Factor 2: These respondents are more than 15 years experience in both industry and academic put together and they are interested in implementing forecasting/ continuous update (refer Table 5 loading score for statement 8). They are positive to view the statements ranked higher in Factor 2 than in other factor array (refer Table 5 loading score for statements 1, 7, 5, and 4). They have positive opinion on implementing the upcoming technology in eliminating inventory, inventory governance, globalization and obsolete products. The negative statements ranked in Factor 2 array than in other factor array (refer Table 5 loading score for the statements7, 6 and 9). They have negative opinion on implementing upcoming technology inventory governance, commercial returns and technology based forecasting. The lowest rank statement is effective collaboration using upcoming technology (refer Table 5 loading score for the statement 3).

CONCLUSION

The demand and supply models in the IJPR for the period of Industrial 3.0 are reviewed elaborately. In Industry 4.0 the demand and supply will get new dimensions because of the upcoming advanced technologies. The Industrial 3.0 issues in demand and supply will be addressed in the Industry 4.0 by using the advanced technologies.

The Q-methodology identified the clusters of the respondents based on the similarity of their opinion about the demand and supply. The findings through Q-methodology are how the practitioners and academicians are evaluated in this study correspond to two distinct groups, each representing on important viewpoint regarding demand and supply from the Industry 4.0 perspective. Each of the two groups is given a descriptive statement based on the most agreed on statements that is forecasting and inventory.

The inventories (Factor 1) were mainly concerned with respect to tracking of transit inventory. The practitioners and academicians are opined that the advanced tracking facilities will enhance the transit reliability in Industry 4.0. The next positive statement of the experts and academicians in the Factor 1 is the effective possible collaboration without relationship during IoT era to deal with uncertainties and complexities in managing the supply and demand and in future advanced technologies will eliminate inventory and enable the companies to meet customers' demand in real-time.

The forecasting (Factor 2) are mainly concerned with real-time sharing, transparency and continuous updating of information will simplify the demand forecasting in the Industry 4.0 era. As per the Factor 2, next positive statement is IoT based demand forecasting which will simplify the planning and reduces complexities in production capacities, improve service level, product management, marketing and product design.

Application of Q-methodology in Industry 4.0 is potentially diverse. The practitioners and academicians attitudes towards demand and supply were discussed. The Q-methodology is a unique research tool that can be applied in other Industry 4.0 applications. This research can be further extended with more number of specific statements with more number of respondents in detailed analysis of each and every aspect of Industry 4.0. This review is considered only IJPR journals, so there is a scope for considering all other reputed journals also. The scope for implementing Industry 4.0 in a county or a region could be easily identified further by using Q-methodology.

REFERENCES

Abdel-Aal, M. A., Syed, M. N., & Selim, S. Z. (2017). Multi-product selective newsvendor problem with service level constraints and market selection flexibility. *International Journal of Production Research, 55*(1), 96–117. doi:10.1080/00207543.2016.1195932

Alfares, H. K., & Attia, A. M. (2017). A supply chain model with vendor-managed inventory, consignment, and quality inspection errors. *International Journal of Production Research, 55*(19), 5706–5727. doi:10.1080/00207543.2017.1330566

Álvarez-Rodríguez, D. A., Normey-Rico, J. E., & Flesch, R. C. C. (2017). Model predictive control for inventory management in biomass manufacturing supply chains. *International Journal of Production Research, 55*(12), 3596–3608. doi:10.1080/00207543.2017.1315191

Arabzad, S. M., Ghorbani, M., & Tavakkoli-Moghaddam, R. (2015). An evolutionary algorithm for a new multi-objective location-inventory model in a distribution network with transportation modes and third-party logistics providers. *International Journal of Production Research, 53*(4), 1038–1050. doi:10.1080/00207543.2014.938836

Bakal, I. S., & Geunes, J. (2010). Order timing strategies in a single-supplier, multi-retailer system. *International Journal of Production Research, 48*(8), 2395–2412. doi:10.1080/00207540902810528

Bandaly, D., Satir, A., & Shanker, L. (2014). Integrated supply chain risk management via operational methods and financial instruments. *International Journal of Production Research, 52*(7), 2007–2025. doi:10.1080/00207543.2013.844376

Battini, D., Gunasekaran, A., Faccio, M., Persona, A., & Sgarbossa, F. (2010). Consignment stock inventory model in an integrated supply chain. *International Journal of Production Research, 48*(2), 477–500. doi:10.1080/00207540903174981

Ben-Daya, M., Hassini, E., Hariga, M., & AlDurgam, M. M. (2013). Consignment and vendor managed inventory in single-vendor multiple buyers supply chains. *International Journal of Production Research, 51*(5), 1347–1365. doi:10.1080/00207543.2012.662725

Berling, P., & Marklund, J. (2014). Multi-echelon inventory control: An adjusted normal demand model for implementation in practice. *International Journal of Production Research, 52*(11), 3331–3347. doi:10.1080/00207543.2013.873555

Bookbinder, J. H., Gümüş, M., & Jewkes, E. M. (2010). Calculating the benefits of vendor managed inventory in a manufacturer-retailer system. *International Journal of Production Research, 48*(19), 5549–5571. doi:10.1080/00207540903095434

Borade, A. B., Kannan, G., & Bansod, S. V. (2013). Analytical hierarchy process-based framework for VMI adoption. *International Journal of Production Research, 51*(4), 963–978. doi:10.1080/00207543.2011.650795

Bräuer, I., & Buscher, U. (2017). A note on 'Pricing and ordering decisions in a supply chain with imperfect quality items and inspection under buyback of defective items'. *International Journal of Production Research*, 1–6.

Brian Hwarng, H., & Wang, Y. (2010). Shift detection and source identification in multivariate autocorrelated processes. *International Journal of Production Research, 48*(3), 835–859. doi:10.1080/00207540802431326

Cannella, S., & Ciancimino, E. (2010). On the bullwhip avoidance phase: Supply chain collaboration and order smoothing. *International Journal of Production Research, 48*(22), 6739–6776. doi:10.1080/00207540903252308

Chakraborty, A., Chatterjee, A. K., & Mateen, A. (2015). A vendor-managed inventory scheme as a supply chain coordination mechanism. *International Journal of Production Research, 53*(1), 13–24. doi:10.1080/00207543.2014.921350

Chan, F. T., & Prakash, A. (2012). Inventory management in a lateral collaborative manufacturing supply chain: A simulation study. *International Journal of Production Research, 50*(16), 4670–4685. doi:10.1080/00207543.2011.628709

Chatfield, D. C., Hayya, J. C., & Cook, D. P. (2013). Stockout propagation and amplification in supply chain inventory systems. *International Journal of Production Research, 51*(5), 1491–1507. doi:10.1080/00207543.2012.697205

Chen, L. T., & Wei, C. C. (2012). Multi-period channel coordination in vendor-managed inventory for deteriorating goods. *International Journal of Production Research, 50*(16), 4396–4413. doi:10.1080/00207543.2011.592159

Chen, S. P., & Cheng, B. H. (2014). Optimal echelon stock policies for multi-stage supply chains in fuzzy environments. *International Journal of Production Research, 52*(11), 3431–3449. doi:10.1080/00207543.2014.885145

Choudhary, D., & Shankar, R. (2015). The value of VMI beyond information sharing under time-varying stochastic demand. *International Journal of Production Research, 53*(5), 1472–1486. doi:10.1080/00207543.2014.951093

Clottey, T. (2016). Development and evaluation of a rolling horizon purchasing policy for cores. *International Journal of Production Research, 54*(9), 2780–2790. doi:10.1080/00207543.2016.1142133

Co, H. C., David, I., Feng, P., & Patuwo, E. (2012). A continuous-review model for dual intercontinental and domestic outsourcing. *International Journal of Production Research, 50*(19), 5460–5473. doi:10.1080/00207543.2011.638941

Danese, P. (2006). Collaboration forms, information and communication technologies, and coordination mechanisms in CPFR. *International Journal of Production Research, 44*(16), 3207–3226. doi:10.1080/00207540600557991

Danese, P. (2011). Towards a contingency theory of collaborative planning initiatives in supply networks. *International Journal of Production Research, 49*(4), 1081–1103. doi:10.1080/00207540903555510

Darvish, M., Larrain, H., & Coelho, L. C. (2016). A dynamic multi-plant lot-sizing and distribution problem. *International Journal of Production Research, 54*(22), 6707–6717. doi:10.1080/00207543.2016.1154623

De Sensi, G., Longo, F., & Mirabelli, G. (2008). Inventory policies analysis under demand patterns and lead times constraints in a real supply chain. *International Journal of Production Research, 46*(24), 6997–7016. doi:10.1080/00207540701528776

Dellino, G., Laudadio, T., Mari, R., Mastronardi, N., & Meloni, C. (2017). A reliable decision support system for fresh food supply chain management. *International Journal of Production Research*, 1–28.

Dolgui, A., Tiwari, M. K., Sinjana, Y., Kumar, S. K., & Son, Y. J. (2018). Optimising integrated inventory policy for perishable items in a multi-stage supply chain. *International Journal of Production Research, 56*(1-2), 902–925. doi:10.1080/00207543.2017.1407500

Fattahi, M., Mahootchi, M., Moattar Husseini, S. M., Keyvanshokooh, E., & Alborzi, F. (2015). Investigating replenishment policies for centralised and decentralised supply chains using stochastic programming approach. *International Journal of Production Research, 53*(1), 41–69. doi:10.1080/00207543.2014.922710

Feng, Y., D'Amours, S., & Beauregard, R. (2010). Simulation and performance evaluation of partially and fully integrated sales and operations planning. *International Journal of Production Research, 48*(19), 5859–5883. doi:10.1080/00207540903232789

Fernandes, R., Gouveia, B., & Pinho, C. (2013). Integrated inventory valuation in multi-echelon production/distribution systems. *International Journal of Production Research, 51*(9), 2578–2592. doi:10.1080/00207543.2012.737947

Fu, D., Ionescu, C. M., Aghezzaf, E. H., & De Keyser, R. (2016). A constrained EPSAC approach to inventory control for a benchmark supply chain system. *International Journal of Production Research, 54*(1), 232–250. doi:10.1080/00207543.2015.1070214

Geunes, J. (2018). Revisiting the two-stage EOQ/EPQ model with inelastic demand: Decentralisation and coordination. *International Journal of Production Research, 56*(1-2), 238–250. doi:10.1080/00207543.2017.1378829

Glock, C. H. (2012). A comparison of alternative delivery structures in a dual sourcing environment. *International Journal of Production Research, 50*(11), 3095–3114. doi:10.1080/00207543.2011.592160

Glock, C. H., & Ries, J. M. (2013). Reducing lead time risk through multiple sourcing: The case of stochastic demand and variable lead time. *International Journal of Production Research, 51*(1), 43–56. doi:10.1080/00207543.2011.644817

Gunasekaran, A., Subramanian, N., & Rahman, S. (2015). Supply chain resilience: Role of complexities and strategies. *International Journal of Production Research, 53*(22), 6809–6819. doi:10.1080/00207543.2015.1093667

Gunasekaran, A., Subramanian, N., & Yusuf, Y. (2018). Strategies and practices for inclusive manufacturing: Twenty-first-century sustainable manufacturing competitiveness. *International Journal of Computer Integrated Manufacturing, 31*(6), 490–493. doi:10.1080/0951192X.2018.1463664

Gupta, M., & Andersen, S. (2012). Revisiting local TOC measures in an internal supply chain: A note. *International Journal of Production Research, 50*(19), 5363–5371. doi:10.1080/00207543.2011.627389

Gupta, M., & Andersen, S. (2018). Throughput/inventory dollar-days: TOC-based measures for supply chain collaboration. *International Journal of Production Research*, 1–17.

Hammami, R., Frein, Y., & Bahli, B. (2017). Supply chain design to guarantee quoted lead time and inventory replenishment: Model and insights. *International Journal of Production Research*, *55*(12), 3431–3450. doi:10.1080/00207543.2016.1242799

Hauke, J., Lorscheid, I., & Meyer, M. (2017). Individuals and their interactions in demand planning processes: An agent-based, computational testbed. *International Journal of Production Research*, 1–15.

Hemmati, M., Fatemi Ghomi, S. M. T., & Sajadieh, M. S. (2017). Vendor managed inventory with consignment stock for supply chain with stock-and price-dependent demand. *International Journal of Production Research*, *55*(18), 5225–5242. doi:10.1080/00207543.2017.1296203

Hennet, J. C. (2009). A globally optimal local inventory control policy for multistage supply chains. *International Journal of Production Research*, *47*(2), 435–453. doi:10.1080/00207540802426458

Hofmann, E., & Rüsch, M. (2017). Industry 4.0 and the current status as well as future prospects on logistics. *Computers in Industry*, *89*, 23–34. doi:10.1016/j.compind.2017.04.002

Huang, J., & Song, J. (2017). Optimal inventory control with sequential online auction in agriculture supply chain: An agent-based simulation optimisation approach. *International Journal of Production Research*, 1–17.

Huang, J. Y., & Yao, M. J. (2013). On the optimal lot-sizing and scheduling problem in serial-type supply chain system using a time-varying lot-sizing policy. *International Journal of Production Research*, *51*(3), 735–750. doi:10.1080/00207543.2012.662604

Hwarng, H. B., Chong, C. S. P., Xie, N., & Burgess, T. F. (2005). Modelling a complex supply chain: Understanding the effect of simplified assumptions. *International Journal of Production Research*, *43*(13), 2829–2872.

Jaber, M. Y., Glock, C. H., & El Saadany, A. M. (2013). Supply chain coordination with emissions reduction incentives. *International Journal of Production Research*, *51*(1), 69–82. doi:10.1080/00207543.2011.651656

Jaber, M. Y., & Goyal, S. K. (2009). A basic model for co-ordinating a four-level supply chain of a product with a vendor, multiple buyers and tier-1 and tier-2 suppliers. *International Journal of Production Research, 47*(13), 3691–3704. doi:10.1080/00207540701805604

Kaku, I., Gong, J., Tang, J., & Yin, Y. (2009). Modeling and numerical analysis of line-cell conversion problems. *International Journal of Production Research, 47*(8), 2055–2078. doi:10.1080/00207540802275889

Kellar, G. M., Polak, G. G., & Zhang, X. (2016). Synchronization, cross-docking, and decoupling in supply chain networks. *International Journal of Production Research, 54*(9), 2585–2599. doi:10.1080/00207543.2015.1107195

Kulkarni, S., Ponnaiyan, S., & Tarakci, H. (2015). Optimal ordering decisions under two returns policies. *International Journal of Production Research, 53*(12), 3720–3734. doi:10.1080/00207543.2014.988883

Kurata, H. (2014). How does inventory pooling work when product availability influences customers' purchasing decisions? *International Journal of Production Research, 52*(22), 6739–6759. doi:10.1080/00207543.2014.916825

Lackes, R., Schlüter, P., & Siepermann, M. (2016). The impact of contract parameters on the supply chain performance under different power constellations. *International Journal of Production Research, 54*(1), 251–264. doi:10.1080/00207543.2015.10 76943

Li, M., & Li, T. (2017). Consumer search, transshipment, and bargaining power in a supply chain. *International Journal of Production Research*, 1–16.

Li, Q., & Disney, S. M. (2017). Revisiting rescheduling: MRP nervousness and the bullwhip effect. *International Journal of Production Research, 55*(7), 1992–2012. doi:10.1080/00207543.2016.1261196

Li, R., & Ryan, J. K. (2012). Inventory flexibility through adjustment contracts. *International Journal of Production Research, 50*(7), 1955–1978. doi:10.1080/00 207543.2011.564667

Li, Y., Guo, H., & Zhang, Y. (2017). An integrated location-inventory problem in a closed-loop supply chain with third-party logistics. *International Journal of Production Research*, 1–20.

Lin, W. J., Jiang, Z. B., Liu, R., & Wang, L. (2014). The bullwhip effect in hybrid supply chain. *International Journal of Production Research, 52*(7), 2062–2084. do i:10.1080/00207543.2013.849013

Liu, X., & Çetinkaya, S. (2011). The supplier–buyer integrated production-inventory model with random yield. *International Journal of Production Research*, *49*(13), 4043–4061. doi:10.1080/00207543.2010.485588

Luthra, S., & Mangla, S. K. (2018). Evaluating challenges to Industry 4.0 initiatives for supply chain sustainability in emerging economies. *Process Safety and Environmental Protection*, *117*, 168–179. doi:10.1016/j.psep.2018.04.018

Ma, Y., Wang, N., Che, A., Huang, Y., & Xu, J. (2013). The bullwhip effect on product orders and inventory: A perspective of demand forecasting techniques. *International Journal of Production Research*, *51*(1), 281–302. doi:10.1080/00207543.2012.676682

Manatkar, R. P., Karthik, K., Kumar, S. K., & Tiwari, M. K. (2016). An integrated inventory optimization model for facility location-allocation problem. *International Journal of Production Research*, *54*(12), 3640–3658. doi:10.1080/00207543.2015.1120903

Martin, P. R., & Patterson, J. W. (2009). On measuring company performance within a supply chain. *International Journal of Production Research*, *47*(9), 2449–2460. doi:10.1080/00207540701725604

Maxwell, A., & Lal, S. (2013). Technological innovations in managing challenges of supply chain management. *Universal Journal of Industrial and Business Management*, *1*(2), 62–69.

Meixell, M. J., & Wu, S. D. (2005). Demand propagation in the extended enterprise: a comparative analysis of product and process design policies. *International Journal of Production Research, 43*(20), 4169-4189.

Merzifonluoglu, Y., & Feng, Y. (2014). Newsvendor problem with multiple unreliable suppliers. *International Journal of Production Research*, *52*(1), 221–242. doi:10.1080/00207543.2013.835497

Monthatipkul, C., & Kawtummachai, R. (2007). Algorithm for constructing a delivery-sequencing/inventory-allocation plan for supply chain control in the operational planning level. *International Journal of Production Research*, *45*(5), 1119–1139. doi:10.1080/00207540600690685

Nabelsi, V., & Gagnon, S. (2017). Information technology strategy for a patient-oriented, lean, and agile integration of hospital pharmacy and medical equipment supply chains. *International Journal of Production Research*, *55*(14), 3929–3945. doi:10.1080/00207543.2016.1218082

Nachiappan, S. P., Gunasekaran, A., & Jawahar, N. (2007). Knowledge management system for operating parameters in two-echelon VMI supply chains. *International Journal of Production Research, 45*(11), 2479–2505. doi:10.1080/00207540601020478

Niranjan, T. T., Wagner, S. M., & Nguyen, S. M. (2012). Prerequisites to vendor-managed inventory. *International Journal of Production Research, 50*(4), 939–951. doi:10.1080/00207543.2011.556153

Otten, S., Krenzler, R., & Daduna, H. (2016). Models for integrated production-inventory systems: Steady state and cost analysis. *International Journal of Production Research, 54*(20), 6174–6191. doi:10.1080/00207543.2015.1082669

Pan, J. H., & Yang, M. F. (2008). Integrated inventory models with fuzzy annual demand and fuzzy production rate in a supply chain. *International Journal of Production Research, 46*(3), 753–770. doi:10.1080/00207540600898072

Persona, A., Grassi, A., & Catena, M. (2005). Consignment stock of inventories in the presence of obsolescence. *International Journal of Production Research, 43*(23), 4969–4988. doi:10.1080/00207540500216631

Pfohl, Y. B., & Kurnaz, T. (2015). *The impact of Industry 4.0. on the supply chain.* Paper presented at the International conference of Logistics (HICL) – 20, Innovations and Strategies for Logistics an Supply Chains.

Rekik, Y., & Sahin, E. (2012). Exploring inventory systems sensitive to shrinkage–analysis of a periodic review inventory under a service level constraint. *International Journal of Production Research, 50*(13), 3529–3546. doi:10.1080/00207543.2012.670919

Ruifeng, C., & Subramaniam, V. (2011). Performance evaluation for tandem multi-factory supply chains: An approximate solution. *International Journal of Production Research, 49*(11), 3285–3305. doi:10.1080/00207541003792243

Salzarulo, P. A., & Jacobs, F. R. (2014). The incremental value of central control in serial supply chains. *International Journal of Production Research, 52*(7), 1989–2006. doi:10.1080/00207543.2013.842014

Sheffi, Y. (2009). Strategy for being Competitive, INCAE. *Business Review (Federal Reserve Bank of Philadelphia), 1/8.*

Shu, J., Wu, T., & Zhang, K. (2015). Warehouse location and two-echelon inventory management with concave operating cost. *International Journal of Production Research, 53*(9), 2718–2729. doi:10.1080/00207543.2014.977456

Sillanpää, V., & Liesiö, J. (2018). Forecasting replenishment orders in retail: Value of modelling low and intermittent consumer demand with distributions. *International Journal of Production Research*, 1–18.

Sirias, D., & Mehra, S. (2005). Quantity discount versus lead time-dependent discount in an inter-organizational supply chain. *International Journal of Production Research*, *43*(16), 3481–3496. doi:10.1080/00207540500117995

Sourirajan, K., Ramachandran, B., & An, L. (2008). Application of control theoretic principles to manage inventory replenishment in a supply chain. *International Journal of Production Research*, *46*(21), 6163–6188.

Staff, M. H. L. (2015, September 16). How Will Industry 4.0 Affect the Supply Chain? *Material Handling & Logistics*.

Stephenson, W. (1935). Correlating persons instead of tests. *Character and Personality*, *4*(1), 17–24. doi:10.1111/j.1467-6494.1935.tb02022.x

Subramanian, P., Ramkumar, N., Narendran, T. T., & Ganesh, K. (2012). A technical note on 'Analysis of closed loop supply chain using genetic algorithm and particle swarm optimisation'. *International Journal of Production Research*, *50*(2), 593–602. doi:10.1080/00207543.2011.593348

Suwanruji, P., & Enns, S. T. (2006). Evaluating the effects of capacity constraints and demand patterns on supply chain replenishment strategies. *International Journal of Production Research*, *44*(21), 4607–4629. doi:10.1080/00207540500494527

Taleizadeh, A. A., Noori-Daryan, M., & Govindan, K. (2016). Pricing and ordering decisions of two competing supply chains with different composite policies: A Stackelberg game-theoretic approach. *International Journal of Production Research*, *54*(9), 2807–2836. doi:10.1080/00207543.2016.1154621

Taleizadeh, A. A., Noori-daryan, M., & Tavakkoli-Moghaddam, R. (2015). Pricing and ordering decisions in a supply chain with imperfect quality items and inspection under buyback of defective items. *International Journal of Production Research*, *53*(15), 4553–4582. doi:10.1080/00207543.2014.997399

Taxakis, K., & Papadopoulos, C. (2016). A design model and a production–distribution and inventory planning model in multi-product supply chain networks. *International Journal of Production Research*, *54*(21), 6436–6457. doi:10.1080/00207543.2016.1158882

Tayur, S. R. (2013). Planned spontaneity for better product availability. *International Journal of Production Research, 51*(23-24), 6844–6859. doi:10.1080/00207543.2013.774507

Thomas, A., Krishnamoorthy, M., Venkateswaran, J., & Singh, G. (2016). Decentralised decision-making in a multi-party supply chain. *International Journal of Production Research, 54*(2), 405–425. doi:10.1080/00207543.2015.1096977

Tsai, D. M. (2011). An optimal production and shipment policy for a single-vendor single-buyer integrated system with both learning effect and deteriorating items. *International Journal of Production Research, 49*(3), 903–922. doi:10.1080/00207540903473375

Tsao, Y. C. (2016). Designing a supply chain network for deteriorating inventory under preservation effort and trade credits. *International Journal of Production Research, 54*(13), 3837–3851. doi:10.1080/00207543.2016.1157272

Pasquali, V. (2015). Industrial Revolution 4.0. *Global Finance, 29*(5), 94.

Vahdani, B., Zandieh, M., & Roshanaei, V. (2011). A hybrid multi-stage predictive model for supply chain network collapse recovery analysis: A practical framework for effective supply chain network continuity management. *International Journal of Production Research, 49*(7), 2035–2060. doi:10.1080/00207540903289748

Vicente, J. J., Relvas, S., & Barbosa-Póvoa, A. P. (2017). Effective bullwhip metrics for multi-echelon distribution systems under order batching policies with cyclic demand. *International Journal of Production Research*, 1–27.

Villegas, F. A., & Smith, N. R. (2006). Supply chain dynamics: Analysis of inventory vs. order oscillations trade-off. *International Journal of Production Research, 44*(6), 1037–1054. doi:10.1080/00207540500338203

Viswanathan, S., Widiarta, H., & Piplani, R. (2007). Value of information exchange and synchronization in a multi-tier supply chain. *International Journal of Production Research, 45*(21), 5057–5074. doi:10.1080/00207540600930057

Wadhwa, S., Bibhushan, & Chan, F. T. S. (2009). Inventory performance of some supply chain inventory policies under impulse demands. *International Journal of Production Research, 47*(12), 3307–3332. doi:10.1080/00207540701689750

Wang, H., Li, S., & Luo, J. (2018). Optimal markdown pricing for holiday basket with customer valuation. *International Journal of Production Research*, 1–15.

Wang, S. J., Huang, C. T., Wang, W. L., & Chen, Y. H. (2010). Incorporating ARIMA forecasting and service-level based replenishment in RFID-enabled supply chain. *International Journal of Production Research, 48*(9), 2655–2677. doi:10.1080/00207540903564983

Wang, S. P., & Lee, W. (2016). A note on 'A two-level supply chain with consignment stock agreement and stock-dependent demand'. *International Journal of Production Research, 54*(9), 2750–2756. doi:10.1080/00207543.2015.1130872

Wong, H., Wikner, J., & Naim, M. (2009). Analysis of form postponement based on optimal positioning of the differentiation point and stocking decisions. *International Journal of Production Research, 47*(5), 1201–1224. doi:10.1080/00207540701549608

Wu, H. H., Huang, H. H., & Jen, W. T. (2012). A study of the elongated replenishment frequency of TOC supply chain replenishment systems in plants. *International Journal of Production Research, 50*(19), 5567–5581. doi:10.1080/00207543.2011.649803

Xu, L. D., Xu, E. L., & Li, L. (2018). Industry 4.0: State of the art and future trends. *International Journal of Production Research*, 1–22.

Yan, H., Tang, S. L., & Yen, G. (2008). Quick response procurement cost control strategy for fabric manufacturing. *International Journal of Production Research, 46*(17), 4769–4784. doi:10.1080/00207540701210037

Yang, D., Xiao, T., Choi, T. M., & Cheng, T. C. E. (2015). Optimal reservation pricing strategy for a fashion supply chain with forecast update and asymmetric cost information. *International Journal of Production Research*, 1–22.

Yang, Y., Pan, S., & Ballot, E. (2017). Mitigating supply chain disruptions through interconnected logistics services in the Physical Internet. *International Journal of Production Research, 55*(14), 3970–3983. doi:10.1080/00207543.2016.1223379

Yi, H., & Sarker, B. R. (2013). An optimal consignment stock production and replenishment policy with controllable lead time. *International Journal of Production Research, 51*(21), 6316–6335. doi:10.1080/00207543.2013.795668

Yin, Y., Stecke, K. E., & Li, D. (2018). The evolution of production systems from Industry 2.0 through Industry 4.0. *International Journal of Production Research, 56*(1-2), 848–861. doi:10.1080/00207543.2017.1403664

Yuan, K. F., & Gao, Y. (2010). Inventory decision-making models for a closed-loop supply chain system. *International Journal of Production Research, 48*(20), 6155–6187. doi:10.1080/00207540903173637

Yuan, K. F., Ma, S. H., He, B., & Gao, Y. (2015). Inventory decision-making models for a closed-loop supply chain system with different decision-making structures. *International Journal of Production Research*, *53*(1), 183–219. doi:10.1080/00207543.2014.946160

Zanoni, S., Ferretti, I., & Tang, O. (2006). Cost performance and bullwhip effect in a hybrid manufacturing and remanufacturing system with different control policies. *International Journal of Production Research*, *44*(18-19), 3847–3862. doi:10.1080/00207540600857375

Zanoni, S., & Jaber, M. Y. (2015). A two-level supply chain with consignment stock agreement and stock-dependent demand. *International Journal of Production Research*, *53*(12), 3561–3572. doi:10.1080/00207543.2014.980012

Zhang, L. H., Li, T., & Fan, T. J. (2018). Inventory misplacement and demand forecast error in the supply chain: Profitable RFID strategies under wholesale and buy-back contracts. *International Journal of Production Research*, 1–18. doi:10.1080/00207543.2018.1530479

Zhao, F., Wu, D., Liang, L., & Dolgui, A. (2016). Lateral inventory transshipment problem in online-to-offline supply chain. *International Journal of Production Research*, *54*(7), 1951–1963. doi:10.1080/00207543.2015.1070971

Compilation of References

Abdallah, T., Diabat, A., & Rigter, J. (2013). Investigating the option of installing small scale PVs on facility rooftops in a green supply chain. *International Journal of Production Economics*, *146*(2), 465–477. doi:10.1016/j.ijpe.2013.03.016

Abdel-Aal, M. A., Syed, M. N., & Selim, S. Z. (2017). Multi-product selective newsvendor problem with service level constraints and market selection flexibility. *International Journal of Production Research*, *55*(1), 96–117. doi:10.1080/00207543.2016.1195932

Abdel-Basset, M., Manogaran, G., El-Shahat, D., & Mirjalili, S. (2018). A hybrid whale optimization algorithm based on local search strategy for the permutation flow shop scheduling problem. *Future Generation Computer Systems*, *85*, 129–145. doi:10.1016/j.future.2018.03.020

Abeyratne, S., & Monfared, R. (2016). Blockchain Ready Manufacturing Supply Chain Using Distributed Ledger. *International Journal of Research in Engineering and Technology*, *05*(09), 1–10. doi:10.15623/ijret.2016.0509001

Agarwal, A., & Vrat, P. (2015). A TISM based bionic model of organizational excellence. *Global Journal of Flexible Systems Management*, *16*(4), 361–376. doi:10.100740171-015-0107-z

Ahmad-Reza Sadeghi, C. W., & Waidner, M. (2016). *Security and Privacy Challenges in*. Fraunhofer Institute for Secure Information Technology, Darmstadt, Germany.

Aksoy, A., Küçükoğlu, İ., Ene, S., & Öztürk, N. (2014). Integrated emission and fuel consumption calculation model for green supply chain management. *Procedia: Social and Behavioral Sciences*, *109*, 1106–1109. doi:10.1016/j.sbspro.2013.12.595

Akter, S., Fosso Wamba, S., & Dewan, S. (2017). Why PLS-SEM is suitable for complex modelling? An empirical illustration in big data analytics quality. *Production Planning and Control*, *28*(11–12), 1011–1021. doi:10.1080/09537287.2016.1267411

Al-Anzi, F. S., & Allahverdi, A. (2005). Using a hybrid evolutionary algorithm to minimize variance in response time for multimedia object requests. *Journal of Mathematical Modelling and Algorithms*, *4*(4), 435–453. doi:10.100710852-005-9027-9

Al-Busaidi, K. A., & Olfman, L. (2017). Knowledge sharing through inter-organizational knowledge sharing systems. *VINE Journal of Information and Knowledge Management Systems*, *47*(1), 110–136. doi:10.1108/VJIKMS-05-2016-0019

Al-Faiz & Mahameda. (2015). GPS-based Navigated Autonomous Robot. *International Journal of Emerging Trends in Engineering Research, 3*(4).

Alfares, H. K., & Attia, A. M. (2017). A supply chain model with vendor-managed inventory, consignment, and quality inspection errors. *International Journal of Production Research*, *55*(19), 5706–5727. doi:10.1080/00207543.2017.1330566

Allahverdi, A. (2000). Minimizing mean flowtime in a two-machine flowshop with sequence-independent setup times. *Computers & Operations Research*, *27*(2), 111–127. doi:10.1016/S0305-0548(99)00010-6

Allahverdi, A., & Al-Anzi, F. S. (2006). A branch-and-bound algorithm for three-machine flowshop scheduling problem to minimize total completion time with separate setup times. *European Journal of Operational Research*, *169*(3), 767–780. doi:10.1016/j.ejor.2004.07.074

Allahverdi, A., Ng, C., Cheng, T. E., & Kovalyov, M. Y. (2008). A survey of scheduling problems with setup times or costs. *European Journal of Operational Research*, *187*(3), 985–1032. doi:10.1016/j.ejor.2006.06.060

Álvarez-Rodríguez, D. A., Normey-Rico, J. E., & Flesch, R. C. C. (2017). Model predictive control for inventory management in biomass manufacturing supply chains. *International Journal of Production Research*, *55*(12), 3596–3608. doi:10.1080/00207543.2017.1315191

Angerhofer, B. J., & Angelides, M. C. (2000). System Dynamics Modelling in Supply Chain Management: Research Review. *Proceedings of the 2000 Winter Simulation Conference, 1*, 342–351. 10.1109/WSC.2000.899737

Ansari, Z. N., & Kant, R. (2017). A state-of-art literature review reflecting 15 years of focus on sustainable supply chain management. *Journal of Cleaner Production*, *142*, 2524–2543. doi:10.1016/j.jclepro.2016.11.023

Arabzad, S. M., Ghorbani, M., & Tavakkoli-Moghaddam, R. (2015). An evolutionary algorithm for a new multi-objective location-inventory model in a distribution network with transportation modes and third-party logistics providers. *International Journal of Production Research*, *53*(4), 1038–1050. doi:10.1080/00207543.2014.938836

Arc-net. (2018). *Arc-net and PwC join forces to use blockchain in combating food fraud*. Available at: https://arc-net.io/blog/arc-net-pwc-blockchain-food-fraud/

Armstrong, J. S., & Overton, T. S. (1977). Estimating Nonresponse Bias in Mail Surveys. *JMR, Journal of Marketing Research*, *14*(3), 396–402. doi:10.1177/002224377701400320

Aste, T., Tasca, P., & Di Matteo, T. (2017). Blockchain Technologies: The Foreseeable Impact on Society and Industry. *Computer*, *50*(9), 18–28. doi:10.1109/MC.2017.3571064

Authenticate. (2018). *Blockchain Technology for the Global Food Supply Network: Transformation or Hype?* Whitepaper.

Autry, C. W., Williams, B. D., & Golicic, S. (2014). Relational and process multiplexity in vertical supply chain triads: An exploration in the U.S. restaurant industry. *Journal of Business Logistics*, *35*(1), 52–70. doi:10.1111/jbl.12034

Bakal, I. S., & Geunes, J. (2010). Order timing strategies in a single-supplier, multi-retailer system. *International Journal of Production Research*, *48*(8), 2395–2412. doi:10.1080/00207540902810528

Bandaly, D., Satir, A., & Shanker, L. (2014). Integrated supply chain risk management via operational methods and financial instruments. *International Journal of Production Research*, *52*(7), 2007–2025. doi:10.1080/00207543.2013.844376

Barbour, R. S. (2008). *Introducing Qualitative Research: A student guide to the craft of doing qualittive research*. London: Sage.

Barney, J. B. (1991). Firm Resources and Sustained Competitive Advantage. *Journal of Management*, *17*(1), 99–120. doi:10.1177/014920639101700108

Battini, D., Gunasekaran, A., Faccio, M., Persona, A., & Sgarbossa, F. (2010). Consignment stock inventory model in an integrated supply chain. *International Journal of Production Research*, *48*(2), 477–500. doi:10.1080/00207540903174981

Baxter, P. (2008). Qualitative Case Study Methodology: Study Design and Implementation for Novice Researcher. *Qualitative Case Study Methodology: Study Design and Implementation*, *13*(4), 544–549.

Bechtsis, D., Tsolakis, N., Vlachos, D., & Srai, J. S. (2018). Intelligent Autonomous Vehicles in digital supply chains: A framework for integrating innovations towards sustainable value networks. *Journal of Cleaner Production*, *181*, 60–71. doi:10.1016/j.jclepro.2018.01.173

Belkhouche, F. (2009). Reactive path planning in a dynamic environment. *IEEE Transactions on Robotics*, *25*(4), 902–911. doi:10.1109/TRO.2009.2022441

Benchoufi, M., Porcher, R., & Ravaud, P. (2017). Blockchain protocols in clinical trials: Transparency and traceability of consent. *F1000 Research*, *6*, 66. doi:10.12688/f1000research.10531.1 PMID:29167732

Ben-Daya, M., Hassini, E., Hariga, M., & AlDurgam, M. M. (2013). Consignment and vendor managed inventory in single-vendor multiple buyers supply chains. *International Journal of Production Research*, *51*(5), 1347–1365. doi:10.1080/00207543.2012.662725

Berling, P., & Marklund, J. (2014). Multi-echelon inventory control: An adjusted normal demand model for implementation in practice. *International Journal of Production Research*, *52*(11), 3331–3347. doi:10.1080/00207543.2013.873555

Biswas, K., Muthukkumarasamy, V., & Tan, W. L. (2017). Blockchain Based Wine Supply Chain Traceability System. *Future Technologies Conference*.

Blowfield, M. (2003). Ethical Supply Chains in the Cocoa, Coffee and Tea Industries. *Greener Management International*, (43).

Bocek. (2017). Blockchains everywhere - a use-case of blockchains in the pharma supply-chain. *IFIP/IEEE Symposium on Integrated Network and Service Management (IM)*. 10.23919/INM.2017.7987376

Bogdan, R. C. (2007). *Qualitative research for education: An introduction to theories and methods* (5th ed.). Boston: Pearson.

Bookbinder, J. H., Gümüş, M., & Jewkes, E. M. (2010). Calculating the benefits of vendor managed inventory in a manufacturer-retailer system. *International Journal of Production Research*, *48*(19), 5549–5571. doi:10.1080/00207540903095434

Borade, A. B., Kannan, G., & Bansod, S. V. (2013). Analytical hierarchy process-based framework for VMI adoption. *International Journal of Production Research*, *51*(4), 963–978. doi:10.1080/00207543.2011.650795

Bounini, F., Gingras, D., & Pollart, H. (2016). Real Time Cooperative Localization for Autonomous Vehicles. *IEEE 19th International Conference on Intelligent Transportation Systems (ITSC),* 1186 – 1191. 10.1109/ITSC.2016.7795707

Bowersox, D.J., & Closs, D.J. (1996). *Logistical Management—The Integrated Supply Chain Process.* McGraw-Hill Companies Inc.

Bräuer, I., & Buscher, U. (2017). A note on 'Pricing and ordering decisions in a supply chain with imperfect quality items and inspection under buyback of defective items'. *International Journal of Production Research*, 1–6.

Braun, V., & Clarke, V. (2006). Using thematic analysis in psychology. *Qualitative Research in Psychology*, *3*(2), 77–101. doi:10.1191/1478088706qp063oa

Brian Hwarng, H., & Wang, Y. (2010). Shift detection and source identification in multivariate autocorrelated processes. *International Journal of Production Research*, *48*(3), 835–859. doi:10.1080/00207540802431326

Brikci, N. (2007, February). *A Guide to Using Qualitative Research Methodology.* Retrieved from Google: https://cloudfront.ualberta.ca/-/media/science/research-and-teaching/teaching/qualitative-research-methodology.pdf

Brody, P. (2017). *How blockchain is revolutionizing supply chain management.* Academic Press.

Brucker, P., Knust, S., & Wang, G. (2005). Complexity results for flow-shop problems with a single server. *European Journal of Operational Research*, *165*(2), 398–407. doi:10.1016/j.ejor.2004.04.010

Bryman. (2004). *Social Research Method* (2nd ed.). Oxford, UK: Oxford University Press.

Bryman, A., & Bell, E. (2011). *Business Research Methods*. New York: Oxford University Press.

Bussmann, O. (2017). *BankThink A public or private blockchain? New Ethereum project could mean both.* Available: https://www.americanbanker.com/opinion/a-public-or-private-blockchain-new-ethereum-project-could-mean-both

Cannella, S., & Ciancimino, E. (2010). On the bullwhip avoidance phase: Supply chain collaboration and order smoothing. *International Journal of Production Research*, *48*(22), 6739–6776. doi:10.1080/00207540903252308

Carter, C. A., Zhong, F., & Zhu, F. (2012). Advances in Chinese agriculture and its global implications. *Applied Economic Perspectives and Policy*, *34*(1), 1–36. doi:10.1093/aepp/ppr047

Carter, C. R., Rogers, D. S., & Choi, T. Y. (2015). Toward the theory of the supply chain. *The Journal of Supply Chain Management*, *51*(2), 89–97. doi:10.1111/jscm.12073

Carvalho, N., Chaim, O., Cazarini, E., & Gerolamo, M. (2018). Manufacturing in the fourth industrial revolution: A positive prospect in Sustainable Manufacturing. *Procedia Manufacturing*, *21*, 671–678. doi:10.1016/j.promfg.2018.02.170

Casey, M. J., & Wong, P. (2017). *Global Supply Chains Are About to Get Better, Thanks to Blockchain*. Available: https://hbr.org/2017/03/global-supply-chains-are-about-to-get-better-thanks-to-blockchain.

Cecere, L. (2014). Supply chain visibility in business networks. *Supply Chain Insights Report*, 1-19.

Cerdas, F., Juraschek, M., Thiede, S., & Herrmann, C. (2017). Life cycle assessment of 3D printed products in a distributed manufacturing system. *Journal of Industrial Ecology*, *21*(S1), S80–S93. doi:10.1111/jiec.12618

Chakraborty, A., Chatterjee, A. K., & Mateen, A. (2015). A vendor-managed inventory scheme as a supply chain coordination mechanism. *International Journal of Production Research*, *53*(1), 13–24. doi:10.1080/00207543.2014.921350

Chan, F. T., & Prakash, A. (2012). Inventory management in a lateral collaborative manufacturing supply chain: A simulation study. *International Journal of Production Research*, *50*(16), 4670–4685. doi:10.1080/00207543.2011.628709

Chatfield, D. C., Hayya, J. C., & Cook, D. P. (2013). Stockout propagation and amplification in supply chain inventory systems. *International Journal of Production Research*, *51*(5), 1491–1507. doi:10.1080/00207543.2012.697205

Chen, Shi, Ren, Yan, & Shi. (2017). A Blockchain-based Supply Chain Quality Management Framework. *The Fourteenth IEEE International Conference on e-Business Engineering*.

Chen, D., Heyer, S., Ibbotson, S., Salonitis, K., Steingrímsson, J. G., & Thiede, S. (2015). Direct digital manufacturing: Definition, evolution, and sustainability implications. *Journal of Cleaner Production*, *107*, 615–625. doi:10.1016/j.jclepro.2015.05.009

Cheng, T., Wang, G., & Sriskandarajah, C. (1999). One-operator–two-machine flowshop scheduling with setup and dismounting times. *Computers & Operations Research, 26*(7), 715–730. doi:10.1016/S0305-0548(98)00087-2

Chen, L. T., & Wei, C. C. (2012). Multi-period channel coordination in vendor-managed inventory for deteriorating goods. *International Journal of Production Research, 50*(16), 4396–4413. doi:10.1080/00207543.2011.592159

Chen, L., Ellis, S. C., & Suresh, N. (2016). A supplier development adoption framework using expectancy theory. *International Journal of Operations & Production Management, 36*(5), 592–615. doi:10.1108/IJOPM-09-2013-0413

Chen, R. Y. (2018). A traceability chain algorithm for artificial neural networks using T–S fuzzy cognitive maps in blockchain. *Future Generation Computer Systems, 80*, 198–210. doi:10.1016/j.future.2017.09.077

Chen, S. P., & Cheng, B. H. (2014). Optimal echelon stock policies for multi-stage supply chains in fuzzy environments. *International Journal of Production Research, 52*(11), 3431–3449. doi:10.1080/00207543.2014.885145

Chen, Y. (2018). Blockchain tokens and the potential democratization of entrepreneurship and innovation. *Business Horizons, 61*(4), 567–575. doi:10.1016/j.bushor.2018.03.006

Chibeles-Martins, N., Pinto-Varela, T., Barbosa-Póvoa, A. P., & Novais, A. Q. (2016). A multi-objective meta-heuristic approach for the design and planning of green supply chains-MBSA. *Expert Systems with Applications, 47*, 71–84. doi:10.1016/j.eswa.2015.10.036

China Ministry of Agriculture. (2017). *China Agricultural Yearbooks*. Beijing, China: China Agriculture Press.

Choi, T. Y., & Dooley, K. J. (2009). Supply networks: Theories and models. *The Journal of Supply Chain Management, 45*(3), 25–26. doi:10.1111/j.1745-493X.2009.03168.x

Choi, T. Y., Dooley, K. J., & Rungtusanatham, M. (2001). Supply networks and complex adaptive systems: Control versus emergence. *Journal of Operations Management, 19*(3), 351–366. doi:10.1016/S0272-6963(00)00068-1

Cho, J.-H., & Kim, Y.-T. (2017). Design of Automated Logistics Transportation Robot System with Fork-Type Lifter. *International Journal of Fuzzy Logic and Intelligent Systems, 17*(3), 177–186. doi:10.5391/IJFIS.2017.17.3.177

Chopra, S., & Sodhi, M. S. (2004). Supply-chain breakdown. *MIT Sloan Management Review, 46*(1), 53–61.

Choudhary, D., & Shankar, R. (2015). The value of VMI beyond information sharing under time-varying stochastic demand. *International Journal of Production Research, 53*(5), 1472–1486. doi:10.1080/00207543.2014.951093

Chou, S. Y., & Pearson, J. M. (2012). Organizational citizenship behaviour in IT professionals: An expectancy theory approach. *Management Research Review, 35*(12), 1170–1186. doi:10.1108/01409171211281282

Clemons, E. K., Jin, F., Wilson, J., Ren, F., Matt, C., Hess, T., & Koh, N. (2013, January). The role of trust in successful ecommerce websites in China: Field observations and experimental studies. In *System Sciences (HICSS), 2013 46th Hawaii International Conference on* (pp. 4002-4011). IEEE.

Clottey, T. (2016). Development and evaluation of a rolling horizon purchasing policy for cores. *International Journal of Production Research, 54*(9), 2780–2790. doi:10.1080/00207543.2016.1142133

Coalition, S. M. L. (2011, June). *Implementing 21st century smart manufacturing*. Workshop summary report.

Co, H. C., David, I., Feng, P., & Patuwo, E. (2012). A continuous-review model for dual intercontinental and domestic outsourcing. *International Journal of Production Research, 50*(19), 5460–5473. doi:10.1080/00207543.2011.638941

Conde, L., Chelim, R., & Nunes, U. (2015). Collaborative Vehicle Self- Localization Using Multi-GNSS Receivers and V2V/V2I Communications. *IEEE 18th International Conference on Intelligent Transportation Systems*, 2525 – 2532. 10.1109/ITSC.2015.406

Corno, F., & Razzak, F. (2012). Intelligent energy optimization for user intelligible goals in smart home environments. *IEEE Transactions on Smart Grid, 3*(4), 2128–2135. doi:10.1109/TSG.2012.2214407

Creasey, S. (2017). *The potential for blockchain in the food industry - just-food global news*. Aroq Limited.

Cui, K., & Shoemaker, S. (2018). *A look at food security in China*. Academic Press.

Dan Craigen, N. D.-T. (2014, October). Defining Cybersecurity. *Science Direct*, 1-9.

Danese, P. (2006). Collaboration forms, information and communication technologies, and coordination mechanisms in CPFR. *International Journal of Production Research*, *44*(16), 3207–3226. doi:10.1080/00207540600557991

Danese, P. (2011). Towards a contingency theory of collaborative planning initiatives in supply networks. *International Journal of Production Research*, *49*(4), 1081–1103. doi:10.1080/00207540903555510

Darvish, M., Larrain, H., & Coelho, L. C. (2016). A dynamic multi-plant lot-sizing and distribution problem. *International Journal of Production Research*, *54*(22), 6707–6717. doi:10.1080/00207543.2016.1154623

Dasgupta & Das. (2015). A Discrete Inter-Species Cuckoo Search for flowshop scheduling problems. *Computers & Operations Research*, *60*, 111-120.

Davies, G. (1992). The two ways in which retailers can be brands. *International Journal of Retail & Distribution Management*, *20*(2). doi:10.1108/09590559210009312

De Sensi, G., Longo, F., & Mirabelli, G. (2008). Inventory policies analysis under demand patterns and lead times constraints in a real supply chain. *International Journal of Production Research*, *46*(24), 6997–7016. doi:10.1080/00207540701528776

DecisionNext. (2018). *The Missing Link in the Food Chain: Blockchain*. Whitepaper.

Declan, E. (2017). *What Jack Ma's New Retail looks like in action*. Available at: https://www.enterpriseinnovation.net/article/what-jack-mas-new-retail-looks-action-967015705

Dellino, G., Laudadio, T., Mari, R., Mastronardi, N., & Meloni, C. (2017). A reliable decision support system for fresh food supply chain management. *International Journal of Production Research*, 1–28.

Deloitte. (2015). *Challenges and Solutions for the Digital Transformation and use exponential technologies*. Zurich: The Creative Studio at Deloitte.

Dickson, B. (2016). *Blockchain has the potential to revolutionize the supply chain*. Aol Tech.

Ding, Lannes, Zhu, Gao, Peng, Song, & Jiang. (2018, March). *Embracing China's New Retail*. Bain report.

Dolgui, A., Tiwari, M. K., Sinjana, Y., Kumar, S. K., & Son, Y. J. (2018). Optimising integrated inventory policy for perishable items in a multi-stage supply chain. *International Journal of Production Research*, *56*(1-2), 902–925. doi:10.1080/00207543.2017.1407500

Dubey, R., Gunasekaran, A., Wamba, S. F., & Bag, S. (2015). Building theory of green supply chain management using Total Interpretive Structural Modeling (TISM). *IFAC-PapersOnLine*, *48*(3), 1688–1694. doi:10.1016/j.ifacol.2015.06.329

Earls, A. R. (2016). *Blockchain not a panacea for supply chain traceability, transparency*. Available: https://searcherp.techtarget.com/feature/Blockchain-not-a-panacea-for-supply-chain-traceability-transparency

Echelmeyer, W., Kirchheim, A., & Wellbrock, E. (2008). Robotics-logistics: Challenges for automation of logistic processes. *2008 IEEE International Conference on Automation and Logistics*, 2099-2103, 10.1109/ICAL.2008.4636510

Edwards, N. (2017). Blockchain meets the supply chain [online]. *MHD Supply Chain Solutions*, *47*(4), 48–50.

European Commission. (2015). *Factories of the Future*. Retrieved from http://ec.europa.eu/research/industrial_technologies/factories-of-thefuture_en.html

FAO. (2011). Global food losses and food waste – Extent, causes and prevention. Rome: FAO.

Fattahi, M., Mahootchi, M., Moattar Husseini, S. M., Keyvanshokooh, E., & Alborzi, F. (2015). Investigating replenishment policies for centralised and decentralised supply chains using stochastic programming approach. *International Journal of Production Research*, *53*(1), 41–69. doi:10.1080/00207543.2014.922710

Fawcett, A. M., Fawcett, S. E., Cooper, M. B., & Daynes, K. S. (2014). *Moments of angst: A critical incident approach to designing customer-experience value systems* (Vol. 21). Benchmarking. doi:10.1108/BIJ-09-2012-0059

Fearne, A., Hornibrook, S., & Dedman, S. (2001). The management of perceived risk in the food supply chain: A comparative study of retailer-led beef quality assurance schemes in Germany and Italy. *The International Food and Agribusiness Management Review*, *4*(1), 19–36. doi:10.1016/S1096-7508(01)00068-4

Feng, Y., D'Amours, S., & Beauregard, R. (2010). Simulation and performance evaluation of partially and fully integrated sales and operations planning. *International Journal of Production Research*, *48*(19), 5859–5883. doi:10.1080/00207540903232789

Fernandes, R., Gouveia, B., & Pinho, C. (2013). Integrated inventory valuation in multi-echelon production/distribution systems. *International Journal of Production Research*, *51*(9), 2578–2592. doi:10.1080/00207543.2012.737947

Fernández-Caramés, T. M., & Fraga-Lamas, P. (2018). A Review on Human-Centered IoT-Connected Smart Labels for the Industry 4.0. *IEEE Access: Practical Innovations, Open Solutions*, *6*, 25939–25957. doi:10.1109/ACCESS.2018.2833501

Fischer, E. A. (2016, August 12). Cybersecurity Issues and Challenges. *Congressional Research Service*, 12.

Fondrevelle, J., Allahverdi, A., Oulamara, A., & Portmann, M. C. (2005). Permutation flowshops with exact time lags to minimize maximum lateness. Nancy, France: Academic Press.

Food Standards Agency. (2018). *FSA trials first use of blockchain*. Available at: https://www.food.gov.uk/news-alerts/news/fsa-trials-first-use-of-blockchain

Ford, S., & Despeisse, M. (2016). Additive manufacturing and sustainability: An exploratory study of the advantages and challenges. *Journal of Cleaner Production*, *137*, 1573–1587. doi:10.1016/j.jclepro.2016.04.150

Fornell, C., & Lacker, D. F. (1981). Evaluating structural equation models with unobservable variables and measurement error. *JMR, Journal of Marketing Research*, *18*(1), 39–50. doi:10.1177/002224378101800104

Fosso Wamba, S., Kamdjoug, K., Robert, J., Bawack, R., & Keogh, J. (2018, June). Bitcoin, Blockchain, and FinTech: A Systematic Review and Case Studies in the Supply Chain Bitcoin, Blockchain, and FinTech : A Systematic Review and Case Studies in the Supply Chain. *Production Planning and Control*.

Fosso-Wamba, S., & Queiroz, M. M. (2018). *Call for Papers for the Theme special section: Blockchain in the Operations and Supply Chain Management*. Retrieved from https://www.journals.elsevier.com/international-journal-of-information-management/call-for-papers/blockchain-in-the-operations-and-supply

Francisco, K., & Swanson, D. (2018). The supply chain has no clothes: Technology adoption of blockchain for supply chain transparency. *Logistics*, *2*(1), 2. doi:10.3390/logistics2010002

Francisco, K., & Swanson, D. (2018). The Supply Chain Has No Clothes: Technology Adoption of Blockchain for Supply Chain Transparency. *Logistics*, *2*(2), 1–13.

Frey, C. B., & Osborne, M. A. (2017). The future of employment: How susceptible are jobs to computerisation? *Technological Forecasting and Social Change*, *114*, 254–280. doi:10.1016/j.techfore.2016.08.019

Fridgen, G., Radszuwill, S., Urbach, N., & Utz, L. (2017). Cross-Organizational Workflow Management Using Blockchain Technology – Towards Applicability, Auditability, and Automation. *51st Annual Hawaii International Conference on System Sciences.*

Fu, D., Ionescu, C. M., Aghezzaf, E. H., & De Keyser, R. (2016). A constrained EPSAC approach to inventory control for a benchmark supply chain system. *International Journal of Production Research, 54*(1), 232–250. doi:10.1080/00207543.2015.1070214

G. K. Saini, M. N. Halgamuge, P. Sharma, and J. Purkis, "A Review on Cyberattacks: Security Threats and Solution Techniques for Different Applications", Secure Cyber-Physical Systems for Smart Cities, IGI Global.

Gao, W., Zhang, Y., Ramanujan, D., Ramani, K., Chen, Y., Williams, C. B., ... Zavattieri, P. D. (2015). The status, challenges, and future of additive manufacturing in engineering. *Computer Aided Design, 69*, 65–89. doi:10.1016/j.cad.2015.04.001

Garey, M. R., Johnson, D. S., & Sethi, R. (1976). The complexity of flowshop and jobshop scheduling. *Mathematics of Operations Research, 1*(2), 117–129. doi:10.1287/moor.1.2.117

Garrone, P., Melacini, M., & Perego, A. (2014). Opening the black box of food waste reduction. *Food Policy, 46*, 129–139. doi:10.1016/j.foodpol.2014.03.014

Gebler, M., Uiterkamp, A. J., & Visser, C. (2014). A global sustainability perspective on 3D printing technologies. *Energy Policy, 74*, 158–167. doi:10.1016/j.enpol.2014.08.033

Geunes, J. (2018). Revisiting the two-stage EOQ/EPQ model with inelastic demand: Decentralisation and coordination. *International Journal of Production Research, 56*(1-2), 238–250. doi:10.1080/00207543.2017.1378829

Ghadge, A., Karantoni, G., Chaudhuri, A., & Srinivasan, A. (2018). Impact of additive manufacturing on aircraft supply chain performance: A system dynamics approach. *Journal of Manufacturing Technology Management, 29*(5), 846–865. doi:10.1108/JMTM-07-2017-0143

Ghodsypour, S. H., & O'Brien, C. (2001). The total cost of logistics in supplier selection, under conditions of multiple sourcing, multiple criteria and capacity constraint. *International Journal of Production Economics, 73*(1), 15–27. doi:10.1016/S0925-5273(01)00093-7

Gibson, I., Rosen, D., & Stucker, B. (2015). *Additive Manufacturing–3D Printing, Rapid Prototyping, and Direct Digital Manufacturing*. Springer.

Gizem Öğütçü, Ö. M. (2016, February). Analysis of personal information security behavior and awareness. *Science Direct, 56*, 83-93. Retrieved from https://www.sciencedirect.com/science/article/pii/S0167404815001406

Glass, C. A., Shafransky, Y. M., & Strusevich, V. A. (2000). Scheduling for parallel dedicated machines with a single server. *Naval Research Logistics, 47*(4), 304–328. doi:10.1002/(SICI)1520-6750(200006)47:4<304::AID-NAV3>3.0.CO;2-1

Glock, C. H. (2012). A comparison of alternative delivery structures in a dual sourcing environment. *International Journal of Production Research, 50*(11), 3095–3114. doi:10.1080/00207543.2011.592160

Glock, C. H., & Ries, J. M. (2013). Reducing lead time risk through multiple sourcing: The case of stochastic demand and variable lead time. *International Journal of Production Research, 51*(1), 43–56. doi:10.1080/00207543.2011.644817

Golicic, S. L., & Mentzer, J. T. (2005). Exploring the drivers of interorganizational relationship magnitude. *Journal of Business Logistics, 26*(2), 47–71. doi:10.1002/j.2158-1592.2005.tb00205.x

Good Logistics. (2017). *Blockchain Next on Food Supply Chain Menu*. Available at: https://www.foodlogistics.com/technology/article/12382466/blockchain-next-on-food-supply-chain-menu

Goutam, R. K. (2015). Importance of Cyber Security. *International Journal of Computers and Applications, 17*. Retrieved from https://pdfs.semanticscholar.org/5cfb/7a5bd2e6c181e8a69ebd49b1dadb795f493b.pdf

gov.uk. (2018). *China lifts ban on exports of British beef*. Available at: https://www.gov.uk/government/news/china-lifts-ban-on-exports-of-british-beef

Govindan, K., Jafarian, A., Khodaverdi, R., & Devika, K. (2014). Two-echelon multiple-vehicle location–routing problem with time windows for optimization of sustainable supply chain network of perishable food. *International Journal of Production Economics, 152*, 9–28. doi:10.1016/j.ijpe.2013.12.028

Govindan, K., Jafarian, A., & Nourbakhsh, V. (2015). Bi-objective integrating sustainable order allocation and sustainable supply chain network strategic design with stochastic demand using a novel robust hybrid multi-objective metaheuristic. *Computers & Operations Research, 62*, 112–130. doi:10.1016/j.cor.2014.12.014

Grabowski, J., & Wodecki, M. (2004, September). A very fast tabu search algorithm for the permutation flow shop problem with makespan criterion. *Computers & Operations Research, 31*(11), 1891–1909. doi:10.1016/S0305-0548(03)00145-X

Graen, G. (1969). Instrumentality theory of work motivation: Some experimental results and suggested modifications. *The Journal of Applied Psychology, 53*(2), 1–25. doi:10.1037/h0027100 PMID:5797825

Grau, G., Kitsomboonloha, R., Swisher, S. L., Kang, H., & Subramanian, V. (2014). Printed transistors on paper: Towards smart consumer product packaging. *Advanced Functional Materials, 24*(32), 5067–5074. doi:10.1002/adfm.201400129

Grawe, S. J., Daugherty, P. J., & Ralston, P. M. (2015). Enhancing dyadic performance through boundary spanners and innovation: An assessment of service provider-customer relationships. *Journal of Business Logistics, 36*(1), 88–101. doi:10.1111/jbl.12077

Griehl, W., & Ruestem, D. (1970). Nylon-12-preparation, properties, and applications. *Industrial & Engineering Chemistry, 62*(3), 16–22. doi:10.1021/ie50723a005

Groenfeldt, T. (2017). *IBM And Maersk Apply Blockchain To Container Shipping*. Available: https://www.forbes.com/sites/tomgroenfeldt/2017/03/05/ibm-and-maersk-apply-blockchain-to-container-shipping/#47cdedb53f05

Gumasta, K., Chan, F. T., & Tiwari, M. (2012). An incorporated inventory transport system with two types of customers for multiple perishable goods. *International Journal of Production Economics, 139*(2), 678–686. doi:10.1016/j.ijpe.2012.06.020

Gunasekaran, A., Subramanian, N., & Rahman, S. (2015). Supply chain resilience: Role of complexities and strategies. *International Journal of Production Research, 53*(22), 6809–6819. doi:10.1080/00207543.2015.1093667

Gunasekaran, A., Subramanian, N., & Yusuf, Y. (2018). Strategies and practices for inclusive manufacturing: Twenty-first-century sustainable manufacturing competitiveness. *International Journal of Computer Integrated Manufacturing, 31*(6), 490–493. doi:10.1080/0951192X.2018.1463664

Gupta, M., & Andersen, S. (2012). Revisiting local TOC measures in an internal supply chain: A note. *International Journal of Production Research, 50*(19), 5363–5371. doi:10.1080/00207543.2011.627389

Gupta, M., & Andersen, S. (2018). Throughput/inventory dollar-days: TOC-based measures for supply chain collaboration. *International Journal of Production Research*, 1–17.

Gupta, V., Kapur, P. K., & Kumar, D. (2016). Modelling and measuring code smells in enterprise applications using TISM and two-way assessment. *International Journal of System Assurance Engineering and Management*, *7*(3), 332–340. doi:10.100713198-016-0460-0

Hackius, N., & Petersen, M. (2017). Blockchain in logistics and supply chain: trick or treat? In *Proceedings of the Hamburg International Conference of Logistics (HICL)* (pp. 3-18). epubli.

Hackman, J. R., & Porter, L. W. (1968). Expectancy theory predictions of work effectiveness. *Organizational Behavior and Human Performance*, *3*(4), 417–426. doi:10.1016/0030-5073(68)90018-4

Haddud, A., DeSouza, A., Khare, A., & Lee, H. (2017). Examining potential benefits and challenges associated with the Internet of Things integration in supply chains. *Journal of Manufacturing Technology Management*, *28*(8), 1055–1085. doi:10.1108/JMTM-05-2017-0094

Hair, J. F., Sarstedt, M., Hopkins, L., & Kuppelwieser, V. G. (2014). Partial least squares structural equation modeling (PLS-SEM): An emerging tool in business research. *European Business Review*, *26*. doi:10.1108/EBR-10-2013-0128

Hair, J. F. Jr, Hult, G. T. M., Ringle, C. M., & Sarstedt, M. (2017). *A Primer on Partial Least Squares Structural Equation Modeling. Long Range Planning* (2nd ed.; Vol. 46). Sage Publications. doi:10.1016/j.lrp.2013.01.002

Hamedi, M., Farahani, R. Z., Husseini, M. M., & Esmaeilian, G. R. (2009). A distribution planning model for natural gas supply chain: A case study. *Energy Policy*, *37*(3), 799–812. doi:10.1016/j.enpol.2008.10.030

Hammami, R., Frein, Y., & Bahli, B. (2017). Supply chain design to guarantee quoted lead time and inventory replenishment: Model and insights. *International Journal of Production Research*, *55*(12), 3431–3450. doi:10.1080/00207543.2016.1242799

Hammerich, T. (2018). *Blockchain and the Internet of Food*. Future of Agriculture.

Han, J., Choi, C. S., Park, W. K., Lee, I., & Kim, S. H. (2014). Smart home energy management system including renewable energy based on ZigBee and PLC. *IEEE Transactions on Consumer Electronics*, *60*(2), 198–202. doi:10.1109/TCE.2014.6851994

Hans de Bruijn, M. J. (2017, January). Building cybersecurity awareness: The need for evidence-based. *Science Direct*, *34*, 1-7. Retrieved from https://www.sciencedirect.com/science/article/pii/S0740624X17300540

Harris, I., Mumford, C. L., & Naim, M. M. (2014). A hybrid multi-objective approach to capacitated facility location with flexible store allocation for green logistics modeling. *Transportation Research Part E, Logistics and Transportation Review*, *66*, 1–22. doi:10.1016/j.tre.2014.01.010

Hasan, S. K., Sarker, R., Essam, D., & Cornforth, D. (2009). Memetic algorithms for solving job-shop scheduling problems. *Memetic Computing*, *1*(1), 69–83. doi:10.100712293-008-0004-5

Hasnah Haron, S. N. (2017). A Hanbook for Business Research Methods. Kuala Lumpur: Pearson Malaysia Sdn Bhd.

Hauke, J., Lorscheid, I., & Meyer, M. (2017). Individuals and their interactions in demand planning processes: An agent-based, computational testbed. *International Journal of Production Research*, 1–15.

Heard, B. R., Taiebat, M., Xu, M., & Miller, S. A. (2018). Sustainability implications of connected and autonomous vehicles for the food supply chain. *Resources, Conservation and Recycling*, *128*, 22–24. doi:10.1016/j.resconrec.2017.09.021

Héctor, Vis, & Roodbergen. (2014). Transport operations in container terminals: Literature overview, trends, research directions and classification scheme. *European Journal of Operational Research, 236*(1), 1-13. doi:10.1016/j.ejor.2013.11.023

Hemmati, M., Fatemi Ghomi, S. M. T., & Sajadieh, M. S. (2017). Vendor managed inventory with consignment stock for supply chain with stock-and price-dependent demand. *International Journal of Production Research*, *55*(18), 5225–5242. doi:10.1080/00207543.2017.1296203

Hennet, J. C. (2009). A globally optimal local inventory control policy for multistage supply chains. *International Journal of Production Research*, *47*(2), 435–453. doi:10.1080/00207540802426458

Heutger, M., & Kückelhaus, M. (2018). *Blockchain in Logistics*. Academic Press.

Hill, M. C., & Ackiss, T. D. (1945). The "Insight Interview" Approach to Race Relations. *The Journal of Social Psychology*, *21*(2), 197–208. doi:10.1080/00224545.1945.9714166

Hobbs, J.E. (2003). *Consumer demand for traceability*. Academic Press.

Hofmann, E., & Rüsch, M. (2017). Industry 4.0 and the current status as well as future prospects on logistics. *Computers in Industry*, *89*, 23–34. doi:10.1016/j.compind.2017.04.002

Holmström, J., & Gutowski, T. (2017). Additive manufacturing in operations and supply chain management: No sustainability benefit or virtuous knock-on opportunities? *Journal of Industrial Ecology*, *21*(1), S21–S24. doi:10.1111/jiec.12580

Hopkinson, N., & Dicknes, P. (2003). Analysis of rapid manufacturing—Using layer manufacturing processes for production. *Proceedings of the Institution of Mechanical Engineers. Part C, Journal of Mechanical Engineering Science*, *217*(1), 31–39. doi:10.1243/095440603762554596

Hosoda. (2011). Autonomous Moving Technology for Future Urban transport. *Hitachi Review, 60*(2).

Huang, J. Y., & Yao, M. J. (2013). On the optimal lot-sizing and scheduling problem in serial-type supply chain system using a time-varying lot-sizing policy. *International Journal of Production Research*, *51*(3), 735–750. doi:10.1080/00207543.2012.662604

Huang, J., & Song, J. (2017). Optimal inventory control with sequential online auction in agriculture supply chain: An agent-based simulation optimisation approach. *International Journal of Production Research*, 1–17.

Huang, Y., Wang, K., Zhang, T., & Pang, C. (2016). Green supply chain coordination with greenhouse gases emissions management: A game-theoretic approach. *Journal of Cleaner Production*, *112*, 2004–2014. doi:10.1016/j.jclepro.2015.05.137

Hung, S.-Y., Tsai, J. C.-A., & Chuang, C.-C. (2014). Investigating primary health care nurses' intention to use information technology: An empirical study in Taiwan. *Decision Support Systems*, *57*, 331–342. doi:10.1016/j.dss.2013.09.016

Hwarng, H. B., Chong, C. S. P., Xie, N., & Burgess, T. F. (2005). Modelling a complex supply chain: Understanding the effect of simplified assumptions. *International Journal of Production Research*, *43*(13), 2829–2872.

I. C. &. Lin. (2017). A Survey of Blockchain Security Issues and Challenges. *International Journal of Network Security*, *19* (5), 653-659.

IBM Blockchain. (2018). *One truth across the supply chain network*. Available at: https://www.ibm.com/blockchain/industries/supply-chain

IBM Institute for Business Value. (2017). *Fast forward Rethinking enterprises, ecosystems and economies with blockchains*. IBM.

IBM. (2016). *An example of using blockchain to increase safety and reveal product provenance in food supply chains*. Retrieved from https://www-03.ibm.com/press/us/en/photo/50817.wss

Ignall, E., & Schrage, L. (1965). Application of the branch and bound technique to some flow-shop scheduling problems. *Operations Research, 13*(3), 400–412. doi:10.1287/opre.13.3.400

Improving efficiency, generating better returns & tackling environmental impacts in beef supply chains. (2017). WRAP. Available at: http://www.wrap.org.uk/sites/files/wrap/Tackling%20hotspots%20in%20the%20beef%20value%20chain_Published%20March%202017%20(2).pdf

Insight Active Learning. (2014, November 26). *Cyber Security and Related Issues: Comprehensive Coverage*. Retrieved from InsightIAS: http://www.insightsonindia.com/2014/11/25/cyber-security-related-issues-comprehensive-coverage/

Iqbal, S. (2016, May 31). On Cloud Security Attacks: A Taxonomy and Instrusion Detection and Preventation as a Service. *Science Direct*, 98-120.

Itten, R., Frischknecht, R., & Stucki, M. (2012). *Life Cycle Inventories of Electricity Mixes and Grid*. Treeze Ltd. Retrieved from: www.lc-inventories.ch

Ivanov, Dolgui, Sokolov, Werner, & Ivanova. (2016). *A dynamic model and an algorithm for short-term supply chain scheduling in the smart factory industry 4.0*. Academic Press.

Ivanov, D., Dolgui, A., Sokolov, B., Werner, F., & Ivanova, M. (2016). A dynamic model and an algorithm for short-term supply chain scheduling in the smart factory industry 4.0. *International Journal of Production Research, 54*(2), 386–402. doi:10.1080/00207543.2014.999958

Jaber, M. Y., Glock, C. H., & El Saadany, A. M. (2013). Supply chain coordination with emissions reduction incentives. *International Journal of Production Research, 51*(1), 69–82. doi:10.1080/00207543.2011.651656

Jaber, M. Y., & Goyal, S. K. (2009). A basic model for co-ordinating a four-level supply chain of a product with a vendor, multiple buyers and tier-1 and tier-2 suppliers. *International Journal of Production Research, 47*(13), 3691–3704. doi:10.1080/00207540701805604

Jaime, C. (2016). The challenges of cybersecurity frameworks to protect data required for the development of advanced maintenance. *Science Direct*, 223.

Jansen, C. (2016). Developing an Operating Industrial Security Services to Mitigate Risks of Digitalization. *Science Direct*, 133-137.

Jansen, D.-I. D.-P. (2016). Developing and operating industrial security services to mitigate risks of digitalization. *Science Direct*, 133-137.

Järvensivu, T., & Törnroos, J. (2010). Case study research with moderate constructionism: Conceptualization and practical illustration. *Industrial Marketing Management*, *39*(1), 100–108. doi:10.1016/j.indmarman.2008.05.005

Jawahar, N., & Balaji, A. (2009). A genetic algorithm for the two-stage supply chain distribution problem associated with a fixed charge. *European Journal of Operational Research*, *194*(2), 496–537. doi:10.1016/j.ejor.2007.12.005

Jay Lee, B. B. (2016). Introduction to cyber manufacturing. *Science Direct*, 13.

Jayalakshmi, B., & Pramod, V. R. (2015). Total interpretive structural modeling (TISM) of the enablers of a flexible control system for industry. *Global Journal of Flexible Systems Management*, *16*(1), 63–85. doi:10.100740171-014-0080-y

Jayant, A., & Azhar, M. (2014). Analysis of the barriers for implementing green supply chain management (GSCM) practices: An interpretive structural modeling (ISM) approach. *Procedia Engineering*, *97*, 2157–2166. doi:10.1016/j.proeng.2014.12.459

Jena, J., Fulzele, V., Gupta, R., Sherwani, F., Shankar, R., & Sidharth, S. (2016). A TISM modeling of critical success factors of smartphone manufacturing ecosystem in India. *Journal of Advances in Management Research*, *13*(2), 203–224. doi:10.1108/JAMR-12-2015-0088

Jofred, P. & Öster, P. (2011). *CO2 Emissions from Freight Transport and the Impact of Supply Chain Management: A case study at Atlas Copco Industrial Technique*. Academic Press.

Johnson, E. (2018). *Blockchain: shaking up logistics*. Epicor.

Johnson, S. M. (1954). Optimal two-and three-stage production schedules with setup times included. *Naval Research Logistics*, *1*(1), 61–68. doi:10.1002/nav.3800010110

Jugder, N. (2016). *The thematic analysis of interview data: an approach used.* University of Leeds. Retrieved from http://hpp.education.leeds.ac.uk/wp-content/uploads/sites/131/2016/02/HPP2016-3-Jugder.pdf

Kakavand, H., Kost De Serves, N., & Chilton, B. (2016). *The Blockchain Revolution: An Analysis Of Regulation And Technology Related To Distributed Ledger Technologies*. Available at: http://www.fintechconnectlive.com/wp- content/uploads/2016/11/Luther-Systems-DLA-Piper-Article-on- Blockchain-Regulation-and-Technology-SK.pdf

Kaku, I., Gong, J., Tang, J., & Yin, Y. (2009). Modeling and numerical analysis of line-cell conversion problems. *International Journal of Production Research*, *47*(8), 2055–2078. doi:10.1080/00207540802275889

Kalyanmoy, D. (2001). *Multi objective optimization using evolutionary algorithms*. John Wiley and Sons.

Kamagaew, A., Stenzel, J., Nettsträter, A., & ten Hompel, M. (2011). Concept of Cellular Transport Systems in facility logistics. *The 5th International Conference on Automation, Robotics and Applications*, 40-45, 10.1109/ICARA.2011.6144853

Kamble, S., Gunasekaran, A., & Arha, H. (2018). Understanding the Blockchain technology adoption in supply chains-Indian context. *International Journal of Production Research*, *0*(0), 1–25. doi:10.1080/00207543.2018.1518610

Kang, H. S., Lee, J. Y., Choi, S., Kim, H., Park, J. H., Son, J. Y., ... Do Noh, S. (2016). Smart manufacturing: Past research, present findings, and future directions. *International Journal of Precision Engineering and Manufacturing-Green Technology*, *3*(1), 111–128. doi:10.100740684-016-0015-5

Katharina, M. D. G. (2015). *Qualification For The Actory Of The Future*. Acta Universitatis Cibniensis – Technical Series.

Kehoe, L., & Gindner, K. (2017). *When two chains combine, supply chain meets blockchain*. Academic Press.

Keliang Zhou, T. L. (2015). Industry 4.0: Towards Future Industrial. *Science Direct*, 2147 - 2152.

Kellar, G. M., Polak, G. G., & Zhang, X. (2016). Synchronization, cross-docking, and decoupling in supply chain networks. *International Journal of Production Research*, *54*(9), 2585–2599. doi:10.1080/00207543.2015.1107195

Kellens, K., Baumers, M., Gutowski, T. G., Flanagan, W., Lifset, R., & Duflou, J. R. (2017). Environmental Dimensions of Additive Manufacturing: Mapping Application Domains and Their Environmental Implications. *Journal of Industrial Ecology*, *21*(1). doi:10.1111/jiec.12629

Khadraoui, A. I. D. (2018). The Security and Traceability of Shared Information in the Process of Transportation of Dangerous Goods. *2018 9th IFIP International Conference on New Technologies, Mobility and Security (NTMS)*.

Khajavi, S. H., Partanen, J., & Holmström, J. (2014). Additive Manufacturing in the Spare Parts Supply Chain. *Computers in Industry*, *65*(1), 50–63. doi:10.1016/j.compind.2013.07.008

Kim, H. J., Lee, J., Park, G. L., Kang, M. J., & Kang, M. (2010). An efficient scheduling scheme on charging stations for smart transportation. In *Security-Enriched Urban Computing and Smart Grid* (pp. 274–278). Berlin, Germany: Springer. doi:10.1007/978-3-642-16444-6_35

Kim, H. M., & Laskowski, M. (2018). Toward an ontology-driven blockchain design for supply-chain provenance. *Intelligent Systems in Accounting, Finance & Management*, *25*(1), 18–27. doi:10.1002/isaf.1424

Knofius, N., van der Heijden, M. C., & Zijm, W. H. M. (2016). Selecting parts for additive manufacturing in service logistics. *Journal of Manufacturing Technology Management*, *27*(7), 915–931.

Kogan, K., & Levner, E. (1998). A polynomial algorithm for scheduling small-scale manufacturing cells served by multiple robots. *Computers & Operations Research*, *25*(1), 53–62. doi:10.1016/S0305-0548(98)80008-7

Kohtala, C. (2015). Addressing sustainability in research on distributed production: An integrated literature review. *Journal of Cleaner Production*, *106*, 654–668. doi:10.1016/j.jclepro.2014.09.039

Koonce, L. (2017). *Blockchains & Food Security in the Supply Chain*. Academic Press.

Korpela & Hallikas. (2017). Digital supply chain transformation toward blockchain integration. *50th Hawaii international conference on system sciences.*

Korpela, K., Hallikas, J., & Dahlberg, T. (2017, January). Digital Supply Chain Transformation toward Blockchain Integration. *Proceedings of the 50th Hawaii International Conference on System Sciences*. 10.24251/HICSS.2017.506

Kshetri, N. (2017). 1 Blockchain's roles in meeting key supply chain management objectives. *International Journal of Information Management*, *39*, 80–89. doi:10.1016/j.ijinfomgt.2017.12.005

Kshetri, N. (2017). Blockchain's roles in strengthening cybersecurity and protecting privacy. *Telecommunications Policy*, *41*(10), 1027–1038. doi:10.1016/j.telpol.2017.09.003

Kulkarni, S., Ponnaiyan, S., & Tarakci, H. (2015). Optimal ordering decisions under two returns policies. *International Journal of Production Research*, *53*(12), 3720–3734. doi:10.1080/00207543.2014.988883

Kumar, H., Singh, M. K., Gupta, M. P., & Madaan, J. (2017). Smart neighbourhood: A TISM approach to reduce urban polarization for the sustainable development of smart cities. *Journal of Science and Technology Policy Management*.

Kurata, H. (2014). How does inventory pooling work when product availability influences customers' purchasing decisions? *International Journal of Production Research*, *52*(22), 6739–6759. doi:10.1080/00207543.2014.916825

Lackes, R., Schlüter, P., & Siepermann, M. (2016). The impact of contract parameters on the supply chain performance under different power constellations. *International Journal of Production Research*, *54*(1), 251–264. doi:10.1080/00207543.2015.1076943

Lambright, K. T. (2010). An update of a classic: Applying expectancy theory to understand contracted provider motivation. *Administration & Society*, *42*(4), 375–403. doi:10.1177/0095399710362714

Lamming, R.C., Caldwell, N., & Phillips, W. (2004). Supply chain transparency. *New et al*, 191-208.

Lane Thames, D. S. (2017). *Cyberrsecurity for Industry 4.0*. Springer Nature. doi:10.1007/978-3-319-50660-9

Lasi, H. F. (2014). Business & Information Systems Engineering. *Industry*, *4*(0), 239–242.

Law, A. (2017). *Smart Contracts and their Application in Supply Chain Management*. Cambridge, MA: Massachusetts Institute of Technology.

Lawrence, A., & Gordon, M. P. (2006). *Managing Cyber Security Resources A cost Benefit Analysis*. McGraw-Hills Companies.

Lee, J. (2013). *Industry 4.0 in Big Data Environment*. German Harting Magazine.

Lee, J., Kao, H. A., & Yang, S. (2014). Service innovation and smart analytics for industry 4.0 and big data environment. *Procedia Cirp*, *16*, 3–8. doi:10.1016/j.procir.2014.02.001

Lee, S. W., Sarp, S., Jeon, D. J., & Kim, J. H. (2015). Smart water grid: The future water management platform. *Desalination and Water Treatment*, *55*(2), 339–346. doi:10.1080/19443994.2014.917887

Lee, S., & Lee, S. (2013). Embedded Visual SLAM: Applications for Low-Cost Consumer Robots. *IEEE Robotics & Automation Magazine*, *20*(4), 83–95. doi:10.1109/MRA.2013.2283642

LeMay, S., Helms, M. M., Kimball, B., & McMahon, D. (2017). Supply chain management: The elusive concept and definition. *International Journal of Logistics Management*, *28*(4), 1425–1453. doi:10.1108/IJLM-10-2016-0232

Levner, E., Kogan, K., & Maimon, O. (1995). Flowshop scheduling of robotic cells with job-dependent transportation and set-up effects. *The Journal of the Operational Research Society*, *46*(12), 1447–1455. doi:10.1057/jors.1995.205

Li, G. J., Cheng, Y., & Hu, Y. (2017). Additive manufacturing technology in spare parts supply chain: A comparative study. *International Journal of Production Research*, *55*(5), 1498–1515. doi:10.1080/00207543.2016.1231433

Li, I.-H., Chen, M.-C., Wang, W.-Y., Su, S.-F., & Lai, T.-W. (2014). Mobile Robot Self-Localization System Using Single Webcam Distance Measurement Technology in Indoor Environments. *Sensors (Basel)*, *14*(2), 2089–2109. doi:10.3390140202089 PMID:24473282

Li, M., & Li, T. (2017). Consumer search, transshipment, and bargaining power in a supply chain. *International Journal of Production Research*, 1–16.

Lin, H.-F. (2017). Antecedents and consequences of electronic supply chain management diffusion: The moderating effect of knowledge sharing. *International Journal of Logistics Management*, *28*(2), 699–718. doi:10.1108/IJLM-01-2016-0023

Lin, W. J., Jiang, Z. B., Liu, R., & Wang, L. (2014). The bullwhip effect in hybrid supply chain. *International Journal of Production Research*, *52*(7), 2062–2084. doi:10.1080/00207543.2013.849013

Li, Q., & Disney, S. M. (2017). Revisiting rescheduling: MRP nervousness and the bullwhip effect. *International Journal of Production Research*, *55*(7), 1992–2012. doi:10.1080/00207543.2016.1261196

Li, R., & Ryan, J. K. (2012). Inventory flexibility through adjustment contracts. *International Journal of Production Research*, *50*(7), 1955–1978. doi:10.1080/00207543.2011.564667

Liserre, M., Sauter, T., & Hung, J. Y. (2010). Future energy systems: Integrating renewable energy sources into the smart power grid through industrial electronics. *IEEE Industrial Electronics Magazine*, *4*(1), 18–37. doi:10.1109/MIE.2010.935861

Liu, P., Huang, S. H., Mokasdar, A., Zhou, H., & Hou, L. (2014). The Impact of Additive Manufacturing in the Aircraft Spare Parts Supply Chain: Supply Chain Operation Reference (SCOR) Model Based Analysis. *Production Planning and Control, 25*(13–14), 1169–1181. doi:10.1080/09537287.2013.808835

Liu, W., Zhu, S., Mundie, T., & Krieger, U. (2017). Advanced Block-Chain Architecture for e-Health Systems. *19th International Conference on E-health Networking, Application & Services (HealthCom): The 2nd IEEE International Workshop on Emerging Technologies for Pervasive Healthcare and Applications (ETPHA 2017).* 10.1109/HealthCom.2017.8210847

Liu, X., & Çetinkaya, S. (2011). The supplier–buyer integrated production-inventory model with random yield. *International Journal of Production Research, 49*(13), 4043–4061. doi:10.1080/00207543.2010.485588

Li, Y., Guo, H., & Zhang, Y. (2017). An integrated location-inventory problem in a closed-loop supply chain with third-party logistics. *International Journal of Production Research*, 1–20.

Li, Z., Wang, W. M., Liu, G., Liu, L., He, J., & Huang, G. Q. (2018). Toward open manufacturing. *Industrial Management & Data Systems, 118*(1), 303–320. doi:10.1108/IMDS-04-2017-0142

Lo. (2017). Evaluating Suitability of Applying Blockchain. *22nd International Conference on Engineering of Complex Computer Systems (ICECCS).*

Lubowe, D., & McDermott, B. (2016). *Trust in trade, towards stronger supply chains.* Academic Press.

Lu, Q., & Xu, X. (2017). Adaptable Blockchain-Based Systems: A Case Study for Product Traceability. *IEEE Software, 34*(6), 21–27. doi:10.1109/MS.2017.4121227

Luthra, S., & Mangla, S. K. (2018). Evaluating challenges to Industry 4.0 initiatives for supply chain sustainability in emerging economies. *Process Safety and Environmental Protection, 117*, 168–179. doi:10.1016/j.psep.2018.04.018

M. &. Iansiti. (2017). The truth about blockchain. *Harvard Business Review, 95*(1), 118–127.

MacLean, L. M., Meyer, M., & Estable, A. (2004). Improving Accuracy of Transcript in Qualitative Research. *Qualitative Health Research, 14*(1), 113–123. doi:10.1177/1049732303259804 PMID:14725179

Mac, T. T., Copot, C., Tran, D. T., & De Keyser, R. (2016). Heuristic approaches in robot path planning: A survey. *Robotics and Autonomous Systems*, *86*, 13–28. doi:10.1016/j.robot.2016.08.001

Maloni, M. J., & Brown, M. E. (2006). Corporate social responsibility in the supply chain: An application in the food industry. *Journal of Business Ethics*, *68*(1), 35–52. doi:10.100710551-006-9038-0

Management, S. (2017a). *Case study: Walmart*. Available at: https://www.cips.org/en/supply-management/analysis/2017/june/case-study-walmart/

Management, S. (2017b). *Six things wrong with blockchain*. Available at: https://www.cips.org/en/supply-management/analysis/2017/december/six-things-wrong-with-blockchain-/

Management, S. (2018). *Blockchain for supply chains: From hype to reality*. Available at: https://www.cips.org/supply-management/analysis/2018/june-/blockchain-hype-reality/

Management, S. (2018). *Unilever and Sainsbury's testing blockchain for supply chain data*. Available at: https://www.cips.org/supply-management/news/2017/december/unilever-and-sainsburys-testing-blockchain-for-small-farmers/

Manatkar, R. P., Karthik, K., Kumar, S. K., & Tiwari, M. K. (2016). An integrated inventory optimization model for facility location-allocation problem. *International Journal of Production Research*, *54*(12), 3640–3658. doi:10.1080/00207543.2015.1120903

Manuel Dominguez, M. A. (2017). Cybersecurity training in control systems using real equipment. *Science Direct*, 12180-12184.

Maria Bada, J. R. (2015). Cyber Security Awareness Campaigns. *Research Gate*, *3*, 15.

Marinagi, C., Trivellas, P., & Sakas, D. P. (2014). The impact of information technology on the development of supply chain competitive advantage. *Procedia: Social and Behavioral Sciences*, *147*, 586–591. doi:10.1016/j.sbspro.2014.07.161

Marjo Heikkila, A. R. (2016). Security Challenges in Small- and Medium-Sized Manufacturing Enterprise. *Science Direct*, 23.

Martin, P. R., & Patterson, J. W. (2009). On measuring company performance within a supply chain. *International Journal of Production Research*, *47*(9), 2449–2460. doi:10.1080/00207540701725604

Material Handling and Logistics (MHL News). (2017). *Blockchain Can Improve Visibility in Food Supply Chain*. Available at: http://www.mhlnews.com/technology-automation/blockchain-can-improve-visibility-food-supply-chain

Maxwell, A., & Lal, S. (2013). Technological innovations in managing challenges of supply chain management. *Universal Journal of Industrial and Business Management*, *1*(2), 62–69.

Ma, Y., Wang, N., Che, A., Huang, Y., & Xu, J. (2013). The bullwhip effect on product orders and inventory: A perspective of demand forecasting techniques. *International Journal of Production Research*, *51*(1), 281–302. doi:10.1080/00207543.2012.676682

McAfee, R. B., Glassman, M., & Honeycutt, E. D. Jr. (2002). the Effects of Culture and Human Resource Management Policies. *Journal of Business Logistics*, *23*(1), 1–18. doi:10.1002/j.2158-1592.2002.tb00013.x

Medium. (2018). *The Amazon of China has Adopted Blockchain to Track high-end Beef Imports*. Available at: https://medium.com/@Michael_Spencer/the-amazon-of-china-has-adopted-blockchain-to-track-high-end-beef-imports-bf927671191a

Mehami, J., Nawi, M., & Ray, Y. (2018). Smart automated guided vehicles for manufacturing in the context of Industry 4.0. *Procedia Manufacturing*, *26*, 1077–1086. doi:10.1016/j.promfg.2018.07.144

Meixell, M. J., & Wu, S. D. (2005). Demand propagation in the extended enterprise: a comparative analysis of product and process design policies. *International Journal of Production Research, 43*(20), 4169-4189.

Mekhilef, S., Saidur, R., & Safari, A. (2011). A review on solar energy use in industries. *Renewable & Sustainable Energy Reviews*, *15*(4), 1777–1790. doi:10.1016/j.rser.2010.12.018

Mena, C., Adenso-Diaz, B., & Yurt, O. (2011). The causes of food waste in the supplier–retailer interface: Evidences from the UK and Spain. *Resources, Conservation and Recycling*, *55*(6), 648–658. doi:10.1016/j.resconrec.2010.09.006

Merzifonluoglu, Y., & Feng, Y. (2014). Newsvendor problem with multiple unreliable suppliers. *International Journal of Production Research*, *52*(1), 221–242. doi:10.1080/00207543.2013.835497

Michael Mylrea, S. N. (2017). An Introduction to Buildings Cybersecurity. *Google Scholar*, 978-982.

Microsoft. (2018). *How block chain will transform the modern supply chain*. Academic Press.

Minggao, Q. (2010). On sustainable coal mining in China. *Journal of China Coal Society*, *35*(4), 529–534.

Min, H. (2018). Blockchain technology for enhancing supply chain resilience. *Business Horizons*. doi:10.1016/j.bushor.2018.08.012

Mishra, N., & Singh, A. (2016). Use of twitter data for waste minimisation in beef supply chain. *Annals of Operations Research*, 1–23.

Mitchell, T. R., & Beach, L. R. (1977). Expectancy theory, decision theory, and occupational preference and choice. In *Human Judgment and Decision Processes in Applied Settings* (pp. 203–226). Elsevier. doi:10.1016/B978-0-12-397240-8.50016-1

Mohammed, A., & Wang, Q. (2017). Developing a meat supply chain network design using a multi-objective possibilistic programming approach. *British Food Journal*, *119*(3), 690–706. doi:10.1108/BFJ-10-2016-0475

Mohanty, S. P., Choppali, U., & Kougianos, E. (2016). Everything you wanted to know about smart cities: The internet of things is the backbone. *IEEE Consumer Electronics Magazine*, *5*(3), 60–70. doi:10.1109/MCE.2016.2556879

Monthatipkul, C., & Kawtummachai, R. (2007). Algorithm for constructing a delivery-sequencing/inventory-allocation plan for supply chain control in the operational planning level. *International Journal of Production Research*, *45*(5), 1119–1139. doi:10.1080/00207540600690685

Moreira, M. R. A., & Alves, R. A. F. S. (2009, May). A methodology for planning and controlling workload in a job-shop: A four-way decision-making problem. *International Journal of Production Research*, *47*(10), 2805–2821. doi:10.1080/00207540701725083

Mosconi, F. (2015). *The new European industrial policy: Global competitiveness and the manufacturing renaissance*. London, UK: Routledge. doi:10.4324/9781315761756

Mutchek, M., & Williams, E. (2014). Moving towards sustainable and resilient smart water grids. *Challenges*, *5*(1), 123–137. doi:10.3390/challe5010123

Mwasilu, F., Justo, J. J., Kim, E. K., Do, T. D., & Jung, J. W. (2014). Electric vehicles and smart grid interaction: A review on vehicle to grid and renewable energy sources integration. *Renewable & Sustainable Energy Reviews*, *34*, 501–516. doi:10.1016/j.rser.2014.03.031

Myers, M. D. (2008). *Qualitative Research in Business & Management*. SAGE Publications.

N. &. Hackius. (2017). Blockchain in logistics and supply chain: trick or treat? *Proceedings of the Hamburg International Conference of Logistics (HICL)*.

Nabelsi, V., & Gagnon, S. (2017). Information technology strategy for a patient-oriented, lean, and agile integration of hospital pharmacy and medical equipment supply chains. *International Journal of Production Research*, *55*(14), 3929–3945. doi:10.1080/00207543.2016.1218082

Nachiappan, S., Gunasekaran, A., & Jawahar, N. (2007). Knowledge management system for operating parameters in two-echelon VMI supply chains. *International Journal of Production Research*, *45*(11), 2479–2505. doi:10.1080/00207540601020478

Naik, S. (2017). Cyber Security - IoT. *IEEE International Conference On Recent Trends in Electronics Information & Communication Technology*, 764.

Nakamoto, S. (2008). *Bitcoin: A Peer-to-Peer Electronic Cash System*. Academic Press. doi:10.100710838-008-9062-0

Nakasumi, M. (2017). Information Sharing for Supply Chain Management based on Block Chain Technology. *IEEE 19th Conference on Business Informatics (CBI)*. 10.1109/CBI.2017.56

Nandi, A., & Rogers, P. (2004, March). Using simulation to make order acceptance/rejection decisions. *Simulation*, *80*(3), 131–142. doi:10.1177/0037549704045046

National center for postsecondary improvement. (2003). *National Center for Postsecondary Improvement*. Retrieved from https://web.stanford.edu/group/ncpi/unspecified/student_assess_toolkit/sampleInterviewProtocol.html

National Food Crime Unit. (2017). *Food Crime Annual Strategic Assessment - A 2016 Baseline*. Food Standards Agency and Food Standards Scotland.

Nawaz, M., Enscore, E. E. Jr, & Ham, I. (1983). A Heuristic Algorithm for the M-Machine, N-Job Flowshop Sequencing Problem. *Omega-International Journal of Management Science*, *11*(1), 91–95. doi:10.1016/0305-0483(83)90088-9

Niranjan, T. T., Wagner, S. M., & Nguyen, S. M. (2012). Prerequisites to vendor-managed inventory. *International Journal of Production Research*, *50*(4), 939–951. doi:10.1080/00207543.2011.556153

Nozick, L. K., & Turnquist, M. A. (2001). Inventory, transportation, service quality and the location of distribution centers. *European Journal of Operational Research, 129*(2), 362–371. doi:10.1016/S0377-2217(00)00234-4

Ntuli, N., & Abu-Mahfouz, A. M. (2016). *Simple security architecture for smart water management system*. Academic Press.

Nurjanni, K. P., Carvalho, M. S., & Da Costa, L. (2014). Green supply chain design with multi-objective optimization. *Proceedings of the 2014 International Conference on Industrial Engineering and Operations Management Bali*, 7-9.

O'Leary, D. E. (2008). Supporting decisions in real-time enterprises: Autonomic supply chain systems. *Information Systems and e-Business Management, 6*(3), 239–255. doi:10.100710257-008-0086-0

O'Leary, D. E. (2017). Configuring blockchain architectures for transaction information in blockchain consortiums: The case of accounting and supply chain systems. *Intell Sys Acc Fin Mgmt, 24*(4), 138–147. doi:10.1002/isaf.1417

Oh, J., & Shong, I. (2017). A case study on business model innovations using Blockchain: Focusing on financial institutions. *Asia Pacific Journal of Innovation and Entrepreneurship, 11*(3), 335–344. doi:10.1108/APJIE-12-2017-038

Oliver, D. G., Serovich, J. M., & Mason, T. L. (2005). Constraints and Opportunties with Interview Transcription: Towards Reflction in Qualitative Research. *Social Forces, 84*(2), 1273–1289. doi:10.1353of.2006.0023 PMID:16534533

Olmsted, B. A., & Davis, M. (Eds.). (2001). *Practical injection molding*. CRC Press. doi:10.1201/9781482294590

Osman, I. H., & Potts, C. N. (1989). Simulated Annealing for Permutation Flowshop Scheduling. *Omega-International Journal of Management Science, 17*(6), 551–557. doi:10.1016/0305-0483(89)90059-5

Otten, S., Krenzler, R., & Daduna, H. (2016). Models for integrated production-inventory systems: Steady state and cost analysis. *International Journal of Production Research, 54*(20), 6174–6191. doi:10.1080/00207543.2015.1082669

Paelke, V. (2014, September). Augmented reality in the smart factory: Supporting workers in an industry 4.0. Environment. In Emerging Technology and Factory Automation (ETFA), 2014 IEEE (pp. 1-4). IEEE.

Panda, S., Modak, N., Basu, M., & Goyal, S. (2015). Channel coordination and profit distribution in a social responsible three-layer supply chain. *International Journal of Production Economics, 168*, 224–233. doi:10.1016/j.ijpe.2015.07.001

Pan, J. H., & Yang, M. F. (2008). Integrated inventory models with fuzzy annual demand and fuzzy production rate in a supply chain. *International Journal of Production Research, 46*(3), 753–770. doi:10.1080/00207540600898072

Panwar, N. L., Kaushik, S. C., & Kothari, S. (2011). Role of renewable energy sources in environmental protection: A review. *Renewable & Sustainable Energy Reviews, 15*(3), 1513–1524. doi:10.1016/j.rser.2010.11.037

Papargyropoulou, E., Lozano, R., Steinberger, J.K., & Wright, N., & bin Ujang, Z. (2014). The food waste hierarchy as a framework for the management of food surplus and food waste. *Journal of Cleaner Production, 76*, 106–115. doi:10.1016/j.jclepro.2014.04.020

Parfitt, J., Barthel, M., & Macnaughton, S. (2010). Food waste within food supply chains: Quantification and potential for change to 2050. *Philosophical Transactions of the Royal Society of London. Series B, Biological Sciences, 365*(1554), 3065–3081. doi:10.1098/rstb.2010.0126 PMID:20713403

Pasquali, V. (2015). Industrial Revolution 4.0. *Global Finance, 29*(5), 94.

Patri, R., & Suresh, M. (2017). Modelling the enablers of agile performance in healthcare organization: A TISM approach. *Global Journal of Flexible Systems Management, 18*(3), 251–272. doi:10.100740171-017-0160-x

Paul van Schaik, D. J. (2017, October). Risk perceptions of cyber-security and precautionary behaviour. *Science Direct, 75*, 547-559. Retrieved from https://www.sciencedirect.com/science/article/pii/S074756321730359X

Peleg-Gillai, B., Bhat, G., & Sept, L. (2006). *Innovators in supply chain security: better security drives business value*. Washington, DC: Manufacturing Institute.

Penang skill development centre. (2017). *Introduction to the Industry 4.0.* Penang: Author.

Persona, A., Grassi, A., & Catena, M. (2005). Consignment stock of inventories in the presence of obsolescence. *International Journal of Production Research, 43*(23), 4969–4988. doi:10.1080/00207540500216631

Pfohl, Y. B., & Kurnaz, T. (2015). *The impact of Industry 4.0. on the supply chain.* Paper presented at the International conference of Logistics (HICL) – 20, Innovations and Strategies for Logistics an Supply Chains.

Pham, Halgamuge, Syed, & Mendis. (2010). Optimizing Windows Security Features to Block Malware and Hack Tools on USB Storage Devices. *The 28th PIERS*, 350-355.

Pham, D. V., Syed, A., & Halgamuge, M. N. (2011, February). Universal Serial Bus Based Software Attacks and Protection Solutions. *Digital Investigation*, *7*(3), 172–184. doi:10.1016/j.diin.2011.02.001

Pilkington, J. H. L. M. (2017). How the Blockchain Revolution Will Reshape the Consumer Electronics Industry [Future Directions]. *IEEE Consumer Electronics Magazine*, *6*(3), 19–23. doi:10.1109/MCE.2017.2684916

Pinedo, M. (2012). *Scheduling: theory, algorithms, and systems* (4th ed.). New York: Springer. doi:10.1007/978-1-4614-2361-4

Pishvaee, M. S., & Razmi, J. (2012). Environmental supply chain network design using multi-objective fuzzy mathematical programming. *Applied Mathematical Modelling*, *36*(8), 3433–3446. doi:10.1016/j.apm.2011.10.007

Podsakoff, P. M., & Organ, D. W. (1986). Self-Reports in Organizational Research: Problems and Prospects. *Journal of Management*, *12*(4), 531–544. doi:10.1177/014920638601200408

Ponelis, S. R. (2015). Using Interpretive Qualitative Case Studies for Exploratory Research in Doctoral Studies: A Case of Information System Research in Small and Medium Enterprise. *International Journal of Doctoral Studies*, *10*, 535–550. doi:10.28945/2339

Pop, P. C., Pintea, C.-M., Sitar, C. P., & Hajdu-Măcelaru, M. (2015). An efficient Reverse Distribution System for solving sustainable supply chain network design problem. *Journal of Applied Logic*, *13*(2), 105–113. doi:10.1016/j.jal.2014.11.004

Portofantwerp.com. (2017). *Antwerp start-up T-Mining develops Blockchain solution for safe, efficient container release.* Available: http://www.portofantwerp.com/en/news/antwerp-start-t-mining-develops-blockchain-solution-safe-efficient-container-release

Prybila, C., Schulte, S., Hochreiner, C., & Weber, I. (2017). Runtime verification for business processes utilizing the Bitcoin blockchain. *Future Generation Computer Systems*. doi:10.1016/j.future.2017.08.024

Qi, W., Liu, J., & Christofides, P. D. (2011). A distributed control framework for smart grid development: Energy/water system optimal operation and electric grid integration. *Journal of Process Control*, *21*(10), 1504–1516. doi:10.1016/j.jprocont.2011.05.010

Queiroz, M. M., & Fosso Wamba, S. (2019). Blockchain adoption challenges in supply chain: An empirical investigation of the main drivers in India and the USA. *International Journal of Information Management, 46*, 70–82. doi:10.1016/j.ijinfomgt.2018.11.021

Radhika, N., & Pramod, V. R. (2014). Modeling structural behaviour of inhibitors of cloud computing: A TISM approach. *Transactions on Networks and Communications, 2*(5), 60–74.

Rahimi-Eichi, H., Ojha, U., Baronti, F., & Chow, M. Y. (2013). Battery management system: An overview of its application in the smart grid and electric vehicles. *IEEE Industrial Electronics Magazine, 7*(2), 4–16. doi:10.1109/MIE.2013.2250351

Rahman, H. F., Sarker, R., & Essam, D. (2015). A genetic algorithm for permutation flow shop scheduling under make to stock production system. Computers & Industrial Engineering, 90, 12-24. doi:10.1016/j.cie.2015.08.006

Rahman, H., Sarker, R., & Essam, D. (2013). Permutation Flow Shop Scheduling with dynamic job order arrival. *Cybernetics and Intelligent Systems (CIS), IEEE Conference on*, 30-35. 10.1109/ICCIS.2013.6751574

Rahman, Sarker, & Essam. (2015). A real-time order acceptance and scheduling approach for permutation flow shop problems. *European Journal of Operational Research, 247*, 488-503.

Rahman, H. F., Sarker, R. A., & Essam, D. L. (2013). A Memetic Algorithm for Permutation Flow Shop Problems. *2013 IEEE Congress on Evolutionary Computation (CEC)*, 1618-1625. 10.1109/CEC.2013.6557755

Rahman, H. F., Sarker, R., & Essam, D. (2018). Multiple-order Permutation Flow Shop Scheduling under Process Interruptions. *International Journal of Advanced Manufacturing Technology, 97*(5-8), 2781–2808. doi:10.100700170-018-2146-z

Rajendran, C., & Ziegler, H. (2004, June 1). Ant-colony algorithms for permutation flowshop scheduling to minimize makespan/total flowtime of jobs. *European Journal of Operational Research, 155*(2), 426–438. doi:10.1016/S0377-2217(02)00908-6

Rajesh, R. (2017). Technological capabilities and supply chain resilience of firms: A relational analysis using Total Interpretive Structural Modeling (TISM). *Technological Forecasting and Social Change, 118*, 161–169. doi:10.1016/j.techfore.2017.02.017

Rau, H., Wu, M.-Y., & Wee, H.-M. (2003). Integrated inventory model for deteriorating items under a multi-echelon supply chain environment. *International Journal of Production Economics*, *86*(2), 155–168. doi:10.1016/S0925-5273(03)00048-3

Reddy, K. B., Narayanan, S. & Pandian, P. (2011). Single-Echelon Supply Chain Two Stage Distribution Inventory Optimization Model for the Confectionery Industry. *Journal of Applied Mathematical Sciences, 5*, 2491-2504.

Rekik, Y., & Sahin, E. (2012). Exploring inventory systems sensitive to shrinkage–analysis of a periodic review inventory under a service level constraint. *International Journal of Production Research*, *50*(13), 3529–3546. doi:10.1080/00207543.2012.670919

René Waslo, T. L. (2017, March 21). *Industry 4.0 and cybersecurity.* Retrieved from Deloitte Insight: https://dupress.deloitte.com/dup-us-en/focus/industry-4-0/cybersecurity-managing-risk-in-age-of-connected-production.html

Richey, R. G., Daugherty, P. J., & Roath, A. S. (2007). Firm Technological Readiness and Complementarity: Capabilities Impacting Logistics Service Competency and Performance. *Journal of Business Logistics*, *28*(1), 195–228. doi:10.1002/j.2158-1592.2007.tb00237.x

Ringle, C. M., Wende, S., & Becker, J.-M. (2015). *SmartPLS 3.* Bönningstedt: SmartPLS.

Ripe.io. (2018). *Ripe – Let's Talk About the Quality of our Food.* Available at: http://www.ripe.io

Risius, M., & Spohrer, K. (2017). A Blockchain Research Framework: What We (don't) Know, Where We Go from Here, and How We Will Get There. *Business & Information Systems Engineering*, *59*(6), 385–409. doi:10.100712599-017-0506-0

Roberts, J. (2012). Organizational ignorance: Towards a managerial perspective on the unknown. *Management Learning*, *44*(3), 215–236. doi:10.1177/1350507612443208

Roca, S. K.-L.-D.-V. (2019). *Cybersecurity Current Challenges and Inria's research directions.* Le Chesnay Cedex, France: Inria.

Rogers, E. (1983). Diffusion of Innovations (3rd ed.). New York: Free Press.

Rogers, E. (2003). Diffusion of Innovations (5th ed.). New York: Free Press.

Rogers, P., & Nandi, A. (2007, October). Judicious order acceptance and order release in make-to-order manufacturing systems. *Production Planning and Control*, *18*(7), 610–625. doi:10.1080/09537280701582422

Rostamzadeh, R., Govindan, K., Esmaeili, A., & Sabaghi, M. (2015). Application of fuzzy VIKOR for evaluation of green supply chain management practices. *Ecological Indicators*, *49*, 188–203. doi:10.1016/j.ecolind.2014.09.045

Ruffo, M., Tuck, C., & Hague, R. (2006). Cost Estimation for Rapid Manufacturing - Laser Sintering Production for Low to Medium Volumes. *Proceedings of the Institution of Mechanical Engineers. Part B, Journal of Engineering Manufacture*, *220*(9), l417–l1427. doi:10.1243/09544054JEM517

Ruggunan, D. S. (2015). Introduction to Qualitative Analysis: Thematic Analysis. *Discipline of Human Resources*, 1-38. Retrieved from http://smitg.ukzn.ac.za/Libraries/General_Docs/Introduction-to-Qualitative-Analysis.sflb.ashx

Ruifeng, C., & Subramaniam, V. (2011). Performance evaluation for tandem multi-factory supply chains: An approximate solution. *International Journal of Production Research*, *49*(11), 3285–3305. doi:10.1080/00207541003792243

Ruiz, R., & Maroto, C. (2005). A comprehensive review and evaluation of permutation flowshop heuristics. *European Journal of Operational Research*, *165*(2), 479–494. doi:10.1016/j.ejor.2004.04.017

Ruiz, R., Maroto, C., & Alcaraz, J. (2006, October). Two new robust genetic algorithms for the flowshop scheduling problem. *Omega-International Journal of Management Science*, *34*(5), 461–476. doi:10.1016/j.omega.2004.12.006

Russell, J. (2017). *Alibaba smashes its Single's Day record once again as sales cross $25 billion.* Available at: https://techcrunch.com/2017/11/11/alibaba-smashes-its-singles-day-record/?guccounter=1

Sabattini, Aikio, M., Beinschob, P., Boehning, M., Cardarelli, E., Digani, V., ... Fuerstenberg, K. (2018). The PAN-Robots Project: Advanced Automated Guided Vehicle Systems for Industrial Logistics. *IEEE Robotics & Automation Magazine*, *25*(1), 55–64. doi:10.1109/MRA.2017.2700325

Saberi, S., Kouhizadeh, M., Sarkis, J., & Shen, L. (2018). Blockchain technology and its relationships to sustainable supply chain management. *International Journal of Production Research*, 1–19.

Sadegheih, A. (2006). Scheduling problem using genetic algorithm, simulated annealing and the effects of parameter values on GA performance. *Applied Mathematical Modelling*, *30*(2), 147–154. doi:10.1016/j.apm.2005.03.017

Sadouskaya, K. (2017). *Adoption of Blockchain Technology in Supply Chain and Logistics*. Academic Press.

Saffar, M., & Razmi, J. (2015). A new multi objective optimization model for designing a green supply chain network under uncertainty. *International Journal of Industrial Engineering Computations*, *6*(1), 15–32. doi:10.5267/j.ijiec.2014.10.001

Sakpere, W., Adeyeye-Oshin, M., & Mlitwa, N. B. W. (2017). A state-of-the-art survey of indoor positioning and navigationsystems and technologies. *South African Computer Journal*, *29*(3), 145–197. doi:10.18489acj.v29i3.452

Salzarulo, P. A., & Jacobs, F. R. (2014). The incremental value of central control in serial supply chains. *International Journal of Production Research*, *52*(7), 1989–2006. doi:10.1080/00207543.2013.842014

Satyavolu, P., & Sangamnerkar, A. (2016). Blockchains Smart Contracts: Driving the Next Wave of Innovation Across Manufacturing Value Chains. *Cognizant 2020 Insignts*.

Saunders, M., & Tosey, P. (2013). *The layers of Research Design*. Academic Press.

Saunders, M., Lewis, P., & Thornhill, A. (2012). *Research methods for business students* (6th ed.). Harlow, UK: Pearson.

Saxena, S. J. S. (2018). *Harness Abundance the Blockchain Way In Supply Chain Management*. Academic Press.

Sazvar, Z., Rahmani, M., & Govindan, K. (2018). A sustainable supply chain for organic, conventional agro-food products: The role of demand substitution, climate change and public health. *Journal of Cleaner Production*, *194*, 564–583. doi:10.1016/j.jclepro.2018.04.118

Schary, P.B., & Skjt-Larsen, T. (2001). *Managing the Global Supply Chain*. Handelsh 鳶skolens forlag.

Schlick, J., Stephan, P., Loskyll, M., & Lappe, D. (2014). Industrie 4.0 in der praktischen Anwendung. In Bauernhansl, Industrie 4.0 in Produktion, Automatisierung und Logistik. Anwendung, Technologien und Migration (pp. 57–84). Academic Press.

Schoenherr, T., Griffith, D. A., & Chandra, A. (2014). Knowledge management in supply chains: The role of explicit and tacit knowledge. *Journal of Business Logistics*, *35*(2), 121–135. doi:10.1111/jbl.12042

Selen, W. J., & Hott, D. D. (1986, December). A mixed-integer goal-programming formulation of the standard flow-shop scheduling problem. *The Journal of the Operational Research Society*, *37*(12), 1121–1128. doi:10.1057/jors.1986.197

Shamsuddoha, M. (2015). Integrated Supply Chain Model for Sustainable Manufacturing: A System Dynamics Approach. *Advances in Business Marketing and Purchasing Sustaining Competitive Advantage Via Business Intelligence, Knowledge Management, and System Dynamics*, 155-399.

Sheffi, Y. (2009). Strategy for being Competitive, INCAE. *Business Review (Federal Reserve Bank of Philadelphia), 1/8.*

Shelzer, R. (2018, April 17). *Cybersecurity and Manufacturing: What's at Risk and How to Protect Your Facility*. Retrieved from Global Electronic Services, Inc: https://blog.gesrepair.com/cybersecurity-and-manufacturing-whats-at-risk-and-how-to-protect-your-facility/

Shen, M., Zhao, D., & Sun, J. (2016). Enhancement of low-cost GNSS localization in connected vehicle networks using Rao-Blackwellized particle filters. *IEEE 19th International Conference on Intelligent Transportation Systems (ITSC),* 834 – 840.

Sherly, J., & Somasundareswari, D. (2015). Internet of things based smart transportation systems. *International Research Journal of Engineering and Technology*, *2*(7), 1207–1210.

Shrouf, F., Ordieres, J., & Miragliotta, G. (2014, December). Smart factories in Industry 4.0: A review of the concept and of energy management approached in production based on the Internet of Things paradigm. In *Industrial Engineering and Engineering Management (IEEM), 2014 IEEE International Conference on* (pp. 697-701). IEEE.

Shu, J., Wu, T., & Zhang, K. (2015). Warehouse location and two-echelon inventory management with concave operating cost. *International Journal of Production Research*, *53*(9), 2718–2729. doi:10.1080/00207543.2014.977456

Siegwart, R., & Nourbakhsh, I. R. (2004). *Introduction to Autonomous Mobile Robots*. MIT Press.

Sillanpää, V., & Liesiö, J. (2018). Forecasting replenishment orders in retail: Value of modelling low and intermittent consumer demand with distributions. *International Journal of Production Research*, 1–18.

Sindhwani, R., & Malhotra, V. (2017). A framework to enhance agile manufacturing system: A total interpretive structural modelling (TISM) approach. Benchmarking. *International Journal (Toronto, Ont.)*, *24*(2), 467–487.

Singh, A. K., & Sushil. (2013). Modeling enablers of TQM to improve airline performance. *International Journal of Productivity and Performance Management, 62*(3), 250–275. doi:10.1108/17410401311309177

Sirias, D., & Mehra, S. (2005). Quantity discount versus lead time-dependent discount in an inter-organizational supply chain. *International Journal of Production Research, 43*(16), 3481–3496. doi:10.1080/00207540500117995

Soda, S., Sachdeva, A., & Garg, R. K. (2018). Green supply chain management drivers analysis using TISM. In *Flexibility in Resource Management* (pp. 113–135). Singapore: Springer. doi:10.1007/978-981-10-4888-3_8

Song, H., Choi, W., & Kim, H. (2016). Robust Vision-Based Relative-Localization Approach Using an RGB-Depth Camera and LiDAR Sensor Fusion. *IEEE Transactions on Industrial Electronics, 63*(6), 3725–3736. doi:10.1109/TIE.2016.2521346

Sourirajan, K., Ramachandran, B., & An, L. (2008). Application of control theoretic principles to manage inventory replenishment in a supply chain. *International Journal of Production Research, 46*(21), 6163–6188.

Sousa, J., Galvão, J., Machado, J., Mendonça, J., Machado, T., & Silva, P. (2018). Modelling and Simulation of a Straight Line Motion Mechanism for Industrial Application. *5th International Conference on Control, Decision and Information Technologies (CoDIT)*, 1104-1109. 10.1109/CoDIT.2018.8394919

Staff, M. H. L. (2015, September 16). How Will Industry 4.0 Affect the Supply Chain? *Material Handling & Logistics*.

Stefansson, G., & Lumsden, K. (2008). Performance issues of smart transportation management systems. *International Journal of Productivity and Performance Management, 58*(1), 55–70. doi:10.1108/17410400910921083

Steiner, J., Baker, J., Wood, G., & Meiklejohn, S. (2015). *Blockchain: the solution for transparency in product supply chains*. Project Provenance Ltd.

Stephenson, W. (1935). Correlating persons instead of tests. *Character and Personality, 4*(1), 17–24. doi:10.1111/j.1467-6494.1935.tb02022.x

Sterman J. D. (2000). *Business dynamics: systems thinking and modelling for a complex world*. McGraw-Hill Education.

Stock, J. R., & Boyer, S. L. (2009). Developing a consensus definition of supply chain management: A qualitative study. *International Journal of Physical Distribution & Logistics Management, 39*(8), 690–711. doi:10.1108/09600030910996323

Stock, T., & Seliger, G. (2016). Opportunities of sustainable manufacturing in industry 4.0. *Procedia Cirp*, *40*, 536–541. doi:10.1016/j.procir.2016.01.129

Su, T. Y., & Lei, X. H. (2008). Research on supply chain cost reduction based on process and time analysis. *IEEE International Conference on Industrial Engineering and Engineering Management,* 1625–1629. 10.1109/IEEM.2008.4738147

Subramanian, P., Ramkumar, N., Narendran, T. T., & Ganesh, K. (2012). A technical note on 'Analysis of closed loop supply chain using genetic algorithm and particle swarm optimisation'. *International Journal of Production Research*, *50*(2), 593–602. doi:10.1080/00207543.2011.593348

Subulan, K., Taşan, A. S., & Baykasoğlu, A. (2015). Designing an environmentally conscious tire closed-loop supply chain network with multiple recovery options using interactive fuzzy goal programming. *Applied Mathematical Modelling*, *39*(9), 2661–2702. doi:10.1016/j.apm.2014.11.004

Su, L.-H., & Chou, F.-D. (2000). Heuristic for scheduling in a two-machine bicriteria dynamic flowshop with setup and processing times separated. *Production Planning and Control*, *11*(8), 806–819. doi:10.1080/095372800750038418

Suwanruji, P., & Enns, S. T. (2006). Evaluating the effects of capacity constraints and demand patterns on supply chain replenishment strategies. *International Journal of Production Research*, *44*(21), 4607–4629. doi:10.1080/00207540500494527

Swan, M. (2015). *Blockchain: Blueprint for a new economy*. O'Reilly Media, Inc.

Taillard, E. (1993, January 22). Benchmarks for Basic Scheduling Problems. *European Journal of Operational Research*, *64*(2), 278–285. doi:10.1016/0377-2217(93)90182-M

Taleizadeh, A. A., Noori-Daryan, M., & Govindan, K. (2016). Pricing and ordering decisions of two competing supply chains with different composite policies: A Stackelberg game-theoretic approach. *International Journal of Production Research*, *54*(9), 2807–2836. doi:10.1080/00207543.2016.1154621

Taleizadeh, A. A., Noori-daryan, M., & Tavakkoli-Moghaddam, R. (2015). Pricing and ordering decisions in a supply chain with imperfect quality items and inspection under buyback of defective items. *International Journal of Production Research*, *53*(15), 4553–4582. doi:10.1080/00207543.2014.997399

Tang, D., Zheng, K., Zhang, H., Sang, Z., Zhang, Z., Xu, C., ... Zechinelli-Martini, J.-L. (2016). Using Autonomous Intelligence to Build a Smart Shop Floor. *Procedia CIRP*, *56*, 354–359. doi:10.1016/j.procir.2016.10.039

Tapscott, D., & Tapscott, A. (2016). *Blockchain revolution: how the technology behind bitcoin is changing money, business, and the world*. Penguin.

Tasgetiren, M. F., Liang, Y. C., Sevkli, M., & Gencyilmaz, G. (2007, March 16). A particle swarm optimization algorithm for makespan and total flowtime minimization in the permutation flowshop sequencing problem. *European Journal of Operational Research*, *177*(3), 1930–1947. doi:10.1016/j.ejor.2005.12.024

Taxakis, K., & Papadopoulos, C. (2016). A design model and a production–distribution and inventory planning model in multi-product supply chain networks. *International Journal of Production Research*, *54*(21), 6436–6457. doi:10.1080/00207543.2016.1158882

Taylor, D. H. (2006). Strategic considerations in the development of lean agri-food supply chains: A case study of the UK pork sector. *Supply Chain Management*, *11*(3), 271–280. doi:10.1108/13598540610662185

Tayur, S. R. (2013). Planned spontaneity for better product availability. *International Journal of Production Research*, *51*(23-24), 6844–6859. doi:10.1080/00207543.2013.774507

Teece, D. J., Pisano, G., & Shuen, A. (1997). Dynamic capabilities and strategic management. *Strategic Management Journal*, *18*(7), 509–533. doi:10.1002/(SICI)1097-0266(199708)18:7<509::AID-SMJ882>3.0.CO;2-Z

Telenko, C., & Seepersad, C. C. (2012). A comparison of the energy efficiency of selective laser sintering and injection molding of nylon parts. *Rapid Prototyping Journal*, *18*(6), 472–481. doi:10.1108/13552541211272018

The Economist. (2015). *Blockchains: The great chain of being sure about things*. Author.

Thiriez, A., & Gutowski, T. (2006). *An Environmental Analysis of Injection Molding*. IEEE International Symposium on Electronics and the Environment, San Francisco, CA.

Thomas, A., Krishnamoorthy, M., Venkateswaran, J., & Singh, G. (2016). Decentralised decision-making in a multi-party supply chain. *International Journal of Production Research*, *54*(2), 405–425. doi:10.1080/00207543.2015.1096977

Tian, F. (2016, June). An agri-food supply chain traceability system for China based on RFID & blockchain technology. In *Service Systems and Service Management (ICSSSM), 2016 13th International Conference on* (pp. 1-6). IEEE.

Traill, B. (1997). Structural changes in the European food industry: Consequences for innovation. In *Products and Process Innovation in the Food Industry* (pp. 38–60). Boston, MA: Springer. doi:10.1007/978-1-4613-1133-1_2

Treiblmaier, H. (2018). The impact of the blockchain on the supply chain: A theory-based research framework and a call for action. *Supply Chain Management, 23*(6), 545–559. doi:10.1108/SCM-01-2018-0029

Tsai, D. M. (2011). An optimal production and shipment policy for a single-vendor single-buyer integrated system with both learning effect and deteriorating items. *International Journal of Production Research, 49*(3), 903–922. doi:10.1080/00207540903473375

Tsao, Y. C. (2016). Designing a supply chain network for deteriorating inventory under preservation effort and trade credits. *International Journal of Production Research, 54*(13), 3837–3851. doi:10.1080/00207543.2016.1157272

Tsao, Y.-C., Thanh, V.-V., Lu, J.-C., & Yu, V. (2018). Designing sustainable supply chain networks under uncertain environments: Fuzzy multi-objective programming. *Journal of Cleaner Production, 174*, 1550–1565. doi:10.1016/j.jclepro.2017.10.272

Tse, E., & Wang, J. (2017). *How China is leading the 'new retail' revolution.* Available at: https://www.scmp.com/comment/insight-opinion/article/2125831/how-china-leading-new-retail-revolution

Tubbs, M. E., Boehne, D. M., & Dahl, J. G. (1993). Expectancy, Valence, and Motivational Force Functions in Goal-Setting Research: An Empirical Test. *The Journal of Applied Psychology, 78*(3), 361–373. doi:10.1037/0021-9010.78.3.361

Turkulainen, V., Roh, J., Whipple, J. M., & Swink, M. (2017). Managing Internal Supply Chain Integration: Integration Mechanisms and Requirements. *Journal of Business Logistics, 38*(4), 290–309. doi:10.1111/jbl.12165

Underwood, S. (2016). Blockchain beyond bitcoin. *Communications of the ACM, 59*(11), 15–17. doi:10.1145/2994581

Urata, T., Yamada, T., Itsubo, N., & Inoue, M. (2015). Modeling and balancing for costs and CO2 emissions in global supply chain network among Asian countries. *Procedia CIRP, 26*, 664–669. doi:10.1016/j.procir.2014.07.107

Vahdani, B., Zandieh, M., & Roshanaei, V. (2011). A hybrid multi-stage predictive model for supply chain network collapse recovery analysis: A practical framework for effective supply chain network continuity management. *International Journal of Production Research*, *49*(7), 2035–2060. doi:10.1080/00207540903289748

Validi, S., Bhattacharya, A., & Byrne, P. (2014). A case analysis of a sustainable food supply chain distribution system—A multi-objective approach. *International Journal of Production Economics*, *152*, 71–87. doi:10.1016/j.ijpe.2014.02.003

Value, I. I. B. (2018). *Blockchain reinvents the consumer experience*. Somers, NY: IBM.

Van Meldert & De Boeck. (2016). *Introducing autonomous vehicles in logistics: a review from a broad perspective*. Working Papers Department of Decision Sciences and Information Management.

Vasiljevic, G., Miklić, D., Draganjac, I., Kovačić, Z., & Lista, P. (2016). High-accuracy vehicle localization for autonomous warehousing. *Robotics and Computer-integrated Manufacturing*, *42*, 1–16. doi:10.1016/j.rcim.2016.05.001

Veal. (2006). *Research Methods for Leisure and Tourism* (3rd ed.). Harlow, UK: Pearson Education.

Vicente, J. J., Relvas, S., & Barbosa-Póvoa, A. P. (2017). Effective bullwhip metrics for multi-echelon distribution systems under order batching policies with cyclic demand. *International Journal of Production Research*, 1–27.

Vidović, M., Ratković, B., Bjelić, N., & Popović, D. (2016). A two-echelon location-routing model for designing recycling logistics networks with profit: MILP and heuristic approach. *Expert Systems with Applications*, *51*, 34–48. doi:10.1016/j.eswa.2015.12.029

Villegas, F. A., & Smith, N. R. (2006). Supply chain dynamics: Analysis of inventory vs. order oscillations trade-off. *International Journal of Production Research*, *44*(6), 1037–1054. doi:10.1080/00207540500338203

Viswanathan, S., Widiarta, H., & Piplani, R. (2007). Value of information exchange and synchronization in a multi-tier supply chain. *International Journal of Production Research*, *45*(21), 5057–5074. doi:10.1080/00207540600930057

Vitasek, K. L., Manrodt, K. B., & Abbott, J. (2005). What makes a lean supply chain? *Supply Chain Management Review*, *9*(7), 39–45.

Vroom. (1964). *Work and motivation*. New York: Wiley.

Wadhwa, S., Bibhushan, & Chan, F. T. S. (2009). Inventory performance of some supply chain inventory policies under impulse demands. *International Journal of Production Research, 47*(12), 3307–3332. doi:10.1080/00207540701689750

Wahlster, W. (2012). From Industry 1.0 to Industry 4.0: Towards the 4th Industrial Revolution. *Forum Business Meets Research.*

Wang, F., Lai, X., & Shi, N. (2011). A multi-objective optimization for green supply chain network design. *Decision Support Systems, 51*(2), 262–269. doi:10.1016/j.dss.2010.11.020

Wang, H., Li, S., & Luo, J. (2018). Optimal markdown pricing for holiday basket with customer valuation. *International Journal of Production Research*, 1–15.

Wang, S. J., Huang, C. T., Wang, W. L., & Chen, Y. H. (2010). Incorporating ARIMA forecasting and service-level based replenishment in RFID-enabled supply chain. *International Journal of Production Research, 48*(9), 2655–2677. doi:10.1080/00207540903564983

Wang, S. P., & Lee, W. (2016). A note on 'A two-level supply chain with consignment stock agreement and stock-dependent demand'. *International Journal of Production Research, 54*(9), 2750–2756. doi:10.1080/00207543.2015.1130872

Wang, S., Wan, J., Zhang, D., Li, D., & Zhang, C. (2016). Towards smart factory for industry 4.0: A self-organized multi-agent system with big data based feedback and coordination. *Computer Networks, 101*, 158–168. doi:10.1016/j.comnet.2015.12.017

Wang, Y. M., Wang, Y. S., & Yang, Y. F. (2010). Understanding the determinants of RFID adoption in the manufacturing industry. *Technological Forecasting and Social Change, 77*(5), 803–815. doi:10.1016/j.techfore.2010.03.006

Wang, Y. Y., Wang, Y. S., & Lin, T. C. (2018). Developing and validating a technology upgrade model. *International Journal of Information Management, 38*(1), 7–26. doi:10.1016/j.ijinfomgt.2017.07.003

Wang, Z., Chen, C., Guo, B., Yu, Z., & Zhou, X. (2016). Internet plus in China. *IT Professional, 18*(3), 5–8. doi:10.1109/MITP.2016.47

Wan, J., Tang, S., Li, D., Imran, M., Zhang, C., Liu, C., & Pang, Z. (2019). Reconfigurable smart factory for drug packing in healthcare industry 4.0. *IEEE Transactions on Industrial Informatics, 15*(1), 507–516. doi:10.1109/TII.2018.2843811

Ward, M. (2018). *Building the Blockchain Logistics Protocol.* Academic Press.

Wells, J. C. (2016). Cyber-physical security challenges in manufacturing systems. *Manufacturing Letters*, 74-77.

Wernerfelt, B. (1984). A resource-based view of the firm. *Strategic Management Journal*, *5*(2), 171–180. doi:10.1002mj.4250050207

Wolf, M. (2017). *How Could Blockchain Be Used With Food*. Smart Kitchen Show.

Wollschlaeger, M., Sauter, T., & Jasperneite, J. (2017). The future of industrial communication: Automation networks in the era of the internet of things and industry 4.0. *IEEE Industrial Electronics Magazine*, *11*(1), 17–27. doi:10.1109/MIE.2017.2649104

Wong, H., Wikner, J., & Naim, M. (2009). Analysis of form postponement based on optimal positioning of the differentiation point and stocking decisions. *International Journal of Production Research*, *47*(5), 1201–1224. doi:10.1080/00207540701549608

Woodward, P. A. (2015, July 20). Cybersecurity vulnerabilities in medical devices: a complex environment and multifaceted problem. *US National Library of Medicines National Institute of Health,* 305-316. Retrieved from https://www.ncbi.nlm.nih.gov/pmc/articles/PMC4516335/

World Health Organization. (2017). *WHO's first ever global estimates of foodborne diseases find children under 5 account for almost one third of deaths*. Available at: http://www.who.int/mediacentre/news/releases/2015/foodborne-disease-estimates/en/

WRAP. (2007). *Understanding the Food Waste*. Research Summary.

Wu, H. H., Huang, H. H., & Jen, W. T. (2012). A study of the elongated replenishment frequency of TOC supply chain replenishment systems in plants. *International Journal of Production Research*, *50*(19), 5567–5581. doi:10.1080/00207543.2011.649803

Wu, H., Li, Z., King, B., Ben Miled, Z., Wassick, J., & Tazelaar, J. (2017). A distributed ledger for supply chain physical distribution visibility. *Information*, *8*(4), 137. doi:10.3390/info8040137

Xu, L. D., Xu, E. L., & Li, L. (2018). Industry 4.0: State of the art and future trends. *International Journal of Production Research*, 1–22.

Xu, L., Chen, L., Gao, Z., Lu, Y., & Shi, W. (2017). CoC: Secure Supply Chain Management System Based on Public Ledger. *26th International Conference on Computer Communication and Networks (ICCCN)*. 10.1109/ICCCN.2017.8038514

Yaaqob, I. (2017). The rise of ransomware and emrging security challenges in the Internet of Things. *Science Direct*, 453-454.

Yadav, D. K., & Barve, A. (2016). Modeling post-disaster challenges of humanitarian supply chains: A TISM approach. *Global Journal of Flexible Systems Management*, *17*(3), 321–340. doi:10.100740171-016-0134-4

Yadav, N. (2014). Total interpretive structural modelling (TISM) of strategic performance management for Indian telecom service providers. *International Journal of Productivity and Performance Management*, *63*(4), 421–445. doi:10.1108/IJPPM-04-2013-0081

Yang, D., Xiao, T., Choi, T. M., & Cheng, T. C. E. (2015). Optimal reservation pricing strategy for a fashion supply chain with forecast update and asymmetric cost information. *International Journal of Production Research*, 1–22.

Yang, Y., Pan, S., & Ballot, E. (2017). Mitigating supply chain disruptions through interconnected logistics services in the Physical Internet. *International Journal of Production Research*, *55*(14), 3970–3983. doi:10.1080/00207543.2016.1223379

Yan, H., Tang, S. L., & Yen, G. (2008). Quick response procurement cost control strategy for fabric manufacturing. *International Journal of Production Research*, *46*(17), 4769–4784. doi:10.1080/00207540701210037

Yeh, W.-C., & Chuang, M.-C. (2011). Using multi-objective genetic algorithm for partner selection in green supply chain problems. *Expert Systems with Applications*, *38*(4), 4244–4253. doi:10.1016/j.eswa.2010.09.091

Yi, H., & Sarker, B. R. (2013). An optimal consignment stock production and replenishment policy with controllable lead time. *International Journal of Production Research*, *51*(21), 6316–6335. doi:10.1080/00207543.2013.795668

Yin, R. K. (2013). *Case study research: Design and methods*. Sage publications.

Yin, R. K. (2015). *Qualitative research from start to finish*. Guilford Publications.

Yin, Y., Stecke, K. E., & Li, D. (2018). The evolution of production systems from Industry 2.0 through Industry 4.0. *International Journal of Production Research*, *56*(1-2), 848–861. doi:10.1080/00207543.2017.1403664

Young, W., Hwang, K., McDonald, S., & Oates, C. J. (2010). Sustainable consumption: Green consumer behaviour when purchasing products. *Sustainable Development*, *18*(1), 20–31.

Yuan, K. F., & Gao, Y. (2010). Inventory decision-making models for a closed-loop supply chain system. *International Journal of Production Research, 48*(20), 6155–6187. doi:10.1080/00207540903173637

Yuan, K. F., Ma, S. H., He, B., & Gao, Y. (2015). Inventory decision-making models for a closed-loop supply chain system with different decision-making structures. *International Journal of Production Research, 53*(1), 183–219. doi:10.1080/00207543.2014.946160

Yulia Cherdantseva, P. B. (2016, February). A review of cyber security risk assessment methods for SCADA systems. *Science Direct, 56*, 1-27. Retrieved from https://www.sciencedirect.com/science/article/pii/S0167404815001388

Zainal, Z. (2007). Case study as a research method. *Jurnal Kemanusiaan*, (9), 1-6.

Zanella, A., Bui, N., Castellani, A., Vangelista, L., & Zorzi, M. (2014). Internet of things for smart cities. *IEEE Internet of Things Journal, 1*(1), 22-32.

Zanoni, S., Ferretti, I., & Tang, O. (2006). Cost performance and bullwhip effect in a hybrid manufacturing and remanufacturing system with different control policies. *International Journal of Production Research, 44*(18-19), 3847–3862. doi:10.1080/00207540600857375

Zanoni, S., & Jaber, M. Y. (2015). A two-level supply chain with consignment stock agreement and stock-dependent demand. *International Journal of Production Research, 53*(12), 3561–3572. doi:10.1080/00207543.2014.980012

Zelbst, P. J., Green, K. W., Sower, V. E., & Reyes, P. M. (2012). Impact of RFID on manufacturing effectiveness and efficiency. *International Journal of Operations & Production Management, 32*(3), 329–350. doi:10.1108/01443571211212600

Zhang, Z. (2016). Social media security and trustworthiness. *Science Direct*, 1-11.

Zhang, C., Gu, P., & Jiang, P. (2015). Low-carbon scheduling and estimating for a flexible job shop based on carbon footprint and carbon efficiency of multi-job processing. *Proceedings of the Institution of Mechanical Engineers. Part B, Journal of Engineering Manufacture, 229*(2), 328–342. doi:10.1177/0954405414527959

Zhang, L. H., Li, T., & Fan, T. J. (2018). Inventory misplacement and demand forecast error in the supply chain: Profitable RFID strategies under wholesale and buy-back contracts. *International Journal of Production Research*, 1–18. doi:10.1080/00207543.2018.1530479

Zhang, Y., Guo, Z., Lv, J., & Liu, Y. (2018). A Framework for Smart Production-Logistics Systems Based on CPS and Industrial IoT. *IEEE Transactions on Industrial Informatics*, *14*(9), 4019–4032. doi:10.1109/TII.2018.2845683

Zhao, F., Wu, D., Liang, L., & Dolgui, A. (2016). Lateral inventory transshipment problem in online-to-offline supply chain. *International Journal of Production Research*, *54*(7), 1951–1963. doi:10.1080/00207543.2015.1070971

Zheng, P., Sang, Z., Zhong, R. Y., Liu, Y., Liu, C., & Mubarok, K. (2018). Smart manufacturing systems for Industry 4.0: Conceptual framework, scenarios, and future perspectives. Academic Press.

Zheng, P., Sang, Z., Zhong, R. Y., Liu, Y., Liu, C., Mubarok, K., ... Xu, X. (2018). Smart manufacturing systems for Industry 4.0: Conceptual framework, scenarios, and future perspectives. *Frontiers of Mechanical Engineering*, 1–14.

Zindler, K., & Geiß, N. (2016). Vehicle Ego-Localization in Autonomous Lane-Keeping Evasive aneuvers. *IFAC-Papers OnLine*, *49*(11), 160–167. doi:10.1016/j.ifacol.2016.08.025

Zobolas, G. I., Tarantilis, C. D., & Ioannou, G. (2009, April). Minimizing makespan in permutation flow shop scheduling problems using a hybrid metaheuristic algorithm. *Computers & Operations Research*, *36*(4), 1249–1267. doi:10.1016/j.cor.2008.01.007

About the Contributors

S. G. Ponnambalam has a Ph.D. degree in Industrial Engineering form Bharathiar University, Coimbatore, India. He is a Full Professor at the Faculty of Manufacturing Engineering at University Malaysia Pahang. Earlier he served as a full Professor at Monash University Malaysia. He is teaching courses in the areas of mechatronics, manufacturing, and manufacturing management, and he is supervising a number of doc students in the area of manufacturing automation and swarm robotics. He is serving as Associate Editor and as editorial board member for many international journals. He has over 300 articles published in various referred journals, refereed conferences and chapters in edited books. His articles are published in different peer reviewed journals those include International Journal of Production Research, International Journal of Advanced Manufacturing Technology, Production Planning and Control, Robotics and Computer-Integrated Manufacturing, Evolutionary Optimization, Journal of Cleaner Production, Computers & Industrial Engineering, Expert Systems with Applications, IEEE Transaction on Cybernetics and Journal of Material Processing Technology.

Nachiappan Subramanian is the Chair of Department of Management and Professor of Operations and Logistics Management and Supply Chains at the University of Sussex UK. Nachi has 20 years of experience including 18 years in academia and 2 years in a consulting company. Nachi has a teaching experience in the UK, China, Australia and India. Prior to post graduate study he worked as consultant and led a software consultancy projects based at India and Japan. He is recognized as the fellow of Chartered Institute of Logistics and Transport (CILT) UK and a Senior Fellow of Higher Education Academy, UK,. Nachi has been recognised for his research and teaching excellence to date he has published 90 peer reviewed refereed research papers. His research areas are sustainable supply chains (environmental and social/humanitarian issues), risk and resilience, technology enabled operations and marketing interface, and in performance measurement. Nachi is he has written several books, book chapters and cases.

* * *

Emily Carey has graduated from Cardiff University with a 2:1 in BSc (Hons) Chemistry with Industrial Experience, with Industrial Placement year at GSK. She has recently finished a MSc in Global Supply Chain and Logistics Management (CIPS and CILT accredited), with a Distinction grade at the University of Sussex. In addition to starting an associate role in the Strategy Team of Product Service & Innovation at Samsung Electronics, as part of the Samsung Emerging Talent Graduate Program.

Atanu Chaudhuri is an Associate Professor in Operations and Supply Chain Management at Aalborg University, Denmark. His research interests are in digital manufacturing and supply chain risk management.

Puteri Fadzline is currently a senior lecturer in Faculty of Industrial Management, Universiti Malaysia Pahang. She graduated from Universiti Technology MARA, Shah Alam, with a PhD in Business Management and from Universiti Sains Malaysia, Penang with BSc and MSc degrees in Bioresources, Paper and Coating Technology under the School of Industrial Technology. Prior to Universiti Malaysia Pahang, she had accumulated industrial experiences in the furniture and wood flooring industry for 4 years. At both companies, she is responsible for matters related to Quality Control and Production processes. Her research interests are Innovation Management, Product Management, and Research Methodology.

Kanagaraj G. obtained his B.E degree in Mechanical Engineering from Anna University affiliated college in 1993 and M.E degree in Engineering Design from Anna University affiliated college in 2000. He also completed Ph.D in Anna University and post doctorate fellowship in Monash University Sunway Campus Malaysia. Currently he is an Associate Professor in the Department of Mechatronics Engineering at Thiagarajar College of Engineering Madurai. His research interests include Meta heuristic and heuristic algorithms, Machine learning algorithms for manufacturing application. He published many national and international research papers and published book under topic Product Design - Reliability Based Total Cost of Ownership Approach. He acts as industrial consultant for many industries and awarded as researcher award by UGC.

Malka N. Halgamuge is a Researcher in the Department of Electrical and Electronic Engineering of the University of Melbourne, and she has also obtained her Ph.D. from the same department in 2007. She is passionate about research and teaching university students (Data Science, Business Intelligence using Big Data, Internet of Things (IoT), Sensor Network, Digital Health, Telemedicine, Bioelectromagnetics, and Hyperthermia). She has published more than 90 peer-reviewed technical articles attracting over 944 Google Scholar Citations with h-index = 16, and her Research Gate RG Score is 33.96. She is currently supervising a PhD student at the University of Melbourne, and 4 PhD students completed their theses in 2013, 2015 and 2018 with her as the principal supervisor. She successfully sought 7 short-term research fellowships at premier Universities in the World. She excels in major commercial work related to the investigation of electromagnetic radiation hazards safety assessment and the provision of extensive technical reports. This work allows her to help organizations and individuals to promote precautionary approaches to the use of technology. This gave her access to contribute in Media (Newspaper Articles and interviews on Radio and Television).

K. Hariharan was born in Tamil Nadu, India. He awarded Ph.D. in 2011 and currently working as an Associate Professor in Department of Electronics and Communication Engineering in Thiagarajar College of Engineering, Tamilnadu, India. He published many research papers in optimisation techniques and analog devices testing applications. He was awarded best project guide Renesas-Japan in 2008.

Pradeep J. is a final year B.E (Mechanical Engineering) student. His areas of interests are Soft computing and optimization in OR problems.

Natarajan Jawahar received his B.E. (Honours) Degree in Mechanical Engineering from University of Madras (GCT, Coimbatore) in 1982, M.E. (Distinction) Degree in Production Engineering from Annamalai University in 1988 and completed his Ph.D. programme in Bharathiar University (PSGCT, Coimbatore) in 1998. He was employed as Engineer Trainee in TI Diamond Chain Ltd., Madras from 1982 to 1983 and Assistant Plant Manager in Neyveli Lignite Corporation Ltd, Neyveli from 1983 to 1989. Since 1989 He has worked in Thiagarajar College of Engineering, Madurai from 1989 to 2018 as Lecturer, Professor of Mechanical Engineering and Dean (Research & Development). Currently he holds the position of Principal Ramco Institute of Technology, Rajapalayam, He has published 6 chapters in edited books, 70 research articles in refereed journals, and 140 technical papers in conferences, and involved with twelve research projects funded by AICTE / DRDO /SAMEER in the field of Automation and Supply Chain Management. He is a reviewer for Int. Journal of IEEE Transactions of Automation Science and Engineering, Int. Journal of Production Research, Int. Journal Rapid Prototyping, European Journal of Operations Research, International Journal of Approximate Reasoning, Int. Journal of Industrial Management and Optimisation, Int. Journal of OR letters, Int. Journal of Robotics and Automation, Int. Journal of Computers and Operations Research, Int. Journal of Asia Pacific Management Review, Int. Journal of Production Economics, Int. Journal of Computers and Industrial Engineering, National Journal of Operation

Research Society of India and National Level Technical Publication ANJAC Journal of Sciences. He has coordinated short-term training programmes in CNC Programming and Supply Chain Management. He has organised National Conference on Operations Research in Information Technology and Supply Chain Management. Eleven scholars have obtained Ph.D. under his guidance and seven more are working for their Ph.D. in different fields of Production and Industrial Engineering. He is a Fellow Member in Institution of Engineers (India), Indian Society of Mechanical Engineers and Indian Institution of Production Engineers, and Member in Indian Society for Technical Education and Operations Research Society of India. Besides, he has been awarded BEST Faculty of TCE - 2007, instituted by CTS and Tamil Nadu State Scientist Award in Engineering & Technology for the year 2012. . His academic and research interests are: Scheduling, Production Planning and Control, Flexible Manufacturing Systems, Process Optimization, Evolutionary Programming, Logistics and Supply Chain Management.

Rahul K. is a final year B.E. (Mechanical Engineering) student. His areas of interests are Soft computing and optimization in OR problems.

Yuhong Li was born in Beijing, China, and finish the Master's degree in University of Sussex, majored in Global supply chain and logistics management. The author has great passion on the transformation of the business world with supply chain management, which pushed him to completed the dissertation of blockchain with supply chain in order to make a difference.

Dennis Massarola is a Masters in Operations and Management Engineering from Aalborg University, Denmark.

Peter Nielsen (PhD) is Associate Professor at Department of Mechanical and Manufacturing Engineering at Aalborg University in Denmark. He is the Coordinating editor for Production and Manufacturing Research journal. He has edited several special issues for reputed international journals such as Production Planning and Control, Annals of Operations Research. His main area of research interests is stochastics in scheduling, inventory management and optimization under stochastic conditions. He has published over 70 articles in reputed journals and conferences.

Mukund Nilakantan (PhD) is a Lecturer in Department of Engineering at University of Leicester in United Kingdom. Prior to this he worked as a Postdoctroal Fellow at Aalborg University, Denmark. He received his PhD from Monash University in 2015 in the area of assembly line balancing and metaheuristics. His main area of interests are metaheuristics, scheduling and manufacturing optimization. He has published over 25 articles in reputed journals and conferences.

Maciel M. Queiroz is a Researcher in Operations and Supply Chain Management at the Naval Architecture and Ocean Engineering Department of the University of São Paulo, Brazil. Maciel holds an MSc and a PhD in Naval Architecture and Ocean Engineering from the University of São Paulo. His current research interests focus on supply chain digital disruptions, digital supply chain capabilities, Industry 4.0, blockchain, big data, IoT, CPS, including the adoption and use of these technologies.

Humyun Fuad Rahman (Ph.D.) is a Research associate at UNSW Australia. Previously he worked as post-doctoral fellow at the Department of Materials and Production at Aalborg University, Denmark. He received his completed his PhD in computer science from University of New South Wales in 2015 in the area of flow shop scheduling. His research mainly covers production scheduling using evolutionary algorithms. Dr. Rahman has published numerous high impact journals and several conference proceedings.

Julius Fusic S. obtained his B.E. degree in Electrical and Electronics Engineering from Anna University affiliated college in 2010 and M.E. degree in Power Electronics& Drives from Anna university affiliated college in 2014. Currently he is an Assistant Professor in the Department of Mechatronics Engineering at Thiagarajar College of Engineering Madurai. His research interests include Autonomous systems, Electrical Machines, Power electronics, Mechatronics systems. He was awarded SAP Award for Excellence from IIT Bombay (Top 253 out of 4051 registered participants).

Bathrinath Sankaranarayanan's areas of interests are Industrial Engineering, Production Scheduling and Optimization, Supply Chain Management, Industrial Safety. He obtained his Post-Doctoral Fellowship in University of Southern Denmark. He has several publications in the area of Industrial Engineering.

S. Saravanasankar is a Professor in the Faculty of Mechanical Engineering at Kalasalingam Academy of Research and Education, India. He has published more than 70 papers in variours International Journals, conferences, book chapters etc. His areas of interests are Industrial Engineering, Production Scheduling and Optimization, Supply Chain Management, Industrial Safety.

Ponnusamy Venkumar, B.E., M.E., Ph.D., M.Sc. (YHE), received his Bachelors in Mechanical Engineering and Masters in Industrial Engineering from Thiagarajar College of Engineering, Madurai, Tamilnadu, India. He had undergone his Doctorate degree in Manonmaniam Sundarnar University, Thirunevelli, India. Moreover, he also did Post Graduation in Yoga for Human Excellence from Tamil Nadu Physical Education and Sports University, Chennai, Tamilnadu, India. He has successfully completed his post-doctoral research in the University of Sussex, Falmer, UK. He is having more than 25 years of teaching experience and one year of industrial experience, presently he is working as a Professor in Mechanical Engineering Department, Kalasalingam Academy of Research and Education, Tamilnadu, India. He has contributed more than 40 research papers in various reputed International and National Journals. His areas of interest include Manufacturing System Design, Optimization, Supply Chain Management, Smart Manufacturing, etc. He has successfully completed a DST, Govt. of India, New Delhi, sponsored research project.

Samuel Fosso Wamba is Full Professor at Toulouse Business School. Prior, He was Associate Professor at NEOMA Business School and Senior lecturer in the School of Information Systems & Technology, University of Wollongong, Australia. Dr Wamba successfully defended on September 2015 his Habilitation of Conducting Research in Management Science at the Telecom Business School, France. He earned his Ph.D. in industrial engineering at the Polytechnic School of Montreal, Canada. He also holds two Masters degrees: one MSc in mathematics, from the University of Sherbrooke in Canada and another MSc in electronic commerce from HEC Montreal, Canada. He is CompTIA RFID+ Certified Professional, Academic Co-Founder of RFID Academia.

Index

M

N

O

P

Q

S

T

W

www.ingramcontent.com/pod-product-compliance
Lightning Source LLC
LaVergne TN
LVHW081345300926
849161LV00008B/2075